FLORIDA STATE
UNIVERSITY LIBRARIES

JUL 18 2000

TALLAHASSEE, FLORIDA

EMERGING TOOLS FOR SINGLE-CELL ANALYSIS

CYTOMETRIC CELLULAR ANALYSIS

Series Editors

J. Paul Robinson
Purdue University Cytometry
 Laboratories
Purdue University
West Lafayette, Indiana

George F. Babcock
Department of Surgery
University of Cincinnati College
 of Medicine
Cincinnati, Ohio

New Volumes in Series

Phagocyte Function: A Guide for Research and Clinical Evaluation
J. Paul Robinson and George F. Babcock, *Volume Editors*

Immunophenotyping
Carlton C. Stewart and Janet K. A. Nicholson, *Volume Editors*

Emerging Tools for Single Cell Analysis: Advances in Optical Measurement Technologies
Gary Durack and J. Paul Robinson, *Volume Editors*

Forthcoming Volume in Series

Cellular Aspects of HIV Infection
Andrea Cossarizza and David Kaplan, *Volume Editors*

EMERGING TOOLS FOR SINGLE-CELL ANALYSIS

ADVANCES IN OPTICAL MEASUREMENT TECHNOLOGIES

Edited by

GARY DURACK
University of Illinois Biotechnology Center
University of Illinois at Urbana-Champaign
Urbana, Illinois

J. PAUL ROBINSON
Purdue University Cytometry Laboratories
Purdue University
West Lafayette, Indiana

A JOHN WILEY & SONS, INC., PUBLICATION

New York • Chichester • Weinheim • Brisbane • Singapore • Toronto

Sci
QH
585.5
.F56
E44
2000

This book is printed on acid-free paper. ∞

Copyright © 2000 by Wiley-Liss, Inc. All rights reserved.

Published simultaneously in Canada.

No part of this publication may be reproduced, stored in a retrieval system or transmitted in any form or by any means, electronic, mechanical, photocopying, recording, scanning or otherwise, except as permitted under Sections 107 or 108 of the 1976 United States Copyright Act, without either the prior written permission of the Publisher, or authorization through payment of the appropriate per-copy fee to the Copyright Clearance Center, 222 Rosewood Drive, Danvers, MA 01923, (978) 750-8400, fax (978) 750-4744. Requests to the Publisher for permission should be addressed to the Permissions Department, John Wiley & Sons, Inc., 605 Third Avenue, New York, NY 10158-0012, (212) 850-6011, fax (212) 850-6008, E-Mail: PERMREQ@WILEY.COM.

For ordering and customer service, call 1-800-CALL-WILEY.

While the authors, editors, and publisher believe that drug selection and dosage and the specification and usage of equipment and devices, as set forth in this book, are in accord with current recommendations and practice at the time of publication, they accept no legal responsibility for any errors or omissions and make no warranty, express or implied, with respect to material contained herein. In view of ongoing research, equipment modifications, changes in governmental regulations, and the constant flow of information relating to drug therapy, drug reactions, and the use of equipment and devices, the reader is urged to review and evaluate the information provided in the package insert or instructions for each drug, piece of equipment, or device for, among other things, any changes in the instructions or indication of dosage or usage and for added warnings and precautions.

Library of Congress Cataloging-in-Publication Data:

Emerging tools for single cell analysis : advances in optical measurement technologies / edited by Gary Durak, J. Paul Robinson.
    p. ; cm. — (Cytometric cellular analysis)
  Includes index.
  ISBN 0-471-31575-3 (alk. paper)
    1. Flow cytometry. 2. Cytophotometry. I. Durack, Gary II. Robinson, J. Paul. III. Series.
    [DNLM: 1. Flow Cytometry—instrumentation. 2. Flow Cytometry—methods. 3. Cell Separation. QH 585.5.F56 E53 2000]
  QH585.5.F56 E44 2000
  571.6'028—dc21
                                                                             99-051911

Printed in the United States of America.
10 9 8 7 6 5 4 3 2 1

Contents

	Preface	vii
	Contributors	ix
1	**Cell-Sorting Technology** Gary Durack	1
2	**High-Speed Cell Sorting** Ger van den Engh	21
3	**Rare-Event Detection and Sorting of Rare Cells** James F. Leary	49
4	**Applications of High-Speed Sorting for CD34+ Hematopoietic Stem Cells** Thomas Leemhuis and David Adams	73
5	**Microfabricated Fluidic Devices for Single-Cell Handling and Analysis** David J. Beebe	95
6	**Single DNA Fragment Detection by Flow Cytometry** Robert C. Habbersett, James H. Jett, and Richard A. Keller	115
7	**Fluorescence Lifetime Imaging: New Microscopy Technologies** Weiming Yu, William W. Mantulin, and Enrico Gratton	139
8	**Fluorescence Lifetime Flow Cytometry** John A. Steinkamp	175
9	**Application of Fluorescence Lifetime and Two-Photon Fluorescence Cytometry** Donald J. Weaver, Jr., Gary Durack, Edward W. Voss, Jr., and Anu Cherukuri	197

10	**Probing Deep-Tissue Structures by Two-Photon Fluorescence Microscopy** *Chen-Yuan Dong, Ki Hean Kim, Christof Buehler, Lily Hsu, Hyun Kim, Peter T. C. So, Barry R. Masters, Enrico Gratton, and Irene E. Kochevar*	**221**
11	**Limits of Confocal Imaging** *James B. Pawley*	**239**
12	**Scanning Near-Field Optical Imaging and Spectroscopy in Cell Biology** *Vinod Subramaniam, Achim K. Kirsch, Attila Jenei, and Thomas M. Jovin*	**271**
13	**White-Light Scanning Digital Microscopy** *J. Paul Robinson and Ben Gravely*	**291**
14	**Illumination Sources** *Howard M. Shapiro*	**307**
15	**Camera Technologies for Cytometry Applications** *Kenneth Castleman*	**323**
	Index	**339**

Preface

This book owes its existence to the recent resurgence of interest in high-resolution evaluation of single-cell properties. One of our goals in assembling this volume was to re-evaluate where current technology stands at the beginning of a new millennium. Further, we felt there was a need for critical discussion of some of these technologies. Take cell sorting, for example. Flow cytometers have been sorting effectively for over 30 years, ever since Fulwyler implemented electrostatic cell sorting in 1965 from inkjet printing technologies developed by Richard Sweet. Sorting in itself, however, did not originate in the sixties—it had been proposed 30 years previously by Moldavin and attempted by several others more or less successfully as technologies matured. In recent years significant gains have been realized. High-speed sorters, once considered useful only for chromosome separation, have supplanted "traditional" sorters in many environments. Now they are used for the isolation of stem cells, mammalian sperm, and a variety of other materials. We have attempted to take a fairly rational approach to the phenomenon of sorting by bringing together in one volume the experts who have played leading roles in the recent revolution. But this was not the only driving force. We wanted a more in-depth discussion of the analytical component of sorting and other technologies. Although biologists frequently veer away from anything containing an equation, we are dependent upon the verification of the mathematical concepts in the implementation process. Thus, this volume attempts to create a domain that successfully integrates engineering and biology.

Now more then ever, there is a blurring of the interface between technologies. Biomedical engineering, and especially one of its component parts, tissue engineering, is the growth business of engineering today. Miniaturization of electronic components has forever changed the nature of computing. Similarly, micro- and now nano-machines capable of performing mechanical action are a reality. Future developments in these areas are sure to impact cytometry in many ways. The success of such technologies will be in a systems approach—combining a good understanding of fluorescence detection with high-speed electronics and miniaturization together with a fundamental grasp of the biology.

Where imaging fits into this picture is anyone's guess, but there is little doubt that imaging is a rapidly advancing component of cellular and molecular biology that has reached previously unattainable heights. New technologies, transformed into usable laboratory instruments, have altered cell biology almost beyond recognition in the last decade of the 20th century. Confocal microscopy, developed nearly 50 years ago but only recently commercialized, gave biologists a unique tool for investigating the inner workings of cell systems. More recently, multiphoton microscopy has taken us a few steps further. The ability to continuously monitor cellular activity in a living system using multiphoton techniques creates a paradigm shift for developmental biologists who can now monitor, in real time, the differentiation processes in embryonic development. Combining these technologies with GFP and other cell-tracking molecules has begun yet another revolution in imaging applications.

An interesting theme that has emerged as this book was being put together is that many of the technologies discussed—sorting, confocal microscopy, color scanning microscopy—all had their roots established well over 50 years ago. Only in recent years have they had a discernable impact, presumably because of parallel technological developments and the more recent commercial commitment by companies that have created off-the-shelf systems usable by scientists who then do not necessarily need to be capable of building such systems themselves.

Our final goal was to compare and contrast totally different technologies that work toward similar goals: evaluating the properties of single cells using optically based measurement systems. Within the biological discovery process, the synergy between cytometry, chemistry, and imaging systems is remarkable. Traditional flow cytometry technologies combined with new imaging and detection technologies have exploded into a plethora of combination systems. It is our hope that this book creates a linkage between the engineering and development of the technologies and the fusion of these technologies into exciting and emerging tools for single cell analysis.

Gary Durack, University of Illinois Biotechnology Center
J. Paul Robinson, Purdue University Cytometry Laboratories

Contributors

David Adams
SyStemix, Inc.
Palo Alto, California

David J. Beebe
University of Wisconsin
Madison, Wisconsin

Christof Buehler
Department of Mechanical Engineering
Massachusetts Institute of Technology
Cambridge, Massachusetts

Kenneth Castleman
Perceptive Scientific Instruments, Inc.
League City, Texas

Anu Cherukuri
Department of Biochemistry
Northwestern University
Evanston, Illinois

Chen-Yuan Dong
Department of Mechanical Engineering
Massachusetts Institute of Technology
Cambridge, Massachusetts

Gary Durack
Biotechnology Center
University of Illinois at Urbana-Champaign
Urbana, Illinois

Enrico Gratton
Laboratory for Fluorescence Dynamics
University of Illinois at Urbana-Champaign
Urbana, Illinois

Ben Gravely
Cosmic Technologies Corporation
Raleigh, North Carolina

Robert C. Habbersett
Cytometry Group
Los Alamos National Laboratory
Los Alamos, New Mexico

Lily Hsu
Department of Mechanical Engineering
Massachusetts Institute of Technology
Cambridge, Massachusetts

Attila Jenei
Department of Molecular Biology
Max Planck Institute for Biophysical
  Chemistry
Goettingen, Germany

James H. Jett
Cytometry Group
Los Alamos National Laboratory
Los Alamos, New Mexico

Thomas M. Jovin
Department of Molecular Biology
Max Planck Institute for Biophysical
  Chemistry
Goettingen, Germany

Richard A. Keller
Cytometry Group
Los Alamos National Laboratory
Los Alamos, New Mexico

Hyun Kim
Department of Mechanical Engineering
Massachusetts Institute of Technology
Cambridge, Massachusetts

Ki Hean Kim
Department of Mechanical Engineering
Massachusetts Institute of Technology
Cambridge, Massachusetts

Achim K. Kirsch
Department of Molecular Biology
Max Planck Institute for Biophysical Chemistry
Goettingen, Germany

Irene E. Kochevar
Wellman Laboratories of Photomedicine
Massachusetts General Hospital
Boston, Massachusetts

James F. Leary
University of Texas Medical Branch
Galveston, Texas

Thomas Leemhuis
Stanford University Medical Center
Stanford, California

William W. Mantulin
Laboratory for Fluorescence Dynamics
University of Illinois at Urbana-Champaign
Urbana, Illinois

Barry R. Masters
University of Bern
Bern, Switzerland

James B. Pawley
Department of Zoology
University of Wisconsin
Madison, Wisconsin

J. Paul Robinson
Purdue University Cytometry Laboratories
Purdue University
West Lafayette, Indiana

Howard M. Shapiro
Howard M. Shapiro, M.D., P.C.
West Newton, Massachusetts

Peter T. C. So
Department of Mechanical Engineering
Massachusetts Institute of Technology
Cambridge, Massachusetts

John A. Steinkamp
Los Alamos National Laboratories
Los Alamos, New Mexico

Vinod Subramaniam
Department of Molecular Biology
Max Planck Institute for Biophysical Chemistry
Goettingen, Germany

Ger van den Engh
University of Washington
Seattle, Washington

Edward W. Voss, Jr.
Department of Microbiology
University of Illinois at Urbana-Champaign
Urbana, Illinois

Donald J. Weaver, Jr.
Department of Microbiology and Immunology
University of North Carolina-Chapel Hill
Chapel Hill, North Carolina

Weiming Yu
Laboratory for Fluorescence Dynamics
University of Illinois at Urbana-Champaign
Urbana, Illinois

1

Cell-Sorting Technology

Gary Durack
University of Illinois at Urbana-Champaign, Urbana, Illinois

INTRODUCTION

Flow-cytometry-based cell sorting was first introduced to the research community more than 20 years ago. It is a technology that has been widely applied in many areas of life science research, serving as a critical tool for those working in fields such as genetics, immunology, molecular biology, and environmental science. Recent advances in underlying technologies, especially digital electronics, have made possible a new generation of high-speed cell-sorting devices. Instruments capable of processing more than 100,000 cells per second are becoming commercially available. Laboratories throughout the world are beginning to find new research and clinical applications for these high-speed cell sorters. On another technology front, rapid advances in MicroElectro-Mechanical Systems (MEMS) design and fabrication techniques are producing *biochips* that can sequentially process cells or other small particles (Masuda et al., 1989; Washizu et al., 1990). These micro–laboratory devices have the potential to revolutionize many cell measurement and processing techniques. The first five chapters of this text describe the state-of-the-art for these rapidly evolving technologies. They cover both theoretical considerations and practical application issues that should be of interest to both instrument designers and end users. This chapter provides a review of cell-sorting technology and serves as an introduction to those not already well acquainted with the field. It is followed by chapters covering

---

Emerging Tools for Single-Cell Analysis, Edited by Gary Durack and J. Paul Robinson.
ISBN 0-471-31575-3 Copyright © 2000 Wiley-Liss, Inc.

high-speed cell sorting (van den Engh), rare-event detection (Leary), clinical applications of high-speed cell sorting (Leemhuis), and MEMS-based cell-handling systems (Beebe). Taken together, this suite of chapters offers a comprehensive presentation of the recent advances in cell-sorting instrumentation and provides some insight into emerging technologies in this important area.

Cell Sorting Versus Cell Separation

A number of excellent texts have been published that exhaustively cover the physical principles that govern cell sorting (Melamed et al., 1990; Shapiro, 1995; Vandilla et al., 1985). The purpose of this chapter is not to duplicate those texts, but rather to provide a concise, comprehensive review of cell-sorting technology for those not already familiar with the field. It will also serve as a reference for the discussion of the emerging high-speed cell sorting, rare-event detection, and MEMS cell-handling technologies that appear in the subsequent chapters.

Unlike bulk cell separation techniques such as immuno-panning or magnetic column separation, flow-cytometry-based cell-sorting instruments measure, classify, and then sort individual cells or particles serially at rates of several thousand cells per second. This rapid "one-by-one" processing of single cells has made flow cytometry a unique and valuable tool for extracting highly pure subpopulations of cells from otherwise heterogeneous cell suspensions. Cells targeted for sorting are usually labeled in some manner with a fluorescent material. The fluorescent probes bound to a cell emit fluorescent light as the cell passes through a tightly focused, high-intensity light beam (usually a laser). A computer records emission intensities for each cell. These data are then used to classify each cell for specific sorting operations. Flow-cytometry-based cell sorting has been successfully applied to hundreds of cell types, cell constituents, and microorganisms as well as to many types of inorganic particles of comparable size.

Types of Cell Sorters

There are two basic types of cell sorters in wide use today: the *droplet cell sorter* and the *fluid-switching cell sorter*. The droplet cell sorter, first reported by Fulwyler (1965), utilizes microdroplets as "containers" to transport selected cells to a collection vessel. The microdroplets are formed by coupling ultrasonic energy to a jetting stream. Droplets containing cells selected for sorting are then electrostatically steered to the desired location, according to a technique first developed by Sweet (1965) for high-speed data recording. A droplet cell sorter can process selected cells at rates of tens of thousands of cells per second, limited primarily by the frequency of droplet generation and the time required for illumination. The weakness of a droplet sorter lies in the stability of the droplet generation system, which can be significantly disrupted by even temporary obstructions in the orifice of the droplet generator. Droplet sorters may also produce aerosols that can present intolerable biohazards when working with pathogenic material (Schmid et al., 1997; Merrill, 1981).

The second type of flow-cytometry-based cell sorter is the fluid-switching cell sorter. Most fluid-switching cell sorters utilize a piezoelectric device to drive a mechanical system that diverts a segment of the flowing sample stream into a collection vessel. These systems were also developed in the late 1970s and early 1980s (Duhnen et al., 1983; Goehde and Shumann, 1987), and several excellent systems are commercially available today. Because they are closed systems, fluid-switching sorters fare better in both their stability of operation and their biosafety characteristics. They also typically employ flow channels with a greater cross-sectional area, which makes them less susceptible to blockage. Compared to droplet cell sorters, fluid-switching cell sorters have a lower maximum cell-sorting rate due to the cycle time of the mechanical system used to divert the sample stream. This cycle time, the time between initial sample diversion and restoration of stable nonsorted flow, is typically significantly greater than the period of a droplet generator on a droplet cell sorter. This longer cycle time limits fluid-switching cell sorters to processing rates of several hundred cells per second. For the same reason, the stream segment switched by a fluid cell sorter is usually at least 10 times the volume of a single microdrop from a droplet generator. This results in a correspondingly lower concentration of cells in the fluid-switching sorter collection vessel as compared to a droplet cell sorter collection vessel.

Not widely used but worthy of note is the *photodamage* technique for cell sorting, often referred to as the cell "zapper." In this case a gated, high-energy laser pulse is used to render the cells not selected for sorting nonviable by inflicting photodamage to DNA (Herweijer and Stokdijk, 1988). Photosensitivity may be intrinsically or selectively induced by adding an appropriate DNA-specific fluorescent probe. This technique was recently described by Los Alamos National Laboratory (Martin et al., 1998) as a means for high-speed chromosome selection. The photodamage method has a much faster cycle time (<1 μs) than the period of a droplet generator. The weakness of this system lies in the expense of the gated pulsed laser system and in the need to calibrate the "death pulse" for the biological target. It also leaves a significant amount of cellular debris in the sorted fraction. At this time photodamage cell sorters are not commercially available.

DROPLET CELL SORTERS

Brief History

A schematic of a typical laser-based droplet cell sorter appears in Figure 1.1. The forerunner of this laser-based cell sorter was a Coulter volume instrument described by Fulwyler (1965), who elegantly combined fundamental flow cytometry with a rapid graph recorder technology developed by Sweet (1965) to produce an instrument that could sort cells based on variations in their volume. Like modern droplet sorters, Fulwyler's sorter measured, classified, and then sorted individual cells by electrostatically steering microdroplets. Bonner et al. (1972) reported on a droplet sorting device that had been constructed in the Herzenberg laboratory and that utilized a modified Crosland-Taylor (1953) coaxial flow system, a 300-mW laser, and fluorescence measurement technology. This instrument incorporated all the major components of today's modern droplet cell sorters. The Stanford Fluorescence Activated

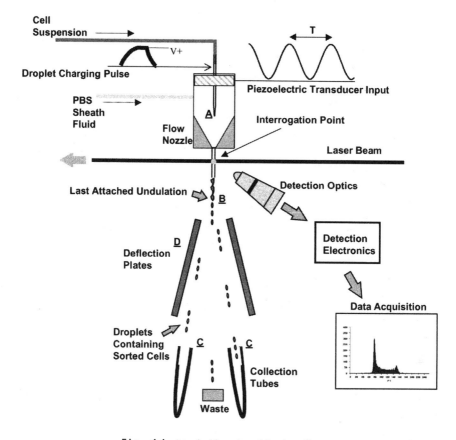

Fig. 1.1. A typical laser-based droplet cell sorter.

Cell Sorter (FACS II) described by Herzenberg et al. (1976) clearly demonstrated the function and utility of droplet cell-sorting technology. Since the FACS II there have been a multitude of technological advancements in optics, electronics, and fluid handling; however, modern droplet cell sorters continue to be built around the fundamental designs developed for these early instruments. It is interesting to note and somewhat ironic that the very first droplet cell sorter provided Coulter volume measurement, yet today, no commercial supplier of droplet cell sorters offers this valuable measurement capability.

Components of Droplet Sorters

Figure 1.1 shows the basic components of a droplet cell sorter. The fluidic design is based on a variation of the Crosland-Taylor coaxial flow system. Figure 1.2 shows a cross section of this coaxial system. The outer stream, or sheath stream, usually consists of phosphate-buffered saline. The inner stream, or core stream, consists of the single-cell suspension to be sorted. The diameter of the core stream is determined by the effective pressure differential between the sheath and the core. The conical geometry of the nozzle causes the sheath stream to undergo acceleration as it converges toward the exit orifice. The single-file stream of cells is introduced by the needle into the center of the sheath flow at the base of this inverted cone (point A in Fig. 1.1). At this point the sheath velocity is low due to the greater cross section of the channel. As flow continues down toward the exit orifice, the converging sheath focuses (narrows) the sample stream, thus accelerating the cells. This acceleration helps to separate the cells spatially along the axis of flow, and the converging sheath confines the cells to a narrow, predictable path at the center of the stream. As long as nonturbulent flow is maintained, the cells will pass through the laser interrogation point at close to constant velocity, reliably positioned near the center of the flow axis. This system of hydrodynamic focusing and coaxial flow serves three primary purposes. First, it creates spatial separation among the individual cells. Second, it aligns the cells at the center of the sheath stream's cross section. Third, it causes the cells to traverse the distance between the measurement and droplet break-off points at very close to a constant velocity, an absolutely essential requirement for accurate cell sorting.

The vehicles that transport the cells to their respective collection vessels are the microdroplets, which ultimately break away from the stream at point B (Fig. 1.1). The droplet formation is based on principles first described by Lord Rayleigh (1879). Stable droplet generation is accomplished by coupling acoustic energy to the jetting stream. Typically, a piezoelectric device, driven by a sinusoidal signal, is used to transfer single-frequency acoustic waves to the exit nozzle tip, where they cause low-level modulation of the stream velocity. This velocity modulation induces periodic variations in the stream diameter. Another approach, often used for sorting large particles (Harkins and Galbraith, 1987), is to actually modulate the stream pressure, and thus its velocity, directly. Either method results in stream undulations that become more defined as the distance from the exit orifice increases. Eventually, the undula-

A Crossection of the Coaxial Flow System

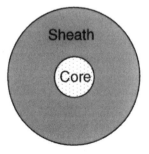

Fig. 1.2. Cross section of the stream for a typical droplet sorter. The ratio of the stream diameter to the core diameter is usually greater than 5 : 1.

Microwell Plate Sorting

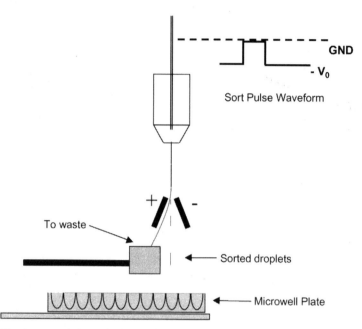

Fig. 1.3A. An approach for sorting into a microwell plate. The waste stream is deflected and the sorted stream drops into the plate. The plate is moved by a computer-controlled X–Y stage.

tions produce uniformly sized microdroplets at a fixed distance from the exit orifice. A stable droplet generator, combined with the previously described flow system, will make the transit time, or delay, between the time a particular cell is measured in the laser beam and the time it arrives at the last attached stream undulation a nearly constant value.

A desired cell is sorted by waiting for the cell to arrive in the last attached undulation and then applying a charge of duration T (the period of the transducer drive signal), synchronized with the droplet formation. If a positive charge is applied, then the undulation, already depleted of electrons, will carry a positive charge on its surface after it has broken free. The charge will reside on the surface of the droplet and will not affect the cell encased in the droplet. The now-free charged droplet then passes through an electric field produced by the potential difference between the deflection plates D (Fig. 1.1). The resulting electromagnetic force steers the droplet along a predictable trajectory based on the geometry of the plates, the value of the potential difference, and the amplitude of the charge applied to the droplet. Care must be taken to eliminate air currents and any other sources of electromagnetic fields in the collection area to ensure proper droplet trajectory. Cells are usually collected into tubes

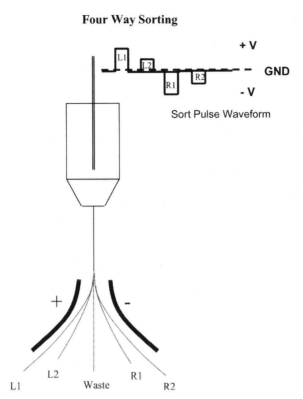

Fig. 1.3B. Four-way sorting configuration. Cell trajectory is controlled by adjusting the amplitude of the charge placed on the stream.

containing appropriate growth media, at collection points C (Fig. 1.1). Unsorted droplets receive no charge and pass straight through to waste or to a collection vessel.

There are several variations on stream deflection based on the collection scheme for the sorted fraction. Figure 1.3A shows a typical orientation for sorting directly into microwell plates. This is the inverse of the traditional bidirectional arrangement previously described. The stream is continuously charged until a droplet is to be sorted. The charge is then removed and the stream returned to the ground potential, causing the sorted microdroplets to fall straight down into the well of the plate. The plate rests on an $X–Y$ stage and indexes its position under computer control. This arrangement is used routinely for selecting cells for cloning and for isolating cells for polymerase chain reaction (PCR) amplification. This method requires a very accurate sorted-droplet trajectory since the diameter of the target in a conical bottom microwell plate can be as small as 1 mm in some cases. Figure 1.3B shows a four-way sorting configuration where the two streams formed on each side are produced by discrete variations in the sort charge amplitude. The advantage of this arrangement is the increased number of sorted fractions that can be collected.

The process of droplet cell sorting can then be summarized by the following sequence of operations: (1) measure the cell; (2) classify the cell as to its sort status;

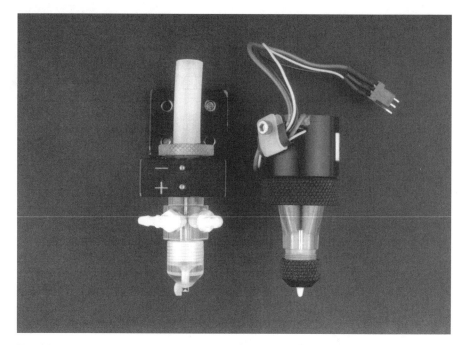

Fig. 1.4. Droplet generator assemblies from two commercial cell sorters: left, Beckman-Coulter instrument; right, Cytomation instrument.

(3) wait for the cell to arrive at the last undulation; (4) properly charge the undulation to accomplish the sort operation; and (5) deflect and then collect the sorted droplets.

Droplet Generation

Proper design and maintenance of the droplet generator are essential for reliable performance of a droplet cell sorter. Droplet generators often consist of a flow cell or nozzle attached in some manner to a piezoelectric device. The droplet generator assemblies for two commercial cell sorters are shown in Figure 1.4. Figure 1.4A (left) shows the generator used on the EPICS Altra™ (Beckman-Coulter) and Figure 1.4B (right) shows the droplet generator used on the MoFlo MLS 2000 (Cytomation). Both generators incorporate a piezoelectric transducer located above their respective flow cell assemblies. For any droplet generator, nozzle diameter, sheath pressure, and droplet generation frequency must all be properly matched to ensure stable droplet generation. For convenience, the relationship and codependencies of these variables are described briefly below. More detailed derivations and discussion of the properties of droplet generators appear in Chapter 2.

As previously discussed, stream undulations are initiated by introducing low-amplitude modulation of the stream velocity. This is accomplished by either coupling acoustic energy to the nozzle tip or directly modulating sheath pressure. The wavelength of these undulations is given by:

Droplet Cell Sorters

$$\lambda = \frac{v}{f}$$

where
λ = undulation wavelength
v = stream velocity
f = modulation frequency

The hydrodynamic properties of the system govern the conditions necessary for stable droplet formation. The details of the fluid dynamics that govern this system are beyond the scope of this text, but the topic is covered in detail elsewhere (Kachel et al., 1977, 1990). In short, droplets form due to forces arising from the surface tension of the stream and perturbations induced by the generator. The range of droplet generation frequencies (Δf) that will result in stable droplet formation is constrained by the diameter of the stream. Kachel et al. (1990) derive that stable droplet break-off will occur only if λ exceeds the circumference of the stream and thus the circumference of the exit orifice. The minimum value for λ (λ_{min}) in terms of the exit orifice diameter d is given by

$$\lambda_{min} = \pi d$$

Kachel et al. proceed to derive the optimal value for λ (λ_{opt}) to be

$$\lambda_{opt} = 4.5d$$

It follows from the previous equations that the maximum and optimal droplet generation frequencies for a droplet sorter are given by

$$f_{max} = \frac{v}{\pi d} \qquad f_{opt} = \frac{v}{4.5d}$$

The optimal droplet generation frequency for a given nozzle diameter is therefore dictated by the jetting stream velocity. In Chapter 2 van den Engh develops the relationship of velocity to sheath pressure, which yields

$$v = \left(\frac{2P}{\rho}\right)^{1/2} \qquad f_{opt} = \frac{(2P/\rho)^{1/2}}{4.5d}$$

where
v = jet velocity
d = jet diameter
ρ = density
P = pressure drop across the exit orifice

For nozzle diameters and pressures typical for cell sorters, the optimal droplet generation frequency is therefore a function of the nozzle diameter and the pressure drop across the exit orifice of the nozzle.

The coaxial flow system employed by droplet cell sorters requires that nonturbulent flow be maintained. This requirement can limit the maximum jet velocity and thereby the maximum droplet generation frequency. For a parabolic velocity profile, this critical operating point can be estimated from the Reynolds number (Re) for the orifice–sheath fluid combination in use. For a flowing stream of phosphate-buffered saline (PBS) with a parabolic velocity profile, the probability that laminar flow may break down is significant for Re > 2300 (Kachel et al., 1990). The equation for the Reynolds number is given as

$$Re = \frac{d\rho v}{\eta}$$

where
- d = nozzle diameter (cm)
- ρ = fluid density (g/cm^3)
- v = average fluid velocity (cm/s)
- η = fluid viscosity (g/cm s)

The short path length in the flow nozzle may prevent full parabolic flow from developing (Lindmo et al., 1990), and therefore nonturbulent flow may be sustainable for higher Re values. However, the impact of temperature on viscosity and ultimately on the maximum allowable jet velocity is important for both large-particle and high-speed cell-sorting applications.

Droplet Sorter Flow Cells

A key component of the droplet generator is the nozzle that produces the jetting stream. It must be large enough to accommodate the largest particles in the sorting suspension, it must maintain the stream position for laser intersection, and it must effectively transfer energy from the generator's transducer to the stream. The diameter of the orifice in a droplet cell sorter nozzle typically ranges 50–150 µm. Cells entering the undulating stream will introduce random perturbations in the stream that may temporarily alter the droplet break-off from the location established without flowing cells. The perturbations become more significant as the cell diameters approach the diameter of the stream (i.e., the orifice diameter). This manifests on a droplet sorter as instability, or fanning of the sorted droplet stream. Observations reported by Bonner et al. (1972) indicate that a 5 : 1 ratio between the sorting orifice diameter and the average cell diameter results in stable droplet generation.

Shown in Figure 1.5 are several different nozzle assemblies used in commercial droplet sorters (5A, Beckman-Coulter; 5B, Becton Dickinson; 5C, Cytomation-Coulter). There are many variations in shape and fabrication materials; however, one major difference among these nozzles is whether the laser intersection occurs in a quartz channel or in the jetting stream in air. The jet-in-air nozzles are relatively easy to construct and offer the simplest transition to laminar flow. Optically, the jetting stream acts as a short-focal-length cylindrical lens. In addition, undulations superimposed on the stream by the droplet generator add to the refractive effect. Flow cells that place the laser intersection in a quartz channel, such as the one shown in Figure 1.5A, offer improved optical performance. The quartz flow cell shown in Figure 1.5A (left) is based on a design described by Watson (1989) and further developed by Larry Arnold (personal communication) at the University of North Carolina. It con-

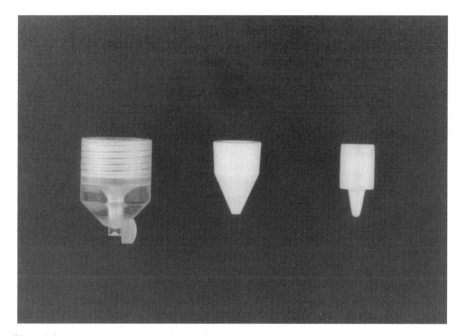

Fig. 1.5. Examples of flow cells used on three commercial sorting systems (left to right): (A) Beckman-Coulter; (B) Becton-Dickinson; (C) Cytomation. Laser intersection for A is in quartz. Laser intersection for B and C is in the jetting stream.

sists of a 1-mm-long, 100-μm-square channel with a lens bonded to one face of the quartz. The integral lens couples light from the laser intersection point to the fluorescence collection system, increasing the effective numerical aperture, thus improving the fluorescence sensitivity of the instrument. This type of flow cell has made possible the use of lower-power lasers for cell sorting. The lensing effect caused by the stream, as described above for jet-in-air systems, is eliminated and the quartz provides a better index-of-refraction match, substantially reducing the scatter at the interface points. Quartz sorting flow cells have been successfully employed and can improve fluorescence collection efficiency and light scatter resolution. On the other hand, they also chip easily, are more difficult to unclog, and are more expensive to replace than their jet-in-air counterparts.

Droplet Charging

The trajectory of the sorted droplets is adjusted on most droplet cell sorters by varying the amplitude of the charging pulse delivered to the stream while maintaining a constant potential difference of 3–5 kV between the deflection plates. Left and right sorting is accomplished by switching the polarity of the charge pulse. Multiple sorted streams can be obtained by establishing a unique charge amplitude–polarity combination for each sorted fraction. It is essential that the charge pulse be applied in phase with droplet generation. The correct phase relationship will produce the maximum droplet deflection, and most droplet sorters provide a method for making this phase adjustment. Figure 1.6 shows the waveform obtained from a commercial cell sorter for a charge pulse applied to sort a single droplet. The pulse is shaped in this manner to correct for charge coupling that occurs between the last undulation and the preceding free charged droplet (Lindmo et al., 1990). Other electromagnetic fields or charged surfaces in the vicinity can affect the trajectory of the charged droplets. Care should be taken to choose collection vessels that will not accumulate a surface charge or to properly ground them.

The arrival times of cells at the laser interrogation point (Fig. 1.1) are expected to be random, independent events and may therefore be described by a Poisson probability distribution. It follows, then, that the relative phase difference between cell arrival times and the synchronous formation of droplets will be uniformly distributed. This means that some droplets will have cells positioned very near the boundaries separating them from adjacent droplets. For cells near the edge of a droplet period when measured, uncertainty regarding the time of flight to the break-off point results in uncertainty as to which of two adjacent free droplets will actually contain the cell. In this case, to achieve a high probability of sorting the cell it is necessary to sort both of the droplets in question, that is, both the droplet that corresponds to the detection time and the adjacent droplet closest (either before or after) to the detection time point. This result has been achieved in some instruments by sorting a three-droplet set. However, sorting three droplets is unnecessary if the sort decision system is designed to use the phase of the cell's arrival time to select the correct droplet pair. Sorting the correct two-droplet pair as described will increase the probability that the identified cell actually makes it into the collection tube. Unfortunately, it also increases the time during which a contaminating cell must be avoided. For this reason, single-droplet sorts will always provide the highest sorting rates.

Cell Queueing and Coincidence

The undulating stream can be thought of as a short queue that advances with the formation of each droplet. As each droplet advances, the sort decision system must

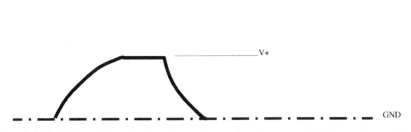

Fig. 1.6. Typical waveform for the voltage pulse used to charge a sorted drop or frame of sorted droplets. This pulse shape corrects for the coupling charge between adjacent droplets.

classify or define the state of the newest queue entry. If we define a contaminating cell as either a cell identified as unselected or a cell detected but unidentified, then the states of interest for a simple one-direction sort would be (1) empty; (2) containing one or more selected cells; (3) containing one or more contaminating cells; or (4) containing one or more selected cells and one or more contaminating cells. When an undulation advances down the queue to the position of last attached undulation, the sorting system performs the prescribed sorting operation for that state. State 4, as described above, is referred to as coincidence. A droplet with state 4 cannot be sorted without also including a contaminating cell. The states become more complicated if we introduce the recognition of the phase location of the cells in the droplet period. States 2, 3, and 4 might be divided into three substates indicating whether the corresponding cells were near the front edge, near the back edge, or centered in the droplet period. In this case, the sort decision system would need to look not only at the state of last attached undulation but also at the states of the previous droplet and the next undulation before deciding on a sort operation.

Electronic Processing of a Cell Measurement

As a labeled cell traverses the focused laser beam, the instrument photodetectors produce electronic pulses proportional to the intensity of the light scatter and fluorescence emissions. Each detector in use produces its own pulse similar to that shown in Figure 1.7. In some cases, analog processing is done on the original pulse to produce secondary pulses having peak amplitudes proportional to the integral, pulse width, or logarithm of the original pulse. After amplification these pulses are used by the sorting electronics to detect the arrival of a cell and then to classify the cell as to its sorting status. Two basic parameters define the theoretical performance of a particular electronic sorting system: the detection time t_{det} and the cycle time t_{cy} of the system. As illustrated in Figure 1.7, the *detection time* starts with the initial detection of cell arrival (pulse amplitude rises above a lower threshold) and ends with the departure of the cell (pulse amplitude passes below a threshold). If another cell arrives during the detection time (as depicted in Fig. 1.7B), it will not be independently recognized. Instead, the electronics will assume that the combination of the cell emissions represent a single cell, and thus the cells will be misclassified. The *cycle time* starts at the same time as the detection time, but it continues until the electronic system is ready, or rearmed to receive the next pulse. A second cell arriving after t_{det} but before t_{cy} (as depicted in Fig. 1.7C) will be detected as present but will remain unclassified because the electronics are busy. The difference between t_{cy} and t_{det} is usually referred to as the system *dead time*.

Detection time (typically 1–2 µs on droplet sorters) is primarily determined by the width of the pulse driving the sorting system. For high-speed applications it is essential to minimize pulse width by not using integral pulses and by reducing the laser beam height to approximately one cell diameter. Cycle time (typically 4–10 µs on droplet sorters) is dictated by the electronics system design. Figure 1.7 represents a relatively simplistic example of event detection; other more complex schemes have

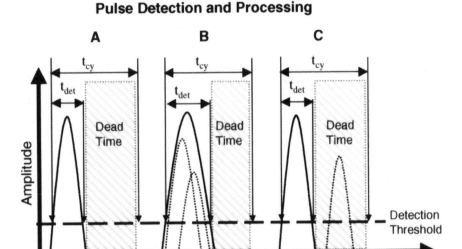

Fig. 1.7. Three examples of pulses detected and then processed by the cell-sorter electronic system. (A) Pulse arrives and is processed with no coincidence. (B) Two pulses arrive (dotted lines) within t_{det}. The pulses cannot be resolved by the electronics and are treated as one longer pulse. This will result in the misclassification of these cells. (C) Another pulse arrives during the dead time (after t_{det} but within t_{cy}). The first pulse is classified. The second pulse is not classified but the coincidence detection system records its presence and can abort the sort decision if necessary. The condition in panel C impacts efficiency and recovery. The condition in panel B can impact purity.

been implemented (van den Engh and Stokodijk, 1989; Dean et al., 1985; Cupp et al., 1984). Given a well-designed, stable droplet generation and sorting system, the performance of a droplet cell sorter will ultimately be limited by its effective t_{det} and t_{cy} values.

Defining Cell Sorter Performance

Cell sorter performance is usually defined in terms of *efficiency, purity, recovery,* and *yield.* These four terms are defined below in terms of the following variables:

- x_{tot} Total number of cells that pass through the laser(s)
- x_{in} Number of cells that should satisfy the sort criteria and that pass through the laser(s)
- x_{srt} Number of cells that the instrument actually attempts to sort
- x_{tub} Number of cells that satisfy the sort criteria and that can be found in the collection tube
- x_{con} Number of contaminating cells or cells which *do not* satisfy the sort criteria and that can be found in the collection tube

Sorting performance terms are defined as:

Sort efficiency (E) in %—sorts attempted as a percent of targeted cells in sample:

$$E = \frac{x_{srt}}{x_{in}} \times 100$$

Sort purity (P) in %—targeted cells collected as a percent of total cells collected:

$$P = \frac{x_{tub}}{x_{tub} + x_{con}} \times 100$$

Sort recovery (R) in %—targeted cells collected as a percent of sorts attempted:

$$R = \frac{x_{tub}}{x_{srt}} \times 100$$

Sort yield (Y) in %—targeted cells collected as a percent of targeted cells in sample:

$$Y = \frac{x_{tub}}{x_{in}} \times 100$$

Predicting Cell Sorter Performance

Accurate predictions of cell sorter performance can be problematic due to the many independent variables involved. Assume a stable droplet generator is properly phased with the sort charging pulses, a fluidic system delivers cells with Poisson-distributed arrival times, and a sample can be reliably classified by fluorescence emission and neglect targeted cells missed due to arrival within t_{det} of another cell. Then the sort efficiency, rate, and maximum purity can be estimated using the following expressions, which are developed in detail by Lindmo et al. (1990) and are presented here as a convenient reference.

Sort efficiency (%):

$$E = \exp[-m(1-a)nT] \times 100$$

Sorting rate (targeted cells per second):

$$R_s = ma \exp(-m(1-a)nT)$$

Maximum purity (%):

$$P_{max} = \frac{1 + amt_{cy}}{1 + mt_{cy}} \times 100$$

where
- m = average analysis rate (cells/s)
- a = x_{in}/x_{tot}
- n = number of droplets sorted
- T = period of droplet generator
- t_{cy} = cycle time

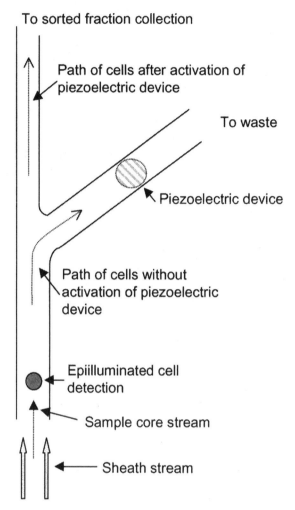

Fig. 1.8A. Representation of the fluidic sorting system that has been implemented in the Partec flow cytometry instruments. The core sample stream and the majority of the sheath normally flow to waste. When a cell targeted for sorting is detected, the piezoelectric device is activated, diverting the core sample stream to sorted fraction collection.

Various numerical and modeling approaches have been used to better predict cell-sorting performance, especially at high speeds (Durack and Durack, 1998; Hoffman and Kouck, 1998; Keij et al., 1991; Rosenblatt et al., 1997). Issues regarding the prediction of these performance parameters for high speed sorting are provided in Chapter 2, and Chapter 3 deals with them in relation to sorting very rare cells.

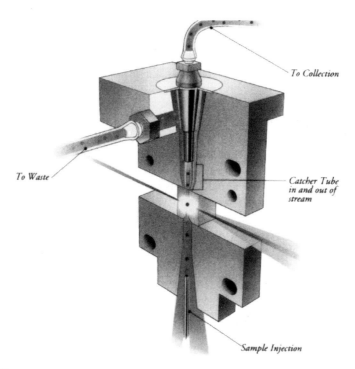

Fig. 1.8B. Fluidic sorting system used in the FACSort™ manufactured by Becton Dickinson Immunocytometry Systems. The catcher tube is extended by a piezoelectric device to collect cells identified for sorting. The tube can cycle up to 500 times per second. *Courtesy of BDIS.*

FLUID-SWITCHING CELL SORTERS

Fluid-switching cell sorters have been commercially available in one form or another for nearly two decades (Duhnen et al., 1983; Goehde and Shumann, 1987). The performance of a fluid-switching cell sorter can be described in the same manner as the droplet cell sorter by substituting the cycle time of the piezoelectric sorting system for the droplet generator period (generally 10–100 times longer). While sorting and analysis rates are much lower than what is possible with droplet cell sorters, fluid-switching cell sorters perform quite reliably and provide a measure of biohazard containment not possible for droplet cell sorters (unless they are enclosed in a biohazard safety cabinet). Commercial bench-top instruments are available that offer a "push-button" approach to cell sorting. Figure 1.8A shows a diagram for the fluidic cell sorter designed by Goehde and Schumann (1987) and implemented in the Partec fluid-switching sorting system. When a desired cell approaches the bifurcation, a valve driven by a piezoelectric device blocks the waste path, temporarily diverting the flow to the sorted cell collection path. Figure 1.8B shows the system implemented by Becton Dickinson Immunocytometry Systems in the FACSort product. In this case a cell-catcher tube is pushed down into the flowing stream by a piezoelectric device to capture cells targeted for sorting. Stream velocities are much lower in these systems, on the order of 2–5 m/s. This coupled with the duty cycle for the sorting device limits cell-sorting rate to less than 1000 cells/s. While not suitable for high-speed sorting applications, these fluid-switching cell sorters are very useful and economical devices for sorting a limited number of cells at a moderate rate.

AN EMERGING TOOL

The invention of the cell sorter is nearly three decades past, yet aspects of this technology are still emerging as important tools for the research laboratory. Today's research community continues to increase its collective expectations for cell-sorting technology, placing heavy demands on the existing instrument systems. There are pressing clinical processing needs that require high-speed, sterile sorting systems for sorting sperm cells and hematopoietic stem cells, as described in Chapter 4. There are demands for higher and higher throughput to meet the challenges of isolating rare cells, as discussed in Chapter 3. There are needs for lower detection limits and broader multicolor capabilities. The technical issues and opportunities relevant to addressing many of these needs are discussed in the following four chapters. One of the newest and most rapidly advancing areas is the cytometry-related MEMS technologies, described in Chapter 5. As more research moves into the microlaboratory environment, there will be a strong demand for cell processing microcytometers to operate there. The "flow cytometer on a chip," predicted by many in the cytometry community, is now an important *emerging tool* almost certain to arrive in our laboratories in the coming decade.

REFERENCES

Bonner WA, Hulett HR, Sweet RG, Herzenberg LA (1972): Fluorescence activated cell sorting. Rev Sci Instrum 43:404–409.

Crosland-Taylor PJ (1953): A device for counting small particles suspended in a fluid through a tube. Nature 171:37–38.

Cupp JE, Leary JA, Cernichiari E, Wood JC, Doherty RA (1984): Rare-event analysis methods for detection of fetal red blood cells in maternal blood. Cytometry 5:138–144.

Dean PN, Merrill T, Pinkel D, Van Dilla M, Gray JW (1985): The LLNL high speed cell sorter: Design features, operational characteristics, and biological utility. Cytometry 6:290–301.

Duhnen J, Stegemann J, Wiezorek C, Mertens H (1983): A new fluid switching flow sorter. Histochemistry 77:117–121.

Durack G, Durack D (1998): Software simulation of cell sorting to predict instrument performance for high speed cell sorting. Cytometry Suppl 9:115.

Fulwyler MJ (1965): Electronic separation of biological cells by volume. Science 150:910–911.

Goehde H, Schumann J (1987): Fluidic cell sorter. European Pat. EP 0177.718.

Harkins KR, Galbraith DW (1987): Factors governing the flow cytometric analysis and sorting of large biological particles. Cytometry 8:60–70.

Herweijer H, Stokdijk WV (1988): High speed photodamage cell selection using bromodeoxyuridine/Hoechst 33342 photosensitized cell killing. Cytometry 9:143–149.

References

Herzenberg LA, Sweet RG, Herzenberg LA (1976): Fluorescence activated cell sorting. Sci Am 234(3):108–117.

Hoffman RA, Kouck DW (1998): High speed sorting efficiency and recovery: Theory and experiment. Cytometry Suppl 9:142.

Kachel V, Fellner-Feldegg H, Menke E (1990): Hydrodynamic properties. In Melamed MR, Lindmo T, Mendelsohn ML (eds). *Flow Cytometry and Sorting,* 2nd ed. New York: Wiley-Liss, pp 27–44.

Kachel V, Kordwig E, Glossner E (1977): Uniform lateral orientation of flat particles in flow through systems caused by flow forces. J Histochem Cytochem 25:774–780.

Keij KL, van Rotterdam A, Groenewegen AC, Stokdijk W, Visser JW (1991): Coincidence in high-speed flow cytometry: Models and measurements. Cytometry 12:398–404.

Lindmo T, Peters DC, Sweet RG (1990): Flow sorters for biological cells. In Melamed MR, Lindmo T, Mendelsohn ML (eds). *Flow Cytometry and Sorting,* 2nd ed. New York: Wiley-Liss, pp 145–170.

Martin JC, Roslaniec MC, Bell-Prince CS, Longmire JL, Jett JH, Cram SL (1998): A photodamage chromosome selector. Cytometry Suppl 9:CT104.

Masuda S, Washizu M, Nanba T (1989): Novel method of cell fusion in field constriction area in fluid integrated circuit. IEEE Trans Ind Applic 25:732–737.

Melamed MR, Lindmo T, Mendelsohn ML (1990): *Flow Cytometry and Sorting,* 2nd ed. New York: Wiley-Liss.

Merrill JT (1981): Evaluation of selected aerosol-control measures on flow sorters. Cytometry 1:342–345.

Lord Rayleigh (1879): On the capillary phenomena of jets. Proc R Soc Lond 29:71–97.

Rosenblatt JI, Hokanson JA, McLaughlin SR, Leary JF (1997): Theoretical basis for sampling statistics useful for detecting and isolating rare cells using flow cytometry and cell sorting. Cytometry 27:233–238.

Schmid I, Nicholson JA, Giorgi JV, Janossy G, Kunkl A, Lopez PA, Perfetto S, Seamer LC, Dean PN (1997): Biosafety guidelines for sorting unfixed cells. Cytometry 28:99–117.

Shapiro HM (1995): *Practical Flow Cytometry,* 3rd ed. New York: Wiley-Liss.

Sweet RR (1965): High frequency recording with electrostatically deflected ink jets. Rev Sci Instrum 36:131–136.

van den Engh G, Stokdijk W (1989): Parallel processing data acquisition system for multi-laser flow cytometry and cell sorting. Cytometry 10:282–293.

Vandilla MA, Dean PN, Laerum OD, Melamed MR (1985): *Flow Cytometry Instrumentation and Data Analysis.* New York: Academic Press.

Washizu M, Nanba T, Masuda S (1990): Handling biological cells using a fluid integrated circuit. IEEE Trans Ind Applic 26:352–358.

Watson J (1989): Flow cytometry chamber with 4 pi light collection suitable for epifluorescence microscopes. Cytometry 10:681–688.

2

High-Speed Cell Sorting

Ger van den Engh
University of Washington, Seattle, Washington

In a cell sorter, cells are injected into a stream of liquid that intersects one or more laser beams. The light signals from the cells identify the different cell types. A vibration causes the stream to separate into discrete drops. An electrical charge is applied to those drops that contain cells of interest, so that these drops can be deflected in an electrical field. The aim is to classify and separate cells accurately, without errors, and as fast as possible.

Speed and precision often pose conflicting requirements. At a low speed, when cells arrive infrequently, it is relatively straightforward to analyze the fluorescence pulses and to charge the few drops that contain cells of interest. The cells dwell in the light spot sufficiently long that their intensities can be quantified accurately without interference from other fluorescent particles. Most drops are empty or contain a single cell and can be selected without ambiguity. However, slow speeds limit the sorter's utility; many experiments require more cells than can be delivered in a single session. For maximum efficiency, we would like to inject as many cells as possible into the instrument, thereby creating conditions under which there is a significant chance of finding more than one cell in the illumination spot or in a deflected droplet. The following sections will show that the highest sorting speeds are reached when every drop, on average, contains a particle. Consequently, high-speed sorting involves more than simply increasing the speed of the instrument's components. High-speed sorting refers to conditions under which all components of a cell sorter are operated

---

Emerging Tools for Single-Cell Analysis, Edited by Gary Durack and J. Paul Robinson.
ISBN 0-471-31575-3 Copyright © 2000 Wiley-Liss, Inc.

near their performance limits, at highest possible efficiency, without loss of yield or precision. High-throughput conditions present novel problems not encountered at slower rates. Unless adequate measures are taken, these complications can lead to impurities and recovery losses that will rapidly negate the advantages of the higher processing speeds of the sorter components.

The perfect cell sorter, one that satisfies all experiments, does not exist. Instruments can be built only to be good enough for specified conditions. We will consider an instrument satisfactory if it meets the following criteria. The measurement errors are small compared to the natural variation in our biological samples. Our ability to differentiate between different cell types is then limited by the quality of cell labeling, not by measurement error. The processing speed is determined by the speed at which we can move cells in and out of the illumination spot, not by the speed of our analysis circuits. Furthermore, the startup and shutdown of the instrument should not significantly add to the length of the experiment. The instrument should be easy to operate and provide clear information about the status of the experiment and the quality of its alignment. Experimental results should be well documented, readily stored, and easily retrieved and compared.

The specific problems that are associated with high processing rates lead to performance specifications for high-throughput, high-duty-cycle sorting. We will see that careful engineering can achieve acceptable compromises between precision and speed. Using current technology, sorters that are capable of reliably analyzing and sorting cells at rates of tens of thousands of cells per second are feasible. My group has designed and built several generations of high-speed sorters. These instruments have been used to sort 40,000 bone marrow cells per second with high purity (Sasaki et al., 1995). These instruments have proven their practical value in sorting chromosomes for the human gene library project of the Department of Energy (DOE) (Van Dilla et al., 1990). Some of this technology has been adopted by commercial manufacturers and is now commercially available. Cytomation's MoFlo is based on the original DOE design. Systemix uses a modified version for the sorting of clinical bone marrow grafts (Sasaki et al., 1995). Cytopeia is preparing a versatile machine for general laboratory use that incorporates new features developed at the University of Washington. The following presents some of the considerations that went into the design of these machines.

DEFINITIONS

Before considering the technology behind high-speed sorting, some terms must be defined. In a sorting experiment, cells are introduced into the machine and processed. The sorted cell fraction is collected. An *event* is any object, be it a cell or a particle, that is registered by the instrument's detectors and processed by its electronic circuitry. The *processing rate* denotes the number of events processed per unit time. The processing rate determines the overall speed. The more events processed per second, the more cells can be collected in a given time interval. Depending on our purpose, we may recognize two *processing rates:* (1) the *peak rate,* or the rate obtained once the instrument is running, and (2) the *effective rate,* or the rate over the whole experiment, including instrument setup time and interruptions for nozzle clogs and realignment. There is often a large difference between these two rates. The engineer who designed the instrument is likely to specify the peak processing rate. The practical biologist is most concerned about the overall effective rate.

The *yield* of an experiment is the amount of material that the investigator takes back to the laboratory. The *recovery* compares the number of cells of a particular type that are collected to the number of these cells that went into the experiment. We may optimize for yield or recovery but can never optimize for both.

Instruments take a certain time to perform an operation. Tasks offered before the machine is finished are ignored or, worse, may corrupt the ongoing process. The time it takes to perform a task and to prepare the machine for the next operation is the *cycle time.* The *dead time* is the portion of the cycle time during which the machine cannot accept a new task because it is still busy with the previous one. For instance, if it takes a circuit 5 μs to analyze a 3–μs-wide pulse, the cycle time is 5 μs with a dead time of 2 μs. Events arriving within one cycle time of each other are said to be coincident. An instrument that is accepting new tasks as soon as a current task is done is said to work at *full duty cycle.* A 50% duty cycle means that an instrument is idle half of the time.

THE WELL-TEMPERED SORTER

Well-built sorters combine a high peak rate with reliability and versatility. In such a machine, the maximum processing speed depends on the properties of the biological sample, that is, cell size, cell fragility, and cell density. The instrument is self-stabilizing and easy to align. Setup time and downtime are relatively short, so that the effective sort rate does not differ much from the theoretical peak performance. The design of the instrument minimizes the chance of failure. Early warning signals indicate when the settings drift too close to the edges of their safe operating zones. Instrument malfunction is detected immediately and normal operation is interrupted automatically. Problems are easy to diagnose and are corrected rapidly. These are the characteristics of a robust design. In cell sorter development, robustness has received little attention. Only recently has the industry started to optimize performance from the perspective of reliability. Huge improvements in sorter design are still possible and are essential if this technology is to be used in clinical applications.

POISSON STATISTICS

Maximum analysis and sort rates are ultimately determined by the frequency of coincident events, either at the point of measurement or in the drops. Therefore, Poisson statistics stand central to the discussion of instrument efficiency. In the case of uncorrelated, stochastic events, Poisson calculates the probability of observing a specific number of events during an interval for which the average event probability is known (Taylor, 1997). If the average number of events in the interval is v, the probability of observing a particular number of μ events is

$$P_v(\mu) = e^{-\mu} \frac{\mu^v}{v!}$$

In our case, the interval of interest most often is a drop, and we need an estimate for the fraction of drops with v cells, when the drops contain μ particles on average. Figures 2.1a and 2.1b show the probability distributions for intervals (drops) with zero and one events for different mean event rates. Figure 2.1c represents the total number of events that are contained in the intervals with two or more events. It should be noted that at low mean event rates ($v < 0.1$) the relationships are simply linear. The largest fraction of intervals with one event (0.37) is observed for $\mu = 1$. In other words, the largest fraction of single events is observed at a high duty cycle, when, on average, all observation intervals contain an event. It is also worth noting that under such conditions multiple events are observed in a substantial fraction of the intervals.

The utility of these graphs is illustrated by considering the distribution of particles (cells) among the drops that are formed by a cell sorter. A convenient unit of time is the period τ of the drop-generation oscillator, which is the inverse of the oscillator's frequency, $f = 1/\tau$. The expected mean number of cells per drop is $\mu = n\tau = n/f$, where n denotes the number of cells per second. In order to obtain pure cells, we deflect only the drops that contain one cell. If a coincident particle is present, the drop will not be sorted. As long as fewer than 1 in 10 drops contain a particle, coincident events are relatively infrequent (less than 10% of drops), and coincident-event rejection will not significantly affect the overall recovery or speed. As the average drop loading approaches one cell per drop, the picture becomes more complex. Figure 2.1a shows that the fraction of empty drops (P_0) at an average occupancy of 1 is approximately 37%. The highest fraction of drops with one cell (Fig. 2.1b) is reached at an average of one cell per drop, where 37% contain one particle. The remaining fraction of drops, $1-P_0-P_1$ (26%), must contain two or more particles. Figure 2.1c plots the fraction of cells that is lost if the drops containing two or more cells are not sorted, which is equal to the total number of events minus the number of single events as a fraction of the total event number, or

$$\frac{v - P_1}{v} = 1 - \frac{P_1}{v}$$

At an average drop loading of one cell per drop, 63% of cells are lost due to coincidence.

It is useful to consider sorting purity without coincident-event rejection. This arrangement may be advantageous when inspecting rare cells that are too valuable to dispose of in the unsorted fraction. Figure 2.2 plots the purity of the sorted sample if all drops with a wanted cell are deflected, regardless of whether the drops contain a coincident cell or not. Even at a high throughput rate, the purity is remarkably high. Up to an occupancy rate of 1.6 cells per drop, more than half of the cells in the sorted sample are of the desired type. When dealing with low-incidence cells, preenrichment by high-throughput sorting without coincidence rejection, followed by a second pass at a low sorting rate, may reduce cell loss and increase overall speed.

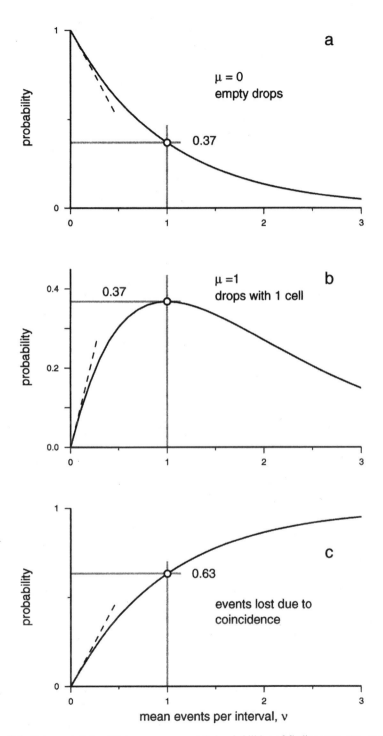

Fig. 2.1. Poisson statistics. The curves represent the probabilities of finding zero, one, or multiple events when the average number of events (ν) is known (see text).

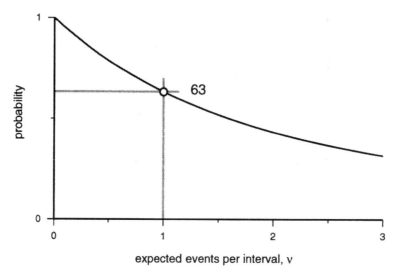

Fig. 2.2. Purity of a cell fraction when all drops that contain wanted cells, including those with multiple cells, are selected.

The Poisson formulas accurately predict overall throughput rates and sort efficiencies only if the events are randomly distributed. The expected distribution of the time intervals between uncorrelated events with a mean event rate of μ events per second is given by

$$P_{time} = \mu e^{-t/\mu}$$

By measuring the event interval distribution, we may determine whether the fluidic system delivers the cells truly randomly or whether the cells arrive in clumped packages (Visser and van den Engh, 1982). The expected interval distribution for random, uncorrelated events is shown in Figure 2.3. If measured intervals between events do not display a relationship as shown in Figure 2.3, Poisson statistics may not apply. In this case, the event spacing is not optimal, and the cell loss due to coincidence rejection will be larger than expected. A redesign of the sample introduction system may improve the event spacing.

MAXIMUM ANALYSIS RATE

Analysis precedes sorting; we cannot sort faster than we can analyze. The cells are measured when they traverse an illuminated spot. Figure 2.4 illustrates that a cell is

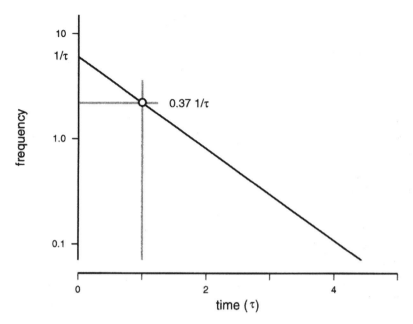

Fig. 2.3. Interval distribution of unrelated events. The mean event rate is μ events per second.

illuminated by the laser light while traveling a distance of the spot plus the cell diameter. When two events are in the illumination spot at the same time, the instrument is likely to have trouble accurately determining the properties of each cell individually. To avoid erroneous classification of cells, the incidence of double occupancy should be kept low. Coincident events, which may cause measurement errors, should be at least an order of magnitude lower than single events. Poisson again serves as a guideline (Fig. 2.1). The average beam occupancy (cycle time) should be below 0.1. Even then, a significant fraction of measurements (again about 0.1) will be affected by coincident events. In uncertain measurement conditions, the instrument should not sort. Therefore, the circuitry should detect coincident events and flag the measurement so that these events can be rejected for sorting by the downstream charge circuitry.

FLUIDICS AND OPTICS

The considerations of the maximum analysis rate lead to the following specifications for the optical and fluid path design. Cells are about 10 µm in diameter. In order to fully illuminate a cell, the laser spot must be somewhat larger. A well-collimated laser and a diffraction-limited lens with a focal length of 100 mm generate a spot about 20 µm in diameter. A jet propelled by 60 psi (1 psi = 0.07 kg/cm^2) moves at ~20 m/s, yielding a cell transit time of ~1.5 µs. Furthermore, some time is needed for the electronics to return to the baseline conditions, a minimum of 1 µs. An illumination duty cycle of 0.1 corresponds to a maximum event rate of 40,000 events/s.

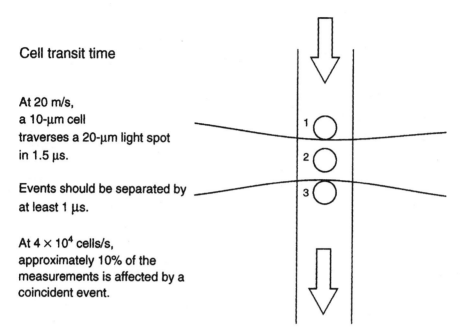

Fig. 2.4. Cell diameter and transit time.

A spot size of 20 μm and fluid pressures of 100 psi or less can be obtained with generally available technology. In a balanced system, the electronic modules must be capable of an analysis rate (and therefore a maximum sort rate) of at least a few 10,000 cells/s. A further increase in speed requires a smaller spot size or a higher jet velocity. A smaller spot size will result in partial illumination of the particle and consequently requires much more complex electronics to obtain an accurate measurement. Higher jet velocities require a significantly higher sheath pressure, complicating the fluid delivery systems. We will limit the current discussion to systems analyzing at 40,000 events/s.

The foregoing identifies the cell transit time as a rate-limiting factor. The importance of the transit time has interesting consequences for sorter design that are not generally appreciated. First, in order to achieve stable jet formation, the pressure drop across the nozzle orifice must be large. Therefore, the velocity of the jet is always considerably faster than the fluid flow in the other parts of the system. Consequently, cuvette systems, in which the cells are measured before they accelerate into the jet, are at least an order of magnitude slower than jet-in-air systems. Second, the number of photons that can be collected from a particle are, to a first approximation, proportional to the illumination time. Short transit times reduce the number of photons collected and therefore the sensitivity. This reduction in photon emission may be compensated to some extent by increasing the illumination intensity (van den Engh and Farmer, 1992). High-speed sorters therefore typically require more powerful lasers than do slow systems.

ELECTRONICS

Electronic circuits accept the signals from the photodetectors (usually photomultipliers), condition the signals, perform analog-to-digital conversion, collate the digital information for event classification, and, after checking for coincident events, issue sort commands for those cells that fall within specified sort windows. The circuits must perform these tasks without affecting the overall speed of sorting.

Experiments with high-speed sorters often search for cells that occur at a very low incidence in the starting sample. In many cases, the unique identification of such rare cells requires determination of a large number of cell characteristics. Throughout the years, the number of fluorescence signals that is considered routine has increased steadily. Recently, Herzenberg's group reported the use of 11 fluorescent tags with 3 excitation lasers for the isolation of hemopoietic progenitor cells (Baumgarth et al., 1998). Even when a limited set of parameters is sufficient to uniquely describe the cell of interest, inclusion of additional parameters may be beneficial. All measurement systems make mistakes occasionally. Erroneous measurements may be confused with a sortable event, leading to sort contamination. In rare-cell sorting, measurement artifacts may become a significant factor. By increasing the number of criteria (parameters) that define a sortable event, we can reduce the chance of an identification error leading to a sorted event. The measurement of redundant properties increases the overall accuracy of our experiments. The electronics for a high-speed sorter should be capable of processing as many parameters as we wish to measure.

High-speed sorters generate fluorescence pulses that are 1–2 µs in width, limiting the average event rate to about 40,000 events/s. The electronic modules should not further constrain this speed. Ideally, the pulse-processing electronics are capable of measuring and digitizing the events with a cycle time of 3 µs or less. The downstream digital acquisition and classification circuits must be able to deal with at least 16 parameters, or 180 ns per parameter. A well-behaved acquisition and event-processing system capable of handling asynchronous events at such speeds requires careful design (van den Engh and Stokdijk, 1989).

A simple test may be performed to determine the cycle time of sorter electronics. A periodic pulse with the duration of a typical fluorescence pulse is offered to the signal amplifiers. The sort windows are set such that all pulses result in a sorted event. Starting at a low repetition rate, the frequency of the pulse generator is gradually increased. At a low rate, the trigger rate and the sort rate will be the same as the frequency of the presented signals. At some point there will be a discrepancy between the number of pulses offered and the number of pulses processed. This frequency marks the cycle time of the instrument.

Two events arriving within one cycle time of each other may not be processed correctly and may lead to measurement errors. The two events are processed as if they represent a single occurrence. The properties reported are a hybrid of the properties of the two original particles. This artifact is known as a correlation error. Suppose we have one cell with fluorescence intensities x_1 and y_2, followed closely by another cell with properties x_2 and y_1 (Fig. 2.5). A correlation error may cause the recording of a nonexistent cell with a signature x_2, y_2. Immunologists might interpret the

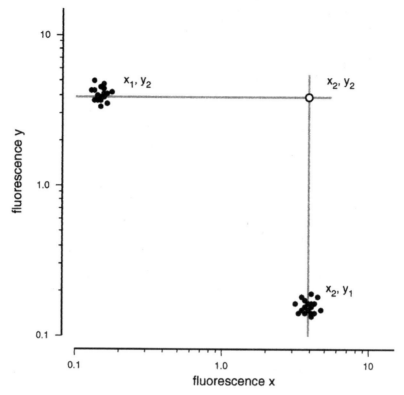

Fig. 2.5. Correlation errors create ghost events. A sample contains two cell types with fluorescent properties x_1, y_2 and x_2, y_1. Two cells that occupy the illumination spot simultaneously may generate the erroneous measurement x_2, y_2, artificially creating a "rare double-positive cell." If a cell population is located orthogonally to two main populations, correlation errors should be considered. The relative frequency of ghost events depends on the cell analysis rate.

measurement error as a double-positive cell. Sorting of events with such properties will deliver a mixed population—both cells will be represented—in which a variety of biological functions may be found. Correlation errors often resemble clusters of rare events at orthogonal positions in a bivariate fluorescence plot. Great care should be exercised when such clusters are selected for sorting. One test for the occurrence of correlation errors is to run a sample at varying rates. If the number of events in an orthogonal cluster is rate dependent, chances are that it represents an electronic artifact rather than a real cell population.

It will be clear that high-speed, multiparameter operation of sorters severely tests the quality of the electronic circuitry. Researchers performing experiments in which purity or accuracy is crucial should check carefully for artifacts. One way to test the error rate of the acquisition electronics is to operate the system near full duty cycle. This test may be done by activating the main trigger circuit with a high-frequency stochastic signal, for instance from a photomultiplier registering dim ambient light or by passing a dense bead suspension through the flow cytometer at a rapid rate. The trigger rate must be such that a significant number of events fall within the cycle time of the instrument. The channels that are not attached to the trigger circuit should sample a simple signal with known characteristics, such as a sine wave from a function generator. Despite the high trigger rate, the measurements should yield plots that are consistent with the simple signals that are offered to the acquisition system. For example, the recorded data, when plotted as a time series, should reconstruct a perfect sine wave or a Lissajoux figure. Sometimes errors happen only at a very precise moment in the trigger cycle, for instance when a second pulse coincides exactly with the transition of a digital component somewhere in the system. When such critical timing conditions are suspected, two digital pulses with a sliding time separation can be offered to the system. By varying the time between the two pulses, the conditions of the rare error can be forced to happen at a higher frequency.

Considering the foregoing, it should be clear that the electronics should be fairly fast for reliable high-speed sorting. Although the speed of even the most advanced cell sorter is modest in terms of modern electronic integrated circuits, a sorter combines many electronic functions that need to work in concert. The asynchronous nature of the events makes error conditions virtually unavoidable. If critical error conditions are possible, one should prepare for their occurrence. A robust acquisition system should have circuitry that interrupts the sort process when conflict conditions are detected. Such circuitry could measure event spacing or could register events that occur within the dead time of the electronics. Modules that interrupt the sorting process need not be limited to monitoring the electronic system. Ideally, a cell sorter has a large number of digital switches, each capable of vetoing the sort process. Such switches could be activated when any relevant parameter (e.g., laser power, drop break-off point, or nozzle temperature) exceeds preset boundaries.

DROP GENERATION

Parallel pulse processing with a cycle time of 3–5 μs (~250,000 events/s) is well within the reach of modern electronics (van den Engh and Stokdijk, 1989). The system that is currently in use in my laboratory handles 32 parameters in 4.5 μs or 16 parameters in 3 μs. Ideally, the frequency of drop formation would match or even exceed the speed of the electronics. As will be shown further on, the generation of 250,000 drops/s requires a jet pressure of about 500 psi, which is much higher than the safe operating pressure for most standard laboratory equipment. Few valves and pressure systems are rated for use above 100 psi. This pressure limits sorters built with readily available components to a drop frequency of ~100 kHz. Unless operating pressures are significantly increased, the rate of cell sorting is determined by the frequency of drop formation.

The discrepancy between the electronic processing rate and the drop frequency is less damaging than it may seem. Unlike the electronic circuitry, the drops can be loaded near full capacity (on average of one cell per drop) without degrading the sort accuracy. With a stable drop generation system and robust deflection electronics, a high average occupancy rate of the drops need not affect sort purity. A high event rate, however, will lead to losses because of coincident drop occupancy. The magnitude of these losses is tightly linked to our ability to predict the position of the measured particles with respect to the drop boundaries. Particles that fall within the error of the drop boundary estimate may end up in the drop that is selected. To avoid unwanted particles in the sorted fraction, we must maintain a safe exclusion zone that is somewhat larger than the area of the drop that we wish to deflect. Any unwanted events that occur inside this buffer zone invalidate a sort decision and lead to the loss of a selected particle. Therefore, the size of the coincidence exclusion zone eventually determines the sort rate. The following section inspects the factors that are important in maintaining stable drop formation and establishing the coincidence exclusion interval.

Excellent and thorough treatments of the fluid physics of jets and drops may be found in Kachel and Menke (1979; unfortunately out of print) and Pinkel and Stovel (1985). The following summarizes the overall conclusions. Fluid flowing slowly from a small, circular aperture will form a growing drop that is attached to the nozzle tip. Surface tension forces the drop into a spherical shape. The drop swells until the surface tension can no longer support its weight, and the drop will separate from the nozzle tip. Under slow-flow conditions, drops form in an irregular, chaotic fashion. At some point, as the fluid velocity is increased, the rate of influx of kinetic energy will exceed the increase of surface energy of the growing drop. The liquid now assumes a cylindrical shape, a jet, with a diameter about the size of the nozzle orifice. The jet is unstable and tends toward reducing its surface energy. Rayleigh (1877) demonstrated that small disturbances along the axis of a cylinder increase the surface area when their wavelengths are shorter than the cylinder's circumference but decrease the surface area when their wavelengths exceed the circumference. Consequently a cylinder of liquid of diameter D will break into drops with a spacing greater than

$$\lambda = \pi D$$

Drops with a spacing much longer than this critical distance tend to partition into smaller drops, posing an upper limit on the drop size. In practice, by impressing small ripples along the surface of a jet, we can break a jet into drops with a spacing of

$$\pi D < \lambda < \sim 4\pi D$$

with an optimum at $\lambda = 4.54D$.

Because there is an optimal spacing of the drops, the drop rate is proportional to the jet velocity (Fig. 2.6). The velocity of a jet itself is proportional to the square root of the jet pressure (Pinkel and Stovel, 1985; Peters et al., 1985). In flow cytometry, the movement of fluid through the tubing is relatively slow, and most of the pressure energy is used in forming the rapidly accelerated jet. Ignoring resistance losses in the tubing, the work performed by the pressure system, $P \times \Delta \text{volume}$, is equal to the kinetic energy of the jet:

$$P \cdot \text{area}_{nozzle} \cdot v_{jet} = 1/2 \, m v_{jet}^2 = 1/2 \, \text{area}_{nozzle} \, \rho v_{jet}^3$$

Drop Generation

where P is the pressure, v_{jet} is the velocity of the jet, m is the mass of the jet, and ρ is the density of the fluid. This formula leads to

$$v_{jet} = \left(\frac{2P}{\rho}\right)^{1/2}$$

Note that the jet velocity is independent of the size of the nozzle orifice. The density of water (ρ_{water} at 20°C is 0.998 kg/m³) is not much affected by temperature; salinity has a small effect ($\rho_{saline,20°C} = \sim 1.005$ kg/m³; Denny, 1993). Saline propelled by a pressure of 1 atm (1 kgf/cm², or 14.22 psi, 9.81 × 10⁴ N/m²) forms a jet with a velocity of

$$(2 \times 9.81 \times 10^4 / 1.005 \times 10^3)^{1/2} = 14 \text{ m/s}$$

The optimal drop rate is related to pressure by

$$f_{optimal} = \frac{v_{jet}}{\lambda} = \frac{(2P/\rho)^{1/2}}{4.54 D_{jet}}$$

Note that the pressure is the pressure drop across the nozzle orifice. In cell sorters, a small but significant amount of pressure is needed to transport the fluid from the sup-

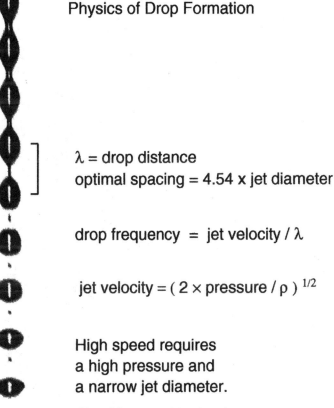

Physics of Drop Formation

λ = drop distance
optimal spacing = 4.54 x jet diameter

drop frequency = jet velocity / λ

jet velocity = (2 × pressure / ρ)$^{1/2}$

High speed requires
a high pressure and
a narrow jet diameter.

Fig. 2.6. Physics of drop formation.

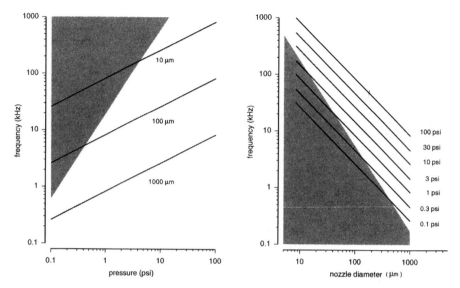

Fig. 2.7. Optimal drop generation frequencies of jets with different velocities and diameters.

ply tank to the nozzle tip. If a 70-μm nozzle receives fluid through 1 m of tubing with a 1.6-mm inside diameter (ID), the pressure drop along the tubing is ~10% of the tank pressure. In-line filters and valves cause additional losses.

These considerations lead to the nomograms of Figure 2.7, which relate pressure and drop frequency for different stream diameters and velocities. The grey zone indicates the area where the fluid drips irregularly from the nozzle because the kinetic energy is insufficient to form a jet.

These formulas and the nomogram do not take into account a number of secondary effects. The flow resistance of the walls of the nozzle orifice has been ignored. Also ignored is the contraction and acceleration of the jet as it leaves the orifice due to the redistribution of kinetic energy from a parabolic into a top-hat flow profile. At some distance from the nozzle tip, jets are somewhat (~0.85 times) narrower than the diameter of the nozzle aperture. Consequently, the velocity is not constant along the length of the jet and corrections must be applied. The nomograms are accurate to a first approximation and may serve as a guideline in generating jets and drops over a wide range of sizes and frequencies. I have found these graphs to be accurate for systems generating drops from 300 Hz at 1.5 m/s to 125 kHz at 100 psi.

FREQUENCY COUPLING

A periodic disturbance, seeded along the surface of a jet at an appropriate spacing, will rapidly grow in amplitude, eventually cleaving the jet into regular drops. A very small acoustic signal, near the optimal frequency, will be sufficient to fix the drop formation. When properly focused, the acoustic signal induces small pressure variations at the nozzle opening that cause variations in the amount of fluid exiting the nozzle aperture. Because the jet, after detachment from the nozzle tip, quickly assumes a uniform speed, the velocity variations in the nozzle orifice result in thinner and fatter regions in the jet's girth. As discussed before, the amount of fluid leaving the nozzle varies with the square root of the pressure. The diameter of the jet in turn varies with the square root of the jet velocity. Hence, the (small) fractional variation in jet diameter is at first approximation

$$\frac{\Delta D_{jet}}{D_{jet}} = 0.25 \frac{\Delta P}{P}$$

Despite the small effect of pressure changes on the jet diameter, minute variations in nozzle pressure accurately determine the location of the drop boundaries. A thin piezoelement driven by 1 V or less, which corresponds to a vibration amplitude of <<100 nm, is sufficient to impose a stable drop formation onto a 70–μm jet (Fig. 2.8).

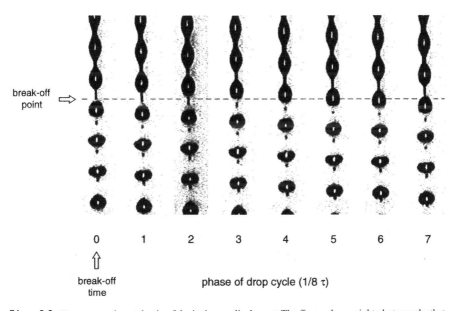

Fig. 2.8. Drop separation at the tip of the jet is a cyclical event. The figure shows eight photographs that are evenly spaced over the drop cycle. The jet moves downward with a constant velocity. The amplitude of a sinusoidal disturbance grows exponentially with time. At some point in space and time, the thin ligament that connects the last drop to the main stream snaps. The separated drop is now electrically isolated from the main stream. Moments later, the ligament separates from the drop that follows forming a microdroplet that is slightly pulled toward the tip of the jet. The microdroplet is soon overtaken by the following drop and merges with it. Note that the drop boundaries are slightly offset from the break-off point. If the stroboscopic flash that freezes the jet image is synchronous with the charge pulse, the image can be used to optimize the timing of the drop charge circuit.

According to Rayleigh (1877; Kachel and Menke, 1979), the amplitude of a surface disturbance of $\lambda > \pi D$ grows at an exponential rate with time,

$$a(t) = a_0 e^{\gamma t}$$

and

$$t(a) = \frac{1}{\gamma} \ln\left(\frac{a}{a_0}\right)$$

The growth rate γ depends on the wavelength and the overall physical characteristics of the fluid, such as viscosity and surface tension, which in turn are temperature sensitive (Denny, 1993). The time it takes for a disturbance to grow from the very small amplitude a_0 at the nozzle tip to the amplitude one-half D_{jet} that partitions the jet diameter is

$$t_{break\text{-}off} = \frac{1}{\gamma} \ln\left(\frac{D_{jet}}{2a_0}\right)$$

The variation in the break-off time as a function of the seed amplitude is

$$\frac{dt_{break\text{-}off}}{da_0} = -\frac{1}{\gamma a_0}$$

and

$$\frac{dt_{break\text{-}off}}{d\gamma} = -\frac{\ln(D_{jet}/2a_0)}{\gamma^2}$$

The stability of the break-off time is inversely related to the amplitude of the seed frequency and the growth rate γ. In order to obtain a stable break-off point, the seed amplitude must be sufficiently large, and the physical properties of the fluid must remain constant. On the other hand, if the seed amplitude is too large, the disturbance will distort the optical properties of the jet, adding noise to the fluorescence and scatter measurements. Setting the amplitude of the seed frequency requires a compromise between jet stability and measurement accuracy.

NUMERICAL EXAMPLE

Experimental observations confirm the relationships described in the previous section. Figure 2.9 presents data measured with a jet generated by 30 psi from a 70-µm nozzle. Assuming a jet contraction of 0.85 and a pressure drop from tank to nozzle of 3 psi, the theoretical optimal drop frequency is ~71.5 kHz. A drop drive frequency of 71.7 kHz was found to yield the shortest jet length (distance from orifice to break-off point). The length of the jet can be expressed as time, length, or drop boundaries. The relationship between the jet length and the piezovoltage follows the theoretical predictions, indicating linearity between the drive voltage and the amplitude of the jet disturbance (Fig. 2.9). Extrapolation predicts that the jet is bisected by an amplitude corresponding to a piezovoltage of 757 V. These numbers yield a surface amplitude of ~40 nm/V. Stable

Construction of the Nozzle Assembly

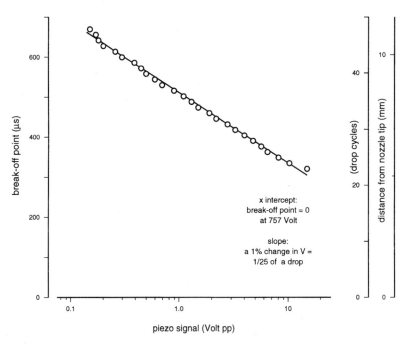

Fig. 2.9. Relationship break-off point and the amplitude of the piezosignal. A piezoelement element driven by an oscillator seeds a periodic vibration onto the jet surface. The break-off point, which can be expressed either as a length (distance from the nozzle tip) or a time (microseconds of drop periods traveled from nozzle tip to break-off point), is proportional to the logarithm of the signal amplitude (see text).

drop formation is already achieved at 150 mV or a surface wave with an amplitude of 6 nm. The graph in Figure 2.9 also demonstrates an increase in jet-length stability with signal amplitude. At a drive voltage of 1 V, a 25-mV variation in the drive signal changes the position of the break-off point by only $1/10$ of a drop. The same change in a drive voltage of 100 mV causes the break-off point to shift one drop.

CONSTRUCTION OF THE NOZZLE ASSEMBLY

The nozzle assembly must impose a small but stable vibration onto the jet. The placement of the vibrating element and the path of the acoustic waves are crucial. A path that leads the waves through the sheath fluid may efficiently guide the energy to the nozzle throat but may render the energy coupling dependent on the properties of the fluid and the presence or absence of air pockets in the fluid cavity. For these reasons, the vibrations are best transmitted through the nozzle body. Resonance frequencies in the nozzle body and associated hardware may be helpful or destructive. A clear resonance frequency may help to preferentially transfer the desired seed frequency while suppressing other parasitic frequencies. On the other hand, resonant modes are difficult to maintain at a constant amplitude, and a small drift in the resonance frequency may cause phase shifts between the vibration driver and the resonant body. Add to all of this that the fluid path must allow laminar flow, deliver the sample precisely, and be easy to sterilize, and it will be clear that the construction of the nozzle assembly and vibration unit is one of the most critical aspects of sorter design.

Figure 2.10 shows an example of a good nozzle and acoustic waveguide combination (van den Engh, 1998, 1999). The fluid and the acoustic waves are guided toward the nozzle tip via separate paths. The presence of air bubbles has minimal effect on the energy transfer. The wetted channels consist of simple straight tubes that can be replaced easily. The annular piezodriver connects to the base of a horn that focuses the energy toward the tip. The horn shape suppresses transverse or higher-order longitudinal resonance modes that may interfere with the vibration signal. A piezosensor near the nozzle tip monitors the acoustic waves in the nozzle body and may be used in a closed-loop oscillator for improved long-term stability of the vibration transfer. This design may be further enhanced with temperature and pressure sensors for stabilizing the physical properties of the fluid and resonator body.

Small acoustic signals guide the drop formation. External noise signals may interfere with the drive signal and temporarily shift the drop boundaries. Therefore, all fluid paths and containers must be acoustically isolated from the outside world. Temperature stability is crucial. Heat- and noise-producing devices should be mechanically isolated from the fluidics and placed well away from the nozzle head and drop deflection system. The trend to build neatly packaged sorters in which lasers, which generate more heat than light, fluidics, pumps, power supplies, and other noisy com-

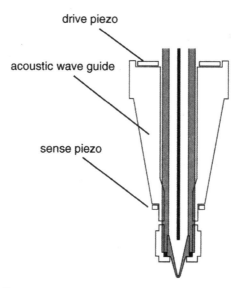

Fig. 2.10. Nozzle with separate acoustic waveguide and sense piezoelement.

ponents are housed in a single enclosure with tubing firmly affixed to an acoustically polluted frame is detrimental for instrument stability. It is remarkable that none of the commercial instruments provides the means to monitor the temperature of the critical sorter components or to thermally stabilize system components other than the sample input and collection path. Apparently the biologists who work with the machines are more concerned about their sample than are the engineers about the stability of the instrument. There is much opportunity to improve sorter performance by paying attention to the thermal and acoustic properties of flow cytometers and cell sorters.

TIMING AND DROP DEFLECTION SYNCHRONIZATION

One of the challenges of cell sorting is to correctly synchronize cell identification with drop deflection and to keep the two in sync over long periods. The drop formation is guided by a periodic vibration that is coupled to the nozzle tip. Drops are deflected by a circuit that applies a charge to the stream in synchrony with the drop separation. The particles to be sorted pass the measurement point at random times. Special circuitry must calculate when these cells are being sequestered into a drop and whether this drop is free of unwanted particles. To deflect the wanted drops, a charge pulse of appropriate polarity and amplitude must be applied during a critical time interval. In current sorters, calibration and adjustment of the timing of these circuits are inaccurate, inconvenient, or both and are surrounded by a certain degree of mystery. Furthermore, the settings tend to drift, becoming less accurate as the experiment gets underway. Temporary disturbances, though easily induced, are not easily detected. These deficiencies rapidly become evident under the high-occupancy conditions associated with high-speed sorting. Often recoveries are less than expected, and contamination is higher than acceptable. Obviously, detailed attention to the timing of the sort events is critical and probably is the aspect with most room for improvement.

In order to appreciate problems that may be encountered in sort timing, some theory is useful. In the following we will revisit the relationship between the jet length, the break-off time, and the amplitude of the vibration that guides the drop formation. This time we will include the discontinuity of drop formation and will determine exactly when a drop separates from the main stream. The piezoelement that causes the vibration is driven by a signal with a period $\tau = 2\pi/\omega$. At the nozzle tip, the vibration is transferred to the jet, where it imprints a very small amplitude on the jet surface:

$$\delta a = \delta A \cos(\omega t)$$

The disturbance moves with the jet at velocity v_j. The magnitude of the disturbance at point x at some distance from the tip is

$$a(x, t) = A(x) \cos\left[\omega\left(t - \frac{x}{v_j}\right)\right]$$

where $A(x)$ represents the local amplitude of the oscillation. In case of a disturbance of suitable wavelength ($\pi D_{jet} < \lambda$), the amplitude of the surface oscillations will grow at an exponential rate as it travels down the jet,

$$A(x) = \delta A\ e^{\gamma t} \quad \text{or} \quad A(x) = \delta A\ e^{\gamma x/v_j}$$

Therefore, at x the jet radius is

$$r_j = r_{av} + a(x) = r_{av} + \delta A\ e^{\gamma x/v_j} \cos\left(\omega t - \frac{\omega x}{v_j}\right)$$

At the break-off point $x_{\text{break-off}}$, the amplitude of the oscillation is just great enough to bisect the jet,

$$dA\ e^{\gamma x_{\text{break-off}}/v_j} = r_{av} \qquad x_{\text{break-off}} = \frac{v_j}{\gamma} \ln\left(\frac{r_{av}}{\delta A}\right)$$

This happens at the time when the waist of the oscillation passes the break-off point:

$$\cos\left[\omega\left(t - \frac{x_{\text{break-off}}}{v_j}\right)\right] = -1 \quad \text{or} \quad \omega\left(t - \frac{x_{\text{break-off}}}{v_j}\right) = \pi + n2\pi$$

where n is an integer. The moment at which a drop separates from the jet is

$$t_{\text{break-off}} = \frac{1}{\gamma \ln(r_{av}/\delta A)} + \frac{\pi}{\omega} + \frac{n2\pi}{\omega}$$

$$= n\tau + \phi_{\text{drop}}$$

The events at the jet tip are of a cyclical nature and are repeated with the periodicity of the drive signal τ. The phase ϕ_{drop} denotes the phase difference between the signal that drives the piezoelement and the drop cycle at the tip of the jet. At a precise moment during each cycle, a drop snaps from the main stream (Fig. 2.8, snap shot 0). Any electrical charges that are present on the drop at that moment will remain after the drop has separated. An electric pulse that is applied to the jet in synchrony with the drop formation will charge the drop that separated during the presence of the charge pulse. At the beginning of each drop cycle, one must decide whether to charge the next drop or not. Drop charging is therefore carried out in two stages. The charging circuit is prepared (armed) during the drop cycle that precedes the wanted drop. The charge circuit is then triggered at the moment that the preceding drop separates from the main stream.

Consider a cell that is contained in the drop being formed during the cycle that is bounded by times $(n-1)\tau + \phi$ and $n\tau + \phi$. The drop boundaries passed the measurement point earlier between $(n-1)\tau + \phi - \Delta t$ and $n\tau + \phi - \Delta t$. In order to deflect the drop that contains the cell, we must arm the charge circuit after a time $\Delta t - \tau$ has elapsed since the cell entered the illumination spot.

ADJUSTMENT OF CHARGE SYNCHRONIZATION

In order to calculate when to apply a charge pulse to the jet to deflect drops that contain desired particles, we need the following information:

1. the status of the drop drive oscillator;
2. the phase difference between the drop formation and the signal to the drive piezo; and
3. the delay time Δt between the measurement point and the break-off point.

The drop drive oscillator is derived from a reference signal in the cell sorter and is therefore known.

The phase of drop formation ϕ can be determined by charging the stream with a pulse that can be varied in phase with respect to the drive signal. The phase setting is optimal when maximal drop deflection is achieved. It is helpful to have a strobe signal that coincides with the charge cycle, rather than with a fixed point of the drive cycle, so that the status of the jet at the onset of charging time may be observed. When adjusted correctly, a strobed video of the jet tip should always show the same image. In a well-tuned system the onset of the charge cycle should coincide with the moment of drop separation (Fig. 2.8, image 0).

The delay between measurement and arming of the charge circuit is determined by sorting test particles onto a microscope slide. The efficiency of the sort is evaluated by counting the number of events in the collected drops. The time delay is set at the value that deposited the largest number of particles onto the slide. This calibration method yields a delay setting with the accuracy of the timing clock.

The delay setting is often determined by the visual method illustrated in Figure 2.11, instead of the empirical method described above. The distance from the measurement point to the break-off point is determined using a microscope with an attached micrometer. The microscope is moved an equal distance downstream from the break-off point. The number of drops along this second trajectory is counted and is used as the number of cycles (minus 1) between measurement and charge notification. This method suffers from the following inaccuracies. The method assumes a constant jet velocity. However, due to jet contraction, some acceleration occurs after the jet leaves the nozzle. In addition, drops will accelerate due to the pull of gravity, inducing further inaccuracy in the measurement. There is also a considerable read error in the determination of the illumination point and the break-off point. There is an error in drop count. If each of these measurements has an error of one-eighth of a drop cycle (the time it takes the particle to travel approximately one-fourth of the jet diameter), the cumulative error can be as high as 0.35 of one drop cycle. The visual method should therefore only be used as a rough indication of the delay time and must always be followed with a test sort in which the delay time is varied until maximum particle recovery is achieved.

The accuracy of the delay time is critical. If the delay time is off by a fraction of the drop cycle, that fraction of events will end up in a drop adjacent to the one that is charged and will be missed. Worse, the lost fraction is replaced by a volume of liquid

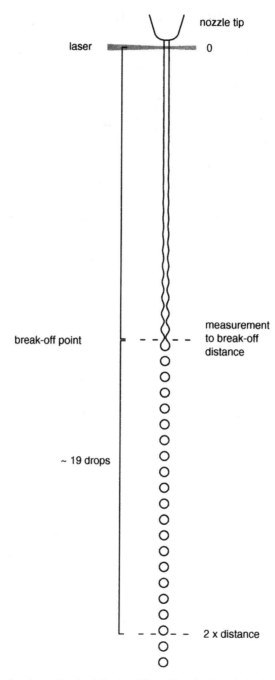

Fig. 2.11. Procedure for setting the deflection delay with a visual method. The distance between the point where the laser intersects the jet of a cell sorter and the break-off point is measured with a microscope. The microscope is then moved an equal distance down from the tip of the jet. The number of drops counted in the second distance is an estimate for the number of drop periods that elapsed between the illumination point and the tip of the jet.

that may contain unwanted particles. Contamination due to charge synchronization errors is a serious problem in high-duty-cycle sorting, often resulting in lower yields and higher contamination with unwanted cells than expected.

The importance of an accurate determination of the delay times is illustrated by the following calculation. Consider a sort in which, on average, one in four drops contains a cell. The delay time is determined with an error of one-eighth of a drop cycle. The yield of wanted particles is $7/8$, or 87%. For every deflected drop there is a chance of $1/8 \times 1/4 = 1/32$ that we collect a contaminating particle. The purity of the sorted fraction is $(7/8) / (7/8 + 1/32) = \sim 96\%$. If the error in the delay time is one-third of a drop, the sort purity will be only ~89%.

The contamination due to inaccuracies in the drop boundary calculation can be avoided by setting an exclusion zone. If any unwanted particles fall in an interval before or after the drop that is calculated to be deflected, the deflection request is ignored. It is prudent to enforce an exclusion zone of twice the timing error on each side of a selected drop. In that case, if the delay time error is one-eighth of a drop, the total area that should be free of coincident events is 1.5 drops. At one cell per four drops, the sort rejection rate will be ~0.35.

Recovery losses that are due to an inaccurate delay setting may be reduced by deflecting more than one drop for each sorted cell. If the drops that precede and follow a selected drop are deflected as well, there is virtually no chance that a cell will be lost because the wrong drop was deflected. However, the coincident rejection zone will obviously increase as the number of sorted drops is increased. A frequently applied strategy is to deflect 1.5 drops. If the cell that is to be selected is calculated to be in the first or last fourth of the drop, the preceding or following drop is deflected as well. Approximately half the sort instructions will result in the deflection of two drops, so that the average number of drops per sorted cell is 1.5. With a further exclusion zone of one-fourth of a drop on either side, the coincident rejection losses at a drop occupancy of 1 in 4 become 0.5. At a drop occupancy exceeding 1 in 20 drops, multiple-drop deflection schemes become counterproductive because the higher recovery of the selected cells is more than negated by the increase in coincident-event rejection.

The foregoing demonstrates that an accurate estimate of the delay time is critical to the cell yield and purity of a sorting experiment. The sorter must have accurate means to determine the proper delay setting at a given break-off point. Once adjusted, the break-off point must remain stable. The next section discusses the monitoring and adjustment of the position of the break-off point.

STABILITY BREAK-OFF POINT

The tip of the jet is observed on a video monitor. During an experiment, a change in conditions may cause the break-off point to drift from its starting position. It is common practice to compensate for changes in the position of the break-off point by adjusting the signal voltage to the piezoelement. If the change in the jet tip is due to a change in efficiency of the coupling of the piezovibration to the nozzle tip, for instance due to the formation of a small air bubble, changing the vibration amplitude of the piezoelement is indeed the proper action. However, not all drifts in the break-off point are associated with a change in the coupling efficiency and therefore do not require adjustment of the vibration energy.

Besides the seed amplitude δA, the position of the break-off point is also dependent on the properties of the sheath fluid, γ, and the velocity of the jet. A change in γ, for instance due to a change in temperature, will change the break-off time. This alteration, like a change in coupling efficiency, should be corrected by adjusting the vibration amplitude.

On the other hand, the theory predicts that a change in position of the tip of the jet that is due to a change in jet velocity will not result in a change of the delay time. A lower jet velocity will move the break-off point up, but the time at which oscillation cleaves the jet remains the same. Experimental observation confirms the notion that the break-off time is relatively insensitive to changes in the jet velocity. A 20-m/s jet vibrating around 70 kHz must be slowed down by approximately five drop cycles in order to change the delay setting one full drop cycle. This concept is important because the velocity of the jet is quite sensitive to environmental conditions and therefore is a likely cause of changes in the position of the break-off point. During the course of an experiment, the level in the sheath fluid reservoir will slowly decrease. This decrease may cause a change in the nozzle pressure large enough to noticeably affect the velocity of the jet. A jet at 30 psi has a break-off point about 8 mm from the nozzle tip. A drop of 10 cm in the fluid level represents a change of 0.5% in pressure. This change will cause a velocity reduction of 0.25%, which brings the break-off point about 20 μm, or $1/_{10}$ of a drop, closer to the nozzle tip but will not change the break-off time. If the change in break-off distance is incorrectly compensated by a decrease in voltage to the piezoelement, an error of ~0.1 drop cycle is introduced in the delay setting, which, as demonstrated before, may cause a significant change in cell recovery and/or purity.

Even greater than the effect of the fluid level on the jet velocity are the effects due to temperature changes of the sheath fluid. Roughly 10–20% of the pressure is needed to overcome the friction of the sheath fluid flowing through narrow tubes. The friction is proportional to a property of the fluid called the kinematic viscosity. The kinematic viscosity is very sensitive to temperature. A change of 10°C is associated with a change of 20–30% in the kinematic viscosity. A few degrees Celsius therefore may change the jet velocity by a few percent.

It is evident that the actions in response to a change in the break-off point must be based on a proper diagnosis of the problem. Considering the inaccuracies of the visual method for estimating the delay time and the possible adverse effects of vibration–amplitude adjustments, systems that automatically adjust the break-off point based on measurements on video images should be used with caution.

DROP DEFLECTION, DROP CHARGE, AND FIELD STRENGTH

The charged drops are deflected in a high-voltage (~4000 V) electrostatic field. The deflection distance over which the drops are deflected depends on their charge-to-mass ratio. The charge of the drop is carried on its surface. At the same voltage, small drops, because of their relatively large surface area, have a larger charge-to-mass ratio than do large drops and hence are deflected over a greater distance. The charge density on the drop surface should not be too large. The surface charges repel each other and counteract the surface tension that keeps the drop together. Drops carrying a high charge become unstable and may fragment into smaller droplets. In addition, surface tension is an important component of γ, the property of the jet that controls the position of the jet's break-off point. A charge pulse of a high voltage may interfere temporarily with the stability of the jet. Information about proper charge pulses and the best methods to apply the charge to the sheath fluid is fragmentary. My personal observation with 70-μm-diameter jets is that these phenomena may cause problems if the charge voltages exceed 80 V. In recurrent stream instabilities, the effect of different charge voltages should be considered. If larger deflection distances are desired, it may be better to extend the length of the deflection plates than to increase the charge voltage.

The surface charge–density of drops may also change by evaporation. In a dry environment, the volume reduction due to evaporation may be significant. Air currents and turbulence from forceful air conditioning equipment that emits alternating blasts of dry and humid air may cause instabilities in the side streams. For long-term stability, the wet bench of the cell sorter should be protected from air currents. The air humidity should be kept constant at a relatively high level, and temperature fluctuations should be avoided. As noted before, the control of temperature (and humidity) is facilitated if heat-producing modules are placed at some distance from the nozzle and sample stream and are thermally isolated from the drop separation and collection area.

MULTIPLE SORT DIRECTIONS AND AUTOMATED CELL COLLECTION

Once stable sort conditions have been obtained, it is possible to apply charge voltages with different amplitudes. Separating a sample into multiple fractions may increase the efficiency of an experiment. Eight-way sorting has been used at the Lawrence Livermore National Laboratory to deposit chromosomes in spots onto hybridization filter (van den Engh et al., unpublished results). One could envision a sorter that would deflect particles with a distance proportional to their fluorescence intensity. An analysis of the gene expression patterns of such a sort could indicate which range of fluorescence intensities is associated with a (DNA) marker with a previously unknown distribution pattern. In this approach the distribution of our marker is determined in a single sort and one test.

In sort experiments, a relatively large amount of time is spent changing sample and collection tubes. These interruptions reduce the effective sort rate. Automated instruments that position tubes, trays, or filters automatically in the collection area or automatically change input samples may greatly increase the overall speed of our experiments. Automated collection equipment may be particularly effective in the collection and presentation of single sorted events. Single-event selection is an important growing area in cell biology. Improvement of collection methods for cell sorters may make a major contribution.

CELL VIABILITY

A discussion of high-speed sorting must include the question of cell viability. Higher pressures and jet velocities increase the risk of cell damage. Fragile cells may be ruptured by high shear forces. Turbulence and rapid acceleration or deceleration of fluid streams should be avoided in all parts of the equipment. The tubing should be short, straight, and without sharp corners and edges. This task is not difficult. The fluid path of the sample can be kept quite simple. Damage due to shear can be kept under control. In the region of highest forces, near the nozzle orifice, the cells follow a trajectory through the center of the stream where the shear forces are smallest. A variety of cell types have been sorted in high-pressure sorters without noticeable detrimental effects (Peters et al., 1985; Sasaki et al., 1995; van den Engh, unpublished observations). There are, however, anecdotal reports of cell damage in systems that use pressures of 30 psi and higher. In reduced recovery of viable cells, the effect of pressure (and sample path modifications) should be considered.

Rapid decompression may cause more of a problem than do shear forces. Oxygen and nitrogen readily dissolve in saline. Under high pressure, significant amounts of gas enter the carrier liquid. In prolonged sort experiments, the content of dissolved gases may change significantly over the course of the experiment. It is possible that high oxygen pressures by themselves are toxic to certain cell types. In passing through the nozzle orifice, the cells undergo a significant and rapid change in pressure that may force the dissolved gases out of solution. Gas bubbles may be an explanation for reduced cell viability. If it is a factor in cell destruction, bubble formation will very much depend on the length of the experiment and the treatment of the sheath fluid before the experiment. Hence the effects of dissolved gases will be erratic in nature. In order to avoid problems with dissolved gases, the use of gas mixtures with a low solubility in water seems worth exploring.

CONCLUSION: PRACTICAL LIMITS OF HIGH-SPEED CELL SORTING

High-speed sorting concerns instruments that generate 10^5 drops/s while processing cells at a rate greater than 10^4 events/s. At such speeds, overall precision and sample purity are determined by coincident events at the illumination site, the maximum frequency of drop formation, multiple cells per drop, and the accuracy of drop boundary calculations. Therefore, high-speed sorting is not merely a matter of increasing the speed of the sorter subsystems. It involves the development of strategies for using the instruments at a high duty cycle, that is, conditions under which the sorter's components operate near their performance limits. High duty cycles present unique problems that the electronic and logical modules must handle unambiguously, without creating measurement artifacts or sorting errors. For overall efficiency, robustness is at least as important as raw speed. Immunity to failure, low-drift mechanical design, closed-loop alignment, and short turnaround times are important in achieving high effective throughput rates.

We have identified factors that constrain the performance of cell sorters. The following conditions present a workable compromise. Using standard fluidic components, an operating pressure to about 690 N/m^2 (100 psi, 7 atm) can be maintained. At that pressure, the jet velocity is 37 m/s. A simple lens system illuminates the jet with a 20-μm light spot. Cells suspended in the jet generate light pulses that are 1–2 μs wide. A 1-μs pulse separation sets the illumination cycle time at 2.5 μs. The number of events that can be handled with confidence (duty cycle of 0.1) is about 4×10^4 events/s. At 37 m/s, a 70-μm jet optimally forms drops at 118 kHz. The break-off time of the drops is approximately 30 drop periods. The break-off point can be kept stable to about one-eighth of a drop. Under these conditions, on average one in three drops contains a cell, with 24% of the drops containing a single event. If all drops with a wanted cell are sorted, accepting drops with multiple events, the instrument delivers samples of 85% purity. Enforcement of a coincidence exclusion zone of 1.5 drops yields better than 99% purity with a yield of 75%. A robust design limits downtime and allows for short start-up and shutdown protocols. Such instruments achieve effective throughput rates of 25,000 cells/s or a little under 10^8 cells/h.

A modest increase in speed of 1.5 times, at an increased risk of nozzle clogging, can be achieved by reducing the diameter of the jet. A jet with a diameter of 50 mm can be made to form drops at 165 kHz. The drop frequency may also be increased by increasing the pressure. However, the square root relationship between pressure and jet velocity limits the practical possibilities. In order to match the drop frequency with the current analysis rate, the pressure must be increased to ~10^4 N/m^2. A pressure of 100 atm can be achieved only with technically sophisticated instrumentation and will negatively affect overall robustness. Furthermore, a higher jet velocity reduces the dwell time of the particle in the laser spot, requiring a higher illumination intensity and much faster electronics. Clearly, the current definition of high-speed sorting represents the practical limits of this technology.

Considering the technical challenges in further increasing sorter speed, parallel operation of several instruments is an attractive alternative. In the parallel approach, technical development concentrates on simplifying sorter design and the sharing of common resources by several instruments. Experimental prototypes of high-speed sorters that are much simpler and much more compact than the current commercial machines have been built in my laboratory. Two compact sorters need not occupy more than a bench that is 1.5 m wide. The instruments can be operated simultaneously by one technician. Two instruments double the effective yield to 2×10^8 cells/h. For critical samples, such as in the preparation of samples for clinical applications, a third sort head could be maintained in a stand-by mode to take over in case of catastrophic failure of one of the operating systems.

ACKNOWLEDGMENTS

This work has been supported by the Department of Energy, grant number DE-FG06-93ER61662, and the Washington Technology Center. Research grants have been supplied by Becton Dickinson, Cytomation, Systemix, Pharmacopeia, and Sumitomo.

Assistance in kind has been provided by V-TEK and Applied Precision. Many students have helped me investigate aspects of flow cytometer design. Ellen Bosgoed, Hugo Gelevert, Maurie Brewer, and Jeannie Fehlauer deserve special mention. Chip Asbury is the senior junior scientist in my laboratory. All new ideas must pass his scrutiny. He has provided valuable help in the preparation of this chapter. The construction of experimental instruments is not possible without the efforts of a dedicated technical crew. Richard Esposito and Brian Wadey have been responsible for much of the design and construction of electronic and mechanical modules. Leah LaTray brought design errors to light, while skillfully sorting cells. I owe much to Barb Trask for a conscientious reading of the manuscript and numerous valuable suggestions.

REFERENCES

Baumgarth N, Parks DR, Bigos M, Stovel RT, Anderson MA, Gerstein RA, King UC, Jager GC, Herman OC, Herzenberg LA, Herzenberg LA (1998): Tracking green fluorescent protein transgene-expressing cells by nine-color FACS. Presentation at the Colorado Springs meeting of the International Society for Analytical Cytology. 28 February–5 March 1998.

Denny MW (1993): *Air and Water.* Princeton, NJ: Princeton University Press.

Gray JW, Dean PN, Fuscoe JC, Peters DC, Trask BJ, van den Engh GJ, Van Dilla MA (1987): High-speed chromosome sorting. Science 238, 323–329.

Kachel V, Menke E (1979): Hydrodynamic properties of flow cytometric instruments. In Melamed MR, Mullaney PF, Mendelsohn ML (eds). *Flow Cytometry and Sorting.* New York: Wiley.

Peters D, Branscomb E, Dean P, Merrill T, Pinkel D, Van Dilla M, Gray JW (1985): The LLNL high-speed sorter: Design features, operational characteristics, and biological utility. Cytometry 6:290–301.

Pinkel D, Stovel R (1985): Flow chambers and sample handling. In Dean PN (ed). *Flow cytometry: Instrumentation and Data Analysis.* New York: Academic Press.

Rayleigh JWS (1877): *Theory of Sound.* London: MacMillan. Reprinted 1945, New York: Dover.

Sasaki DT, Tichenore EH, Lopez F, Combs J, Uchida N, Smith C, Stokdijk W, Vardenega M, Buckle A-M, Chen B, Tushinski R, Tsukamoto AS, Hoffman R (1995): Development of a clinically applicable high-speed flow cytometer for the isolation of transplantable human hematopoietic stem cells. J Hematother 4:504–514.

Taylor JR (1997): *An Introduction to Error Analysis.* Sausolito, CA: University Science Books.

van den Engh GJ (1998): Particle separating apparatus and method, US Pat. 5,819,948.

van den Engh GJ (1999): Particle separating apparatus and method, US Pat. 6,003,678.

van den Engh GJ, Farmer C (1992): Photo-bleaching and photon saturation in flow cytometry. Cytometry 13:669–677.

van den Engh GJ, Stokdijk W (1989): Parallel processing data acquisition system for multi-laser flow cytometry and cell sorting. Cytometry 10, 282–293.

Van Dilla MA, Deaven LL, Albright KL, Allen NA, Bartholdi MF, Brown NC, Campbell EW, Carrano AV, Christensen M, Clark LM, Cram LS, Dean PN, de Jong P, Fawcett JJ, Fuscoe JC, Gray JW, Hildebrand CE, Jackson PJ, Jett JH, Killa S, Longmire JL, Lozes CR, Luedemann ML, McNinch JS, Mendelsohn ML, Meyne J, Meincke LJ, Moyzis RK, Mullikin J, Munk AC, Perlman J, Pederson L, Peters DC, Silva AJ, Trask BJ, van den Engh G (1990): The national laboratory gene library project. In Gray JW (ed). *Flow Cytogenetics.* London: Academic Press, pp 257–274.

Visser J, van den Engh G (1982): Immunofluorescence measurements by flow cytometry. In Wick G, Trail KN, Schauenstein K (eds). *Immunofluorescence Technology.* Amsterdam: Elsevier.

3

Rare-Event Detection and Sorting of Rare Cells

James F. Leary
University of Texas Medical Branch, Galveston, Texas

INTRODUCTION

Why Is the Analysis and Isolation of Rare Cells Important?

Just as in physics, where many of the simpler problems have been solved, biology is now entering a similar era. This is not to say that the major problems have been solved. Indeed, with the sequencing portion of the Human Genome Project about to be completed, a new era is about to begin. But many of the new biological problems of importance will involve looking beyond the simpler analyses of the past. For example, many of the new basic and clinical problems in biology will involve the detection, isolation, and molecular characterization of rare cells.

Rare cells can be defined as those of less than 0.1% frequency. Ultrarare cells can be defined as those of less than 0.001% frequency. These are terms for ease of discussion and are not universally agreed upon. In terms of flow cytometry and cell sorting, rare-cell applications not only push the limits of the technology but also require levels of staining specificity beyond those assumed by most biologists. It is important for engineers to understand that good technology can be made irrelevant by bad cell staining and preparation. The greatest difficulty with rare-cell applications is not the immediate technological constraints but the requirement that there be no weak links in the

Emerging Tools for Single-Cell Analysis, Edited by Gary Durack and J. Paul Robinson.
ISBN 0-471-31575-3 Copyright © 2000 Wiley-Liss, Inc.

experimental methodology anywhere in the process from cell preparation, flow cytometry/cell sorting, and data analysis or subsequent analysis of isolated cells. Each step in the methodology must be excellent, and the entire process must be thought through to eliminate or deal with the weaker links of a given rare-cell application.

There are many rare-cell applications of importance to basic or clinical research. The ones discussed in this chapter help show the range not only of the applications but also of the technological challenges to engineers working in this field. Some basic research applications include (1) isolation of rare-cell clones with specific mutations or transfected genes, (2) isolation of clones with combinatorial libraries of inserted genes, and (3) studies of environmentally induced mutations in human cells. Some clinical applications include (1) isolation and sequence analysis of rare metastatic cells with mutated tumor suppressor genes for minimal residual disease monitoring and (2) isolation of rare human stem/progenitor cell subsets for autologous transplantation.

One of the major impediments of flow cytometry/cell-sorting technology is that biologists are unaware that many of the limitations of the currently available commercial instrumentation are due more to the lack of markets for manufacturers than to true technological limitations. Also, many biologists are unaware of what can be accomplished technologically with the instrumentation that they already have. Indeed, lack of a true understanding of the technology by many biology researchers has been one of the major reasons why many of the more challenging biological applications either are not being done or are being done poorly. Thus engineers may be puzzled or frustrated by the fact that even the present technology is not being fully utilized. While there are some fundamental technical limitations, as there are in all fields, few of these limits are as yet being reached by current commercial instrumentation. New technology must be "smart technology" that does not require considerable expertise of users who are frequently not well trained in the physical sciences. So the challenges to engineers are not only to push the technology but also to make something complex appear simple because it is implemented with smart technology that does not depend more than is absolutely necessary on the skill of the cell sorter operator. Unfortunately, due to the lack of smart technology in cell sorting, cell sorting remains an "art" that is not widespread. It has also led to the popularity of a number of alternative "simpler" cell separation technologies that appeal to users due to their simplicity, even though flow cytometry/cell sorting is considerably more powerful if it is properly used.

Why Are the Analysis and Isolation of Rare Cells Difficult?

There are a number of reasons why the analysis and isolation of rare cells are difficult. First, at the level of rare cells, few, if any, individual markers/probes for rare cells have adequate signal-to-noise (S/N) ratios. The main limitations for rare-cell analysis and sorting are not the S/N ratio of the instrument but rather the S/N ratio of the cell sample. Sometimes it is not possible to construct cell probes that have the specificity required by a given rare-cell application. But since flow cytometry and cell sorting approach the problem from the vantage point of individual cells, there are major possibilities for enhancement of S/N ratios by eliminating cells not-of-interest from the analysis. This can lead to overall S/N ratios that are many thousands of times better on single cells than they would be on populations of cells. For example, a probe may do a good job of labeling the rare cells of interest, but it may also stain another cell type not-of-interest. However, if the cells not-of-interest can be labeled with another probe that makes it possible to exclude them from the analysis, the first probe may be able to allow successful isolation of the rare cell when used in a Boolean logic of measured probe levels in cells, which eliminates the cells not-of-interest. Thus it is at least as important to have good negative markers/probes to eliminate as many as possible of the non–rare cells not-of-interest. As the cells become more rare, more markers/probes are frequently required, leading to the requirement for additional flow cytometric parameters. The true power of this technology is its ability to look at multiple probes on single cells. It is very unlikely for a single probe to be both sensitive and specific enough to detect rare cells, and almost never does a single probe work for ultrarare cells. Second, as cells become more rare, the sampling statistics problem can become more and more difficult. On the one hand, if total sample size is limited as in some applications, the researcher is forced to look for only a few cells in the entire sample. If, on the other hand, the total sample size is very large but the frequency of the rare cells is very low, a huge number of total cells must be examined. Unless the flow cytometer/cell sorter is very fast, the time required may be totally impractical or, in the case of the analysis of living cells, beyond the time limits imposed by either the stability of the parameter being measured or the viability of the cells. Third, due to the limitations of the preceding problems, there are necessarily trade-offs between yield and purity of the analyzed or sorted sample. This problem is starting to be addressed in commercial instrumentation, but only in some simple ways.

WHY IS IT IMPORTANT TO CONSIDER THEORETICAL ASPECTS OF THE PROBLEM?

It is important to consider as many of the theoretical aspects as possible to avoid getting into impractical or impossible situations. Unfortunately many engineers forget to take into account fundamental requirements of cells, particularly living cells, in their designs of flow cytometers/cell sorters. In this chapter many of these limitations, including those of the cells, will be taken into account to produce some technical specifications for the "ideal" instrument. Likewise, many biologists are so eager to perform a rare-cell analysis or sort that they fail to consider theoretical limitations imposed by instrument design and basic physics or by the rare-cell sampling statistics imposed by the given application. The problems of rare-cell analysis and cell sorting cover a large range of disciplines. Failure to address the full range of problems frequently results in failure of the particular application. The most difficult thing about rare-cell analysis and sorting is that everything has to work very well at the same time. If any aspect of the process is weak, the total effort frequently results in failure. Hence many of the partial or only average-quality solutions of non-rare-cell applications do not work for rare-cell applications. For example, small amounts of nonspecific staining are frequently not important for non-rare-cell applications. But a 1% or even a 0.1% nonspecific staining background (in terms of numbers of false-positive staining cells) in an application involving rare- or ultra-rare-cell subpopulations can make the application impossible unless ways can be found to reduce this nonspecific staining level to some acceptable level lower than the specific staining frequency (the frequency of true-positive cells).

SOME FUNDAMENTAL PROBLEMS IN RARE-CELL DETECTION AND ISOLATION

Statistical Sampling Limits

In the case of limited total sample size, there must be a sufficient number of cells analyzable to obtain statistically or biologically meaningful results. For example, if total sample size is limited and frequencies of rare cells that are analyzable are low, it may not be possible to analyze enough cells to be meaningful at either the statistical or biological level. Since most rare-cell experiments are difficult and costly, this should be taken into account before plunging mindlessly into an attempt to analyze or sort rare cells. Mathematical and statistical modeling of the situation is wise to do beforehand, as it may change the fundamental approach to the problem. When the number of rare cells is very small, it is important to use combinatorial rather than Poisson statistics as Poisson statistics are only an approximation that is not particularly stable below about 50 rare cells (Rosenblatt et al., 1997). For example, to ensure with 95% confidence that the single rare (10^{-6} frequency) cell has been sorted actually requires that a minimum of nearly three million cells must be sampled. Since the probability of successfully isolating the rare cell of interest above false-positive background events becomes more difficult for rare cells, it is important to sample an even higher number than this because it is highly likely that there will be false-positive cells sorted. The times needed to isolate a given number of desired rare cells as a function of sorting rate are illustrated in Table 3.1 (from Rosenblatt et al., 1997).

If rare single cells are to be isolated for either cloning or single-cell polymerase chain reaction (PCR), the proper approach is to take these rare-cell sampling statistics into account and perform single-cell, multitube sorting. An important principle of rare-cell sorting is that at very rare cell levels perhaps no cell markers, for example, monoclonal antibodies, will be good enough to pull out pure populations of rare cells from a noisy background. But if rare cells are isolated by single-cell sorting, 100% pure rare cells can always be isolated. If the S/N ratio is very poor, the number of tubes of single isolated cells becomes impractical. But with modern molecular biology methods such as PCR, even single-copy DNA or small numbers of messenger RNA (mRNA) sequences from a single cell can be quickly expanded into 10^6–10^8 copies (Zhang et al., 1992). So the principle is to recover a single copy of the DNA or mRNA template from a sorted single cell and then make many, many copies of that template much more rapidly than can be accomplished even by high-speed cell sorting. The basic idea of single-cell, multitube cell sorting is shown in Figure 3.1. Obvi-

TABLE 3.1. Time Required to Collect Desired Cells as Function of Sorting Rate[a]

Desired Number of Cells with Selected Properties	Total Number of Sort Decisions	Time (s) Required to Collect Desired Cells by Sorting Rate[b]					
		2,500	5,000	10,000	20,000	50,000	100,000
1	2,990,000	1,196	598	299	150	60	30
10	15,705,214	6,282	3,141	1,570	785	314	157
100	116,997,126	46,799	23,399	11,700	5,850	2,340	1,170
1000	1,052,577,091	421,031	210,515	105,258	52,629	21,052	10,526

[a]Frequency of cells with selected characteristics = 10^{-6} (0.95 level of assurance).
[b]In sort decisions per second.

ously cells can be sorted into tube arrays, or multiwell plates. The flow cytometric information associated with each sorted cell can remain associated with all the subsequent molecular measurements by a method known in the field as *indexed cell sorting* (Stovel and Sweet, 1979). We have combined flow cytometric indexed cell-sorting data with molecular data in hybrid files that we have then used for subsequent analysis with multivariate statistical packages such as S-Plus for Windows.

Our practical experience agrees quite well with these theoretical predictions. When we have attempted to isolate rare (10^{-6} frequency) cells (labeled with very good probes!), we have found that we usually get 2 to 3 rare cells successfully isolated per 10 single-cell sorts, as depicted in Figure 3.1. Similarly, we have had applications where the probes are not very good and the combination of limited sampling statistics due to low numbers of cells available and poor probes have made impossible the successful isolation of less rare cells of a frequency of 10^{-5}. So one must pay careful attention to whether enough cells are available to reach proper sampling statistics and whether the probes are good enough to practically isolate the rare cells of interest.

Rare-Event Detection Is Usually Only Possible with Multiple Probes

While many markers/probes look very good in non-rare-cell applications, few markers/probes look good at the rare-cell level and very few at the level of ultrarare cells. It is important to have some idea of the S/N ratio of the marker/probe system being used before wasting considerable energies trying to do an impossible application. Simple mathematical modeling of data using each probe can be used to predict the S/N ratio of that probe. In most cases one should expect to use multiple probes to detect rare cells (Ryan et al., 1984). Ultra-rare-cell applications usually require a minimum of three to four parameters and frequently cocktails of probes designed to minimize the number of required acquisition parameters and to reduce data complexity. The *receiver operating characteristic* (ROC; Swets and Pickett, 1982; Beck and Shultz, 1986; Hanley, 1989; Choi, 1998) long familiar to many engineers but only beginning to be used in flow cytometry and cell sorting (Leary et al., 1997,

CONCEPT OF SINGLE-CELL / MULTI-TUBE PCR ANALYSIS OF RARE SORTED CELLS

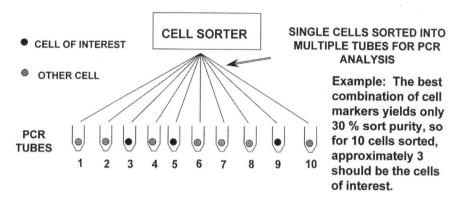

Then, use "multiplex PCR" with one set of primers to confirm the cells of interest and another set of primers to look for specific gene sequences.

Fig. 3.1. General principle of single-cell, multitube PCR analysis of sorted single cells is to obtain one or more copies of a DNA or mRNA template from a single sorted cell. Some of the singly sorted cells will be false positives due to the inability of specific marker probes to have a high enough S/N ratio above false-positive staining background.

1998a, b, 1999; Hokanson et al., 1999) can be used to predict the power of each probe, singly and in combination. A discussion of ROC analysis and an example appear later in this chapter.

Analysis Speeds Must Be Fast Enough to Be Practical

When the total sample size is large and the rare-cell frequency small, there may be some fundamental limits imposed by instrument stability/operation, marker stability, and cell viability. Thus there is usually a time window imposed by these variables that should be carefully considered when either designing an instrument for an application or attempting an application on an existing instrument. In most cases this time window will require an analysis speed in excess of 10,000 cells/s for rare-cell applications and an analysis speed in excess of 40,000 cells/s for ultra-rare-cell applications. In terms of fundamental limitations it is possible to build instruments capable of analysis speeds in excess of 100,000 cells/s, if properly designed. But to do efficient high-speed sorting of rare cells, it is wise to do the fundamental cell classification/sort decision in two stages. The first stage should classify most of the cells with easy decisions so that the second stage can spend more time making the tough decisions. Most of the time we know that the cells are not of interest; the decision is straightforward and can usually be accomplished by some simple Boolean combination of simple gates of the original parameters. But there will always be some potential false-positive cells that must be distinguished from the true-positive rare cells of interest. Making these tougher choices may require more elaborate gating and perhaps mathematical combinations of parameters (e.g., discriminant functions) that are more complicated and require more time to process the signals. One should also remember that mindless high-speed digitization of signals merely leads to a river of digital data that must be processed at least partially in real time. Otherwise, one would have to deal with the problem of storing and processing many gigabytes of data per sample.

Limits in Sorting Speeds and Purities

Sorting is fundamentally more difficult and slower than analysis for a number of reasons. First, sorting requires that the rare cells be properly classified, which itself is difficult. Cell sorting should really be considered as real-time data analysis. The calculations and classifications must be done fast enough to reach a sort decision, typically less than a millisecond after excitation of the cell. Fortunately there are new methods available to accomplish this feat (Corio and Leary, 1993, 1996; Leary et al., 1997, 1998, 1999). By using one or more linked, high-speed lookup tables, one can perform complex linear and nonlinear "calculations" at memory speeds, enabling their use for real-time statistical classifications suitable for cell-sorting decisions. The lookup tables are precalculated for a range of useful values based on data obtained from an aliquot of the cell sample, as is typically done for any cell-sorting decision. The incoming data are then compared to lookup table values that then can be used to issue a sort/no-sort decision.

Second, as any flow cytometrist knows, the Achilles heel of a flow cytometer is the flow; that is, the stability of the fluidics is fragile and difficult to maintain. Since high-speed fluidic switching is still under development and has not yet been applied to cell sorting, droplet sorting is presently the method of choice, although this may change in the not-too-distant future. While fluidic switching has problems of its own, it has many potential advantages over droplet sorting, not the least of which is the safety issue of sorting biohazardous materials in a closed system rather than in an open system with aerosol generation. Advances in microfluidics may well change this situation.

Not only is the break-off position of a cell sorter difficult to maintain, but as the number of cells per second is increased for sorting of rare cells, the stability becomes less reliable. The stickiness of the cells makes clumping a problem at high cell concentrations (e.g., greater than 5×10^7 cells/ml). In terms of the fluidics, the cells themselves start to contribute to the viscosity of the sample stream, making transport difficult and less predictable. As discussed later in more detail, a high-speed cell sorter for rare cells should be able to measure the arrival statistics of the cells to determine if the cell preparation is adequate to permit isolation of these rare cells. As discussed later in this chapter, cell arrival statistics is a critically important measure of cell sorter performance during actual cell sorting of rare cells. If the fluidics are working properly, cell arrival statistics should be random. Cell arrival statistics can be described by simple queuing theory (Gross and Harris, 1985) familiar to many engineers, and discussed in the flow cytometry/cell-sorting field in a number of papers by different groups (Lindmo et al., 1981; van Rotterdam et al., 1992; Leary, 1994).

IMPORTANCE OF MODELING IN RARE-CELL ANALYSIS

Simple One-Parameter Modeling

The simplest modeling involves either constructing actual data mixtures of negative, non–rare cells and positive, rare cells or constructing mathematical models of these subpopulations with mathematical curves with known properties (Cupp et al., 1984). If pure populations of each cell type are available, it is possible to "tag" the data points so that when a data mixture is made, the yield and purity of rare cells in a sort gate can be modeled. If such data are not available, the cell subpopulations can be parametrically modeled, most easily but not always correctly, with Gaussians. This mathematical approach leads to easy statistical analysis of yields and purities, including probabilities of misclassification, a subject to be explored in greater depth later in this chapter.

Multiparameter Modeling and the Use of "Tagged" Data Mixtures

As the data become more complex, methods have been developed that permit easy modeling of multiparameter flow cytometric data (Redelman and Coder, 1994; Leary et al., 1996). Basically, each cell in a listmode flow cytometry data file is tagged with an additional classification parameter that is not used in the subsequent analysis process other than testing the validity of the classification process at the very end. An example of this cell-tagging process in a datafile is shown in Figure 3.2 (Leary et al., 1997).

Testing classifiers with listmode data

listmode flow cytometry data file

Identifier	<--Raw Parameters-->			<--Principal Components-->			<-- Classifiers -->			<--Actual-->
cell #	Par 1	Par 2 ...	Par 4	PC1	PC2	PC3	Discrim 1	K Means	BrPlot	TrueClass
1	122	451 ...	2352	1126	3222	1765	1	1	2	1
2	1333	3111 ...	455	3033	1055	891	3	3	3	3
3	1777	2881 ...	622	2698	899	787	3	2	2	2
.	.	.	.	.	.	.	.	.	.	.
.	.	.	.	.	.	.	.	.	.	.
10000	89	444 ...	2111	1023	3091	1555	1	1	1	1

Fig. 3.2. Tagged parameter is an additional correlated listmode parameter that is not used directly in the data analysis routine being used but instead serves as a truth table for the correct identity of each cell.

The actual use of such a tagged datafile to test cell classifier systems is shown in Figure 3.3. The principle is to run purified populations of cells separately on the flow cytometer. Then an additional correlated parameter (a "tag") is added to the data but not used during the actual data analysis or sort classification process. But at the end of that process, the classifier tag serves as a way to have a *truth table* to evaluate the efficiency of the sort. Since cell sorting is just a *real-time data analysis,* sorting algorithms for rare cells labeled with different probes can be quickly tested in software without having to resort to costly and laborious testing of each possibility by actual cell sorting. Defined data mixtures simulating different rare-cell frequencies can be very rapidly modeled such that several weeks worth of actual sorting can be simulated in a single day. Then the best method for isolating the rare cells by actual cell sorting can be chosen. The ROC analysis can be used to estimate the performance of various cell classifiers both singly and in combination (Leary et al., 1998; Hokanson et al., 1999), as shown in Figure 3.4. In Figure 3.4A we see the general theory of ROC analysis. In Figure 3.4B we see ROC analysis applied to a rare-cell classification problem, in this case the problem of rare metastatic tumor cell analysis and purging (negative sorting to remove these cells) from stem cells to prevent their being cotransplanted back into a patient—a potentially very important future clinical application.

One of the most critical problems of rare-event analysis is the correct classification of the cells. To determine the correctness, one must have a truth standard; that is, one must be able to unequivocally determine the identity of each cell in a training set. Obviously in the new test sample there will be incoming data that reside in a classification "gray zone" where different cell subpopulations overlap. But if the training set is done properly, every cell can at least have a probability of assignment calculated on a cell-by-cell basis. This probability becomes an important part of a determination of sort boundaries to allow optimization of sorted cell yield and purity. In the case of cells that truly have partial membership in two or more classes (e.g., in normally differentiating cells or hematopoietic cancers such as leukemias), fuzzy logic can be used to provide fuzzy rather than crisp classifiers.

Multivariate Statistical Classification of Rare Cells

In some rare-cell applications it is important to know the probability of misclassification for analysis or sorting of rare cells. An example of an application for which this is important is the isolation of stem/progenitor cells free from contaminating rare tumor cells in autologous transplants of cancer patients (Schultze et al., 1997). Gene-marking studies have shown that reinfusion of contaminating rare tumor cells can lead to the growth of serious new tumors in the patients (Brenner et al., 1993). One way of estimating the probability of misclassification of these very rare tumor cells in this application is through the use of discriminant function analysis (DFA; Lachenbruch, 1975; Klecka, 1980; McLachlan, 1992; Kleinbaum et al., 1992) recently applied to flow cytometry and cell sorting (Hokanson et al., 1999). An example of the theory and use of DFA in these studies is shown in Figure 3.5.

By moving the classification line we can choose a point of "acceptable" costs of misclassification, where *acceptable* is defined as presenting the patient with a tumor

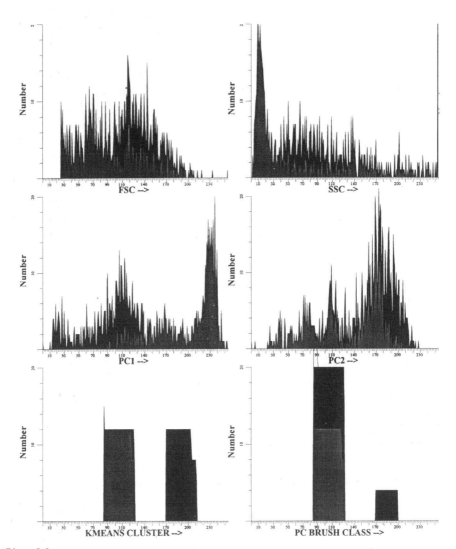

Fig. 3.3. Use of listmode data to test classifiers. Actual listmode data mixtures of human bone marrow cells and MCF-7 human breast cancer cells were analyzed using the methods described in Fig. 3.5. In addition to the raw flow cytometric parameters P1–P4, the first three principal components PC1–PC3

load that clinicians estimate the immune system of the patient can handle. While DFA is a relatively simple multivariate statistical classification method, it does make two important assumptions. First, the data are assumed to be multivariate normally distributed. This assumption is clearly violated by virtually all data, but the method is still relatively robust to violations of this assumption. A second assumption is that

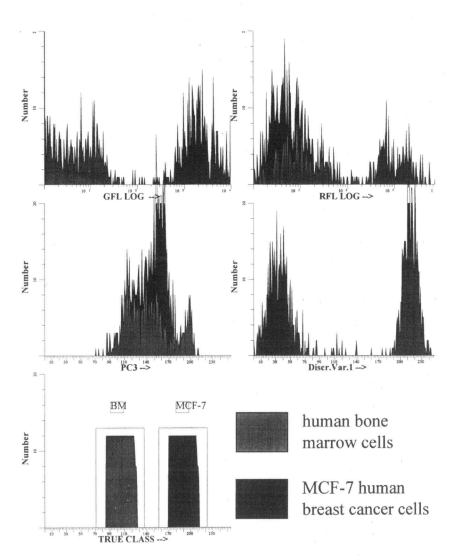

and a discriminant function D1 were calculated. Cells were classified using *K*-means clustering or brush plotting through principal-component space. Data were gated on the true classifiers, which show how well they separated the two general types of cells.

there are only two groups of cells. Each group can consist of several subpopulations, but there can be only two groups. If a third group of cells is required for the analysis, DFA is insufficient. In the case of multivariate statistical classification of more than two groups, logistic regression techniques should be used. Logistic regression, a relatively new multivariate statistical technique, makes no assumptions about the distributional characteristics of the data (they need not be assumed multivariate normal) and there is no limit to the number of groups that can be classified.

Real-Time Multivariate Statistical Classification of Rare Cells for Cell Sorting

The pioneering work of Bartels showed that multivariate statistical classification of cells analyzed by image analysis could be a powerful tool for cell classification (Bartels, 1980a, b). Interestingly, it is possible to implement real-time multivariate statistical classification of cells for cell sorting using high-speed lookup tables that can compute linear or even nonlinear mathematical functions at memory speeds (Leary et al., 1997, 1998). One of the biggest problems in sorting rare cells has been the relatively crude, "seat-of-the-pants" choice of sort boundaries. The most common method used by most researchers is to set boundaries for "positive" cells "out there, somewhere" on one or more parameters. This is frequently done because the researchers were either unable or unwilling to sample enough cells to see where the rare cells are in multidimensional space. Another problem is that few researchers have ac-

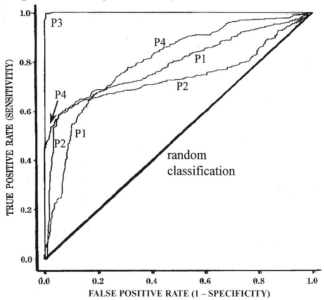

Fig. 3.4. In this ROC plot of actual flow cytometric parameters P1–P4, we see that P2 (side scatter) is similar in performance as a classifier to P1 (forward scatter). A fluorescent monoclonal antibody against antigens found on breast cancer cells, P3, is a very good positive selection probe and classifier. A fluorescent monoclonal antibody against CD45 found on blood cells but not on breast cancer cells, P4, is a negative selection probe and is needed when the breast cancer cells become very rare in blood.

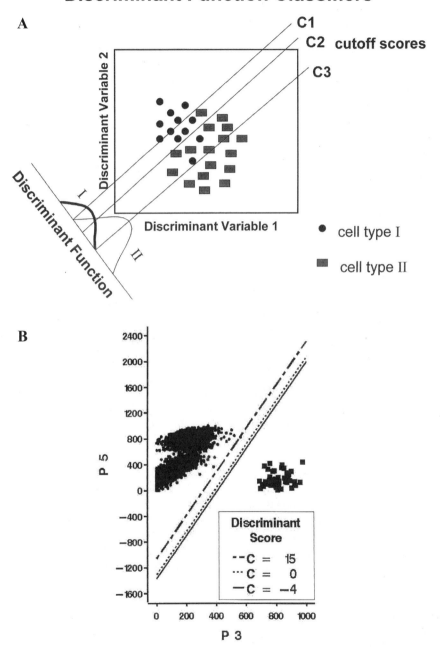

Fig. 3.5. (A) Stem cells and rare tumor cells are separated by DFA with different cutoff scores to exclude varying percentages of tumor cells from stem cells. (B) DFA data classification with costs of misclassification for normal human bone marrow and rare MCF-7 breast cancer cells where discriminant scores are in units of standard deviations (from Hokanson et al., 1999).

cess to software to allow interactive three-dimensional visualization of either the raw parameter data space or better yet some multidimensional projection method (e.g., principal-component analysis). "Blind" sorting of cells from a multidimensional data space is extremely dangerous, as many or most of those positive cells out there, somewhere are actually false-positive background from the cells not of interest.

Most experimenters choose sort boundaries by human pattern recognition and arbitrarily drawn bitmap sorting boundaries. As the data required to find rare cells become more multidimensional, more complicated techniques are required to permit human visualization of the data. As rare cell sorting enters the clinic (e.g., in human stem cell sorting for subsequent transplantation), such arbitrariness is extremely unwise and will probably someday encounter Food and Drug Administration (FDA) regulations in the United States. Clinicians may someday have to provide for quality assurance as to why such sort boundaries were chosen and what the probabilities are for contamination that could adversely affect the patient. This is an area of rare-cell analysis and isolation that is critically important and needs more research and development.

DEFINING AND MEASURING INSTRUMENT PERFORMANCE FOR RARE-CELL SORTING BY QUEUING THEORY

There are a number of considerations that should be taken into account in assessing instrument performance, including cell–cell coincidence in the excitation source and/or in the sorting unit, instrument deadtime, and biological limitations imposed by the requirements of sorting rare cells. Many of these properties that involve mathematical modeling of random cell arrival can be easily evaluated using queuing theory (Gross and Harris, 1985).

Coincidence

There are at least two important types of coincidence that must be taken into account by both instrument users and designers. First, there is the coincidence, or probability, that two or more cells lie within the excitation source such that the signals from these multiple cells cannot be separated from one another. Second, there is coincidence of two or more cells within the sorting unit such that all of these cells must be either sorted in a contaminated sort or rejected (using "anticoincidence" circuitry and logic) with a subsequent loss of cell yield. The second situation is frequently encountered by virtually all experimenters because the probability for cell coincidence in the sort unit is significant on most instruments even at rates of 5000 cells/s, as shown in Figure 3.6 (from Leary, 1994), which shows cell loss as a function of cell analysis rate and the number of droplets per sorting unit.

Cell–cell coincidence in the excitation source is less often seen by most experimenters unless cells are analyzed at rates in excess of about 20,000 cells/s on most instruments. The loss of cells not seen by the system due to cell coincidence is shown in Figure 3.7 (from Leary, 1994).

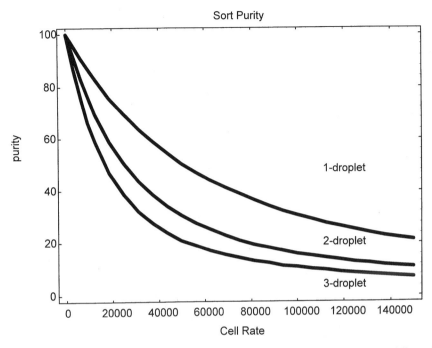

Fig. 3.6. Queuing theory model of sort purity statistics produced by mathematical modeling using Mathematica (Wolfram, 1996). Sort purity decreases with the number of cells per second processed since these cells arrive randomly and must be contained within a fixed number of sorting units per second that are equally spaced. As the number of droplets per sorting unit increases, the queue length increases, leading to increased coincidence of multiple cells within a given sort unit.

In the high-speed enrichment sorting of rare cells, there will be multiple cells in each sort unit (typically one, two, or three droplets). For enrichment sorting it is important to know the identity of each cell and to have software algorithms capable of deciding whether adding that sort unit to the total sorted cell container is going to satisfy the requirements of yield versus purity for that application. Rare-cell sampling statistics and the endpoint use of the sorted rare cells must be taken into account in the sorting decision. Hence sorting decisions should be "weighted" according to experimental considerations that are not simple sort decisions on the flow cytometric parameters alone (cf. patents by Corio and Leary, 1993, 1996). This then leads to the subject of cost of misclassification or cost of classification/sort impurities, a subject to be discussed later in this chapter.

Deadtime

One of the most important characteristics of any instrument is its deadtime, that is, the time period that the instrument takes to process one cell and be ready for the next cell. It is a more complex characteristic than just signal processing time (e.g., analog-to-digital conversion) because signals might have to be further processed, stored in a

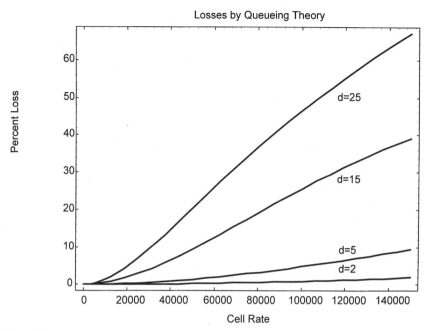

Fig. 3.7. Mathematica (Wolfram, 1996) -generated graph of the percent loss in the number of observed cells per second compared to the expected number of cells per second as a function of instrument deadtime based on queuing theory and instrument deadtimes.

buffer, buffers periodically emptied, and so on. A better way of taking all these many, and sometimes hidden, sources of deadtime in a system is to experimentally determine the *functional deadtime,* that is, the time required to process all the cells over the period of time required by the particular application. Depending on the instrument, more parameters or multiple excitation beams may introduce more deadtime. Since many systems acquire data directly into computer memory buffers, the time to empty those buffers as they fill during an experiment should be taken into account. The effect of instrument deadtime on instrument performance as measured by the numbers of cells "missed" during the deadtime of the instrument is shown graphically in Figure 3.8 (from Leary, 1994).

Empirically obtaining an estimate of the functional deadtime of a particular instrument is relatively simple and has been previously described (Leary, 1994). Briefly, you will typically need a series of cell samples of concentrations from about 5×10^6 to about 1×10^8 cells/ml. Most flow cytometers use about 30–60 μl/min of sample, but this may vary widely with the instrument model. Then simply record the total number (n) of cells per second counted by your instrument over that number of minutes and compare with the number (N) of cells per second that should have passed through the instrument based on the product of the number of cells per milliliter times the sample volume used. Then use the equation below to calculate the deadtime. Rearranging the equation yields the familiar relationship

$$d = \frac{N-n}{Nn}$$

If N and n are in cells per second, the deadtime d is in seconds. For example, if on the basis of cell concentration and volume throughput you expected 20,000 cells/s with your system and you observed only 15,000 cells/s, your instrument deadtime (under those conditions!) would be

$$d = \frac{20{,}000 - 15{,}000}{20{,}000 \times 15{,}000} = 16.7 \times 10^{-6}, \text{ or } 16.7\ \mu s$$

Were it not for the effects on subsequent sorting, it is well known how to account for cell–cell coincidence in the excitation source. By running known numbers of cells per second through an instrument, one can easily calculate the rate of coincidences and automatically compensate mathematically to get true total counts. Since rare cells are by definition rare, the probability of a rare-cell/rare-cell coincidence is extremely low. Hence, by correctly counting all rare cells, their frequency in the total population can be easily calculated. For this reason, obtaining rare-cell frequencies is not very difficult even at very high (e.g. 100,000 cells/s) cell analysis rates.

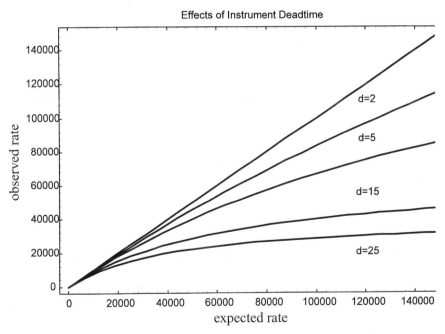

Fig. 3.8. Observed versus expected numbers of cells parameterized according to instrument deadtime as predicted by queuing theory modeling using Mathematica (Wolfram, 1996). Early generation commercial flow cytometers had deadtimes in the 15–20-µs range, while more recent commercial higher speed flow cytometers are now in the 5–10-µs range. Some experimental flow cytometers have been in the 2–3-µs deadtime range for over 10 years.

Limitations Imposed by Live-Cell Requirements

It is frequently forgotten that many aspects of instrument performance are determined by the cells that are to be processed through those instruments. Fixed cells, while not indestructible, are nonetheless very hardy and can be subjected to relatively high shear, explosive decompression, and deceleration (on impact) forces. Many of these parameters were studied in one of the first high-speed cell sorters developed at Lawrence Livermore National Laboratory in the early 1980s (Peters et al., 1985). They and their probes being measured also tend to be more stable over time, thereby greatly reducing the constraints of experiment time required for analysis or sorting. Interestingly, the new technology of PCR has greatly affected rare-cell sorting by reducing the requirements for numbers of sorted cells needed to as few as one since the PCR reaction can amplify DNA or RNA sequences from a single copy to many millions in a few hours.

If live rare cells must be analyzed or sorted, many serious constraints are then placed on instrument design and performance. First, since the cells are frequently dying over a matter of hours, the cells must be processed as quickly as possible. Second, since live-cell probes may be changing over a short time span, that time interval of relative stability sets a major constraint on the instrument performance characteristics. Third, since live cells are relatively fragile, they cannot be exposed to excessive shear and deceleration forces or to explosive decompression extremes. Since cell flow rates and droplet generation frequencies are a function of system pressures and flow cell orifice diameters, this sets some upper limits to the rates at which cells can be processed. There is a natural engineering trade-off between the flow rates and the intercell spacing that then affects instrument analysis and sorting deadtimes and coincidences. The intercell spacing can be easily calculated by queuing theory (Leary, 1994).

DESIGNING THE IDEAL RARE-CELL SORTER INSTRUMENT

Designing an ideal rare-cell sorter involves a number of factors, some of which lead to performance trade-offs, others of which set absolute constraints. The trick for instrument designers is to keep in mind that flow cytometers/cell sorters are tightly coupled in terms of their engineering parameters. When one engineering parameter is changed, the effect frequently propagates throughout the system, sometimes in unexpected ways.

Minimal Cell–Cell Coincidence in Excitation Source While Maintaining Sufficient Overall Sensitivity

The trade-off here is minimizing the queue length in the direction of flow to minimize cell–cell coincidence in the beam, while providing enough excitation energy and fluorescence "duty cycles" to have sufficient sensitivity. This dictates a focused laser excitation spot that is ellipsoidal in structure with a semiminor axis length comparable to the diameter of the cells being analyzed and a semimajor axis of length sufficient to allow the cells to receive a very similar amount of excitation energy without variations produced by variations in cell trajectory through the focused spot.

Fast Sensors and Preamplifiers, Good Sensitivity

The sensors must be fast and sensitive enough that there is little or no loss in signal thresholding for proper triggering or loss of fluorescence signal due to slew rate limits of the preamplifiers and amplifiers connected to the detectors. The sensors should have good spectral and quantum efficiency so that the S/N ratio is as good as possible.

Minimal Instrument Deadtime

The instrument deadtime must be minimized. For rare-cell analysis and sorting the instrument deadtime should be no more than about 5 µs for most systems and preferably closer to 2 µs. Obviously this deadtime depends on a number of factors, including cell velocities. To determine the effects of deadtime, the signal processing should be modeled by queuing theory rather than simplistic calculations of instrument deadtime that fail to take into account random arrival statistics. This has been described earlier in this chapter.

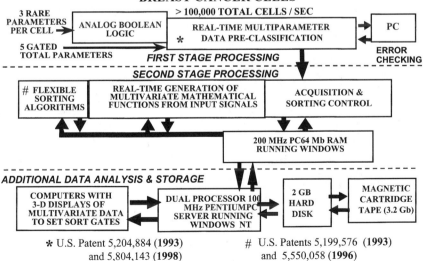

Fig. 3.9. For high-speed flow cytometry and cell sorting the cell classification process is more efficiently handled in a two-step process. The first stage classifies most of the cells as "not-of-interest" and then passes those remaining cells either likely to be "of-interest" or "not-sures" to a secondary stage classifier that can use far more sophisticated classification algorithms to improve the real-time classification process.

Multistage, Multiqueuing Signal Processing

As cells in close proximity are analyzed for high-speed flow cytometry and cell sorting, there comes a point where the major limitation is the signal processing deadtime. Perhaps the best way to improve signal processing throughput under these conditions is to use either multistage or multiqueuing signal processing. Multistage signal processing has the advantage of preprocessing a significant fraction of the incoming signals so that more careful signal processing can be used for the signals that matter. This has been previously implemented in a high-speed, two-stage processing, double-queue digital signal-processing system as part of a high-speed flow cytometer/cell sorter (cf. patents by Corio and Leary, 1993, 1998; and Leary et al., 1993 and 1998. Fig. 3.9). This method is relatively inexpensive and easy to implement on new high-speed flow cytometer/cell sorters and can be easily retrofitted to most existing commercial instruments.

Another way of accomplishing this multistage signal processing is to perform high-speed digitization of the incoming signals and then discard unneeded data along a digital pipeline (van den Engh and Stokdijk, 1989). Such a system has been successfully implemented on the commercially available Mo-Flo instrument (Cytomation, Fort Collins, CO).

Discriminant Function Cell Sorting

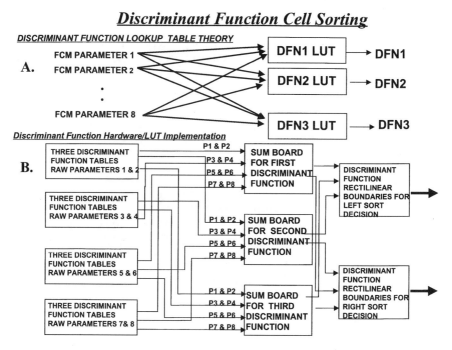

Fig. 3.10. The fundamental idea is that for a cell sorter to perform real-time statistical functions, all input parameters must be linked in an architecture similar to that of a single-layer neural network. To satisfy the mathematical requirements, a combination of multiply and add operations must be performed at high speeds. One implementation of such an architecture is shown in this figure as described previously (Leary et al., 1996, 1997, 1998a, 1998b, 1999).

Linked, Transformed Parameters

As the parameters required to separate rare cells from background become more numerous, the data obviously become more complex. As described earlier in this chapter, there are a number of multivariate statistical techniques useful for improved classification of rare cells. Such multivariate statistical functions such as principal-component analysis, discriminant functions as pioneered by Bartels (Bartels et al., 1980), or more recently logistic regression analysis (as shown by Hokanson et al., 1999) all require weighted linear combinations of most or all of the input parameters. While this is easy to accomplish for rare-cell data analysis off-line, it is much more difficult to implement for real-time situations such as cell sorting. To generate such multivariate statistical classification functions for rare cells in real time requires that all input parameters be hardwire linked to each other in a structure analogous to a single-layer neural network as has been previously implemented and as shown in Figure 3.10.

NEED FOR SOPHISTICATED, "FLEXIBLE," SORTING ANTICOINCIDENCE

For high-speed sorting of rare cells there needs to be sophisticated anticoincidence hardware and software. Anticoincidence systems are used to determine which cells being analyzed will occur in a given sort unit. If one knows the sample fluid flow rate

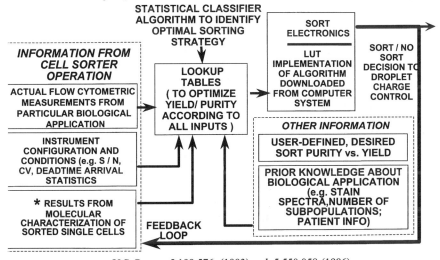

U.S. Patents 5,199,576 (1993) and 5,550,058 (1996)

Fig. 3.11. Flexible sorting algorithms controlling the sort decision essentially change the weighting factors applied to multiple sources of information, both real-time quantitative from each cell and other non-real-time information containing prior knowledge. The resulting algorithms, which can be optimized for each application, can provide for improved yield and purity as well as applying cost of misclassification to the sort decision; that is, how many "good" cells are we willing to throw away to avoid sorting "bad" cells.

and cell concentration (or rate of cells per second), the intercell spacing can be predicted by queuing theory (Leary, 1994). This intercell spacing relationship is built into the circuitry such that if two cells arrive within a given interval that would cause them to occur in the same sort unit, there is the option to either accept or reject this situation. If the situation is accepted, the two cells are sorted in the same sort unit. If the situation is rejected, both cells are rejected and not sorted. To really do this properly, one must consider not only the cell following a given cell but also the cell preceding a given cell. The situation becomes even more complicated during high-speed sorting of rare cells because there may be multiple cells preceding or postceding a given cell. The identity of each cell can be stored and weighted according to the desirability/undesirability of that cell in a register so that this process can be performed continuously and in a manner "flexible" to the needs and knowledge of the experimenter. This flexible sorting, which encompasses not only programmable anticoincidence strategies but also other factors, has been implemented in a high-speed cell sorter for isolation of rare cells (Corio and Leary, 1993, 1996; Leary et al., 1996, 1999), as shown in Figure 3.11.

Flexible, Modular Design for Optimization to Particular Application

If we put together all these requirements for cell sorting, we produce a system similar to what has been produced in one research high-speed cell sorter optimized for the analysis and isolation of rare cells (Leary et al., 1996, 1997, 1998; Fig. 3.12). It

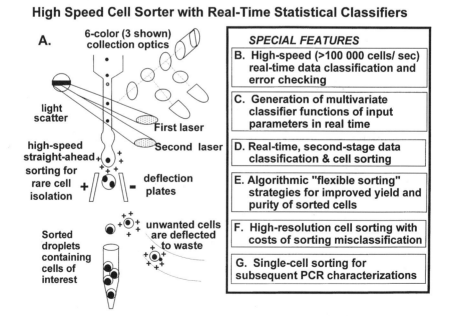

Fig. 3.12. A possible implementation of an "ideal," high-speed cell sorter for rare cells. While a number of specific features remain to be optimized, the most basic necessary features are shown.

is essential for the engineer to understand that while the design should be modular, it is the nature of cell sorting that changes in any part of the design tend to propagate through the entire design. By its nature, and particularly for high-speed sorting of rare cells, a cell sorter is tightly designed.

REFERENCES

Bartels PH (1980a): Numerical evaluation of cytologic data. VI. Multivariate distributions and matrix notation. Anal Quant Cytol 2(3):155–160.

Bartels PH (1980b): Numerical evaluation of cytologic data. IV. Discrimination and classification. Anal Quant Cytol 2(1):19–24.

Beck JR, Shultz EK (1986): The use of relative operating characteristic (ROC) curves in test performance evaluation. Arch Pathol Lab Med 110:13–20.

Choi BCK (1998): Slopes of a receiver operating characteristic curve and likelihood ratios for a diagnostic test. Am J Epidemiol 148(11):1127–1132.

Corio MA, Leary JF (1993 and 1996): A system for flexibly sorting particles, U.S. Pat. 5,199,576 and 5,550,058 (International patents pending in Europe, Canada, and Japan).

Cupp JE, Leary JF, Cernichiari E, Wood JC, Doherty RA (1984): Rare-event analysis methods for detection of fetal red blood cells in maternal blood. Cytometry 5:138–144.

Gross D, Harris CM (1985): *Fundamentals of Queuing Theory,* 2nd ed. New York: Wiley.

Hanley JA (1989): Receiver operating characteristic (ROC) methodology: The state of the art. Crit Rev Diagn Imag 29(3):307–335.

Hokanson JA, Rosenblatt JI, Leary JF (1999): Some theoretical and practical considerations for multivariate statistical cell classification useful in autologous stem cell transplantation and tumor cell purging. Cytometry 36: 60–70.

Klecka WR (1980): *Discriminant Analysis.* Sage University Paper Series Quantitative Applications in the Social Sciences, Series 07-019.

Kleinbaum DG, Kupper LL, Miller KE (1992): *Applied Regression and Other Multivariate Methods.* Boston: PWS-Kent, pp 560–594.

Lachenbruch PA (1975): *Discriminant Analysis.* New York: Hafner Press, pp 128–204.

Leary JF, McLaughlin SR, Hokanson JA, Rosenblatt JI (1998a): High speed real-time data classification and cell sorting using discriminant functions and probabilities of misclassification for stem cell enrichment and tumor purging. SPIE (J Opt Soc Am) 3260:274–281.

Leary JF, McLaughlin SR, Hokanson JA, Rosenblatt JI (1998b): New high speed cell sorting methods for stem cell sorting and breast cancer cell purging. SPIE (J Opt Soc Am) 3259:114–121.

Leary JF, McLaughlin SR, Reece LN, Rosenblatt JI, Hokanson JA (1999): Real-time multivariate statistical classification of cells for flow cytometry and cell sorting: a data mining application for stem cell isolation and tumor purging. SPIE 3604:158–169.

Leary JF, Hokanson JA, McLaughlin SR (1997): High speed cell classification systems for real-time data classification and cell sorting. SPIE (J Opt Soc Am), 2982:342–352.

Leary JF, McLaughlin SR, Kavanau K (1996): New methods for detection, analysis and isolation of rare cell populations. SPIE (J Opt Soc Am) 2678:240–253.

Leary JF (1994): Strategies for rare cell detection and isolation. In Darzynkiewicz Z, Robinson JP, Crissman H (ed). *Methods in Cell Biology: Flow Cytometry,* Vol 42. New York: Academic Press, pp 331–358.

Leary JF, Corio MA, McLaughlin SR (1993 and 1998): System for high-speed measurement and sorting of particles. U.S. Pat. 5,204,884 and U.S. Pat. 5,804,143.

Lindmo T, Fundingsrund K (1981): Measurement of the distribution of time intervals between cell passages in flow cytometry as a method for the evaluation of sample preparation procedures. Cytometry 2:151–154.

McLachlan GJ (1992): *Discriminant Analysis and Statistical Pattern Recognition.* New York: Wiley, pp 7–9.

Peters D, Branscomb E, Dean P, Merrill T, Pinkel D, Van D, Gray J (1985): The LLNL high-speed sorter: Design features, operational characteristics, and biological utility. Cytometry 6:290–301.

Redelman D, Coder DM (1994): Cell subset (CS) parameter to record the identities of individual cells in flow cytometric data. Cytometry 18(2):95–102.

Rosenblatt JI, Hokanson JA, McLaughlin SR, Leary JF (1997): A theoretical basis for sampling statistics appropriate for the detection and isolation of rare cells using flow cytometry and cell sorting. Cytometry 26:1–6.

Ryan DH, Mitchell SJ, Hennessy LA, Bauer KD, Horan PK, Cohen HJ (1984): Improved detection of rare CALLA-positive cells in peripheral blood using multiparameter flow cytometry. J Immunol Methods 74:115–128.

Schultze R, Schultz M, Wischnik A, Ehnle S, Doukas K, Behr W, Ehret W, Schlimok G (1997): Tumor cell contamination of peripheral blood stem cell transplants and bone marrow in high-risk breast cancer patients. Bone Marrow Transplant 19(12):1223–1228.

Stovel RT, Sweet RG (1979): Individual cell sorting. J Histochem Cytochem 27:284–288.

Swets JA, Pickett RM (1982): *Evaluation of Diagnostic Systems Methods from Signal Detection Theory.* New York: Academic Press.

van den Engh G, Stokdijk W (1989): Parallel processing data acquisition system for flow cytometry and cell sorting. Cytometry 10:282–293.

van Rotterdam A, Keij J, Visser JW (1992): Models for the electronic processing of flow cytometric data at high particle rates. Cytometry 13:149–154.

Wolfram S (1996): *Mathematica,* 3rd ed. New York: Addison-Wesley.

Zhang L, Cui X, Schmitt K, Hubert R, Navidi W, Arnheim N (1992): Whole genome amplification from a single cell: Implications for genetic analysis. Proc Natl Acad Sci 89:5847–5851.

4

Applications of High-Speed Sorting for CD34⁺ Hematopoietic Stem Cells

Thomas Leemhuis* and David Adams
SyStemix, Inc., Palo Alto, California

INTRODUCTION

Flow cytometry is the premier cell selection technology available for isolating very pure populations of rare-cell types for research purposes. Great advances have been made with this technology in the past decade, such that the potential applications are now quite diverse, provided that a single-cell suspension is possible and relatively small cell numbers are needed. In particular, the ability to quantitate and select rare populations of cells based upon multiple antigen–antibody combinations (and cell cycle status) simultaneously has enabled great advances in understanding hematopoietic function. Most clinical and experimental hematology laboratories now have flow cytometry capability and most publications in the field reference the technology either for analysis or cell selection purposes. The best example to date may be the way flow cytometry has contributed to understanding the pathogenesis of human immunodeficiency virus (HIV) and its effects on the immune system. The ability to test the viral effects on purified cell populations was (is) extremely useful. The real fascina-

---

*Currrent affiliation: Stanford University Medical Center, Stanford, California.

Emerging Tools for Single-Cell Analysis, Edited by Gary Durack and J. Paul Robinson.
ISBN 0-471-31575-3 Copyright © 2000 Wiley-Liss, Inc.

tion with this technology for us, however, lies in the potential for purifying rare-cell populations for in vivo study; therefore, this chapter will focus on high-speed sorting for hematopoietic stem and progenitor cells for use in human clinical trials.

Human hematopoietic stem cells (HSCs) are rare cells in bone marrow (<1%) and cytokine-mobilized peripheral blood (<5%) that are capable of self-renewal and differentiation into multiple hematopoietic lineages. The precise cell surface phenotype of these cells is widely debated among both clinical and experimental hematologists. Most, but not all, agree that hematopoietic stem and progenitor cells express the cell surface antigen known as CD34 (Strauss et al., 1986). Bone marrow cells expressing CD34 comprise only approximately 1% of harvested marrow but are still a functionally heterogeneous population of cells. Only a fraction of CD34$^+$ cells are believed to be responsible for long-term engraftment following bone marrow transplantation (BMT). High-speed cell sorting is ideally suited, in principle, to the purification of these rare cells, but the practical aspects of achieving a sufficient cell dose in a safe, reproducible, and cost-efficient manner are daunting.

This chapter has three purposes. First, a review of the potential clinical benefits of using high speed cell sorting to isolate hematopoietic stem cells for transplantation will be presented. A brief review of the different cell surface markers associated with hematopoietic reconstitution potential will be followed by a summary of those diseases that could, at least theoretically, be treated with primitive stem-cell populations. Second, we will present the unique instrument design and methodology requirements that must be taken into account when isolating cells for human clinical trials approved by the Food and Drug Administration (FDA). "Manufacturing" cells for use in phase I/II autologous or allogeneic transplantation studies in the United States requires compliance with current FDA regulations governing industry-sponsored cellular therapy trials. This chapter will present a brief review of those regulations and how they should translate into good quality-assurance practices. Several pages are devoted to a discussion of the SyStemix high-speed cell-sorting process that is currently approved by the FDA for production of clinical trial material. Although the focus is on hematopoietic cells, the same principles would apply to using flow cytometry for production of other cellular therapy products as well. Third, the chapter concludes with preliminary data from SyStemix's most recent clinical trials using high speed cell sorting to isolate CD34$^+$Thy-1$^+$ stem cells for the treatment of hematopoietic and nonhematopoietic malignancies.

CD34 AND THE STEM CELL PHENOTYPE

The CD34 antigen was first described in 1984 (Civin et al., 1984). It is a 120-kD cell surface protein expressed on immature hematopoietic cells and endothelial cells. Like most cell surface proteins, there is both a cytoplasmic and a cell-surface domain, and since several protein kinase C activators stimulate phosphorylation of the cytoplasmic domain, the antigen likely serves a signaling function (Barclay et al., 1993). The precise function of the molecule is unknown, but it is considered to be the best marker available to identify precursors of colony-forming cells in human bone marrow (Andrews et al., 1990). Extensive in vitro culture data support the understanding that CD34$^+$ cells are a heterogeneous population of primitive hematopoietic cells that vary in their ability to proliferate and differentiate into functional lineages. Clinical trial data showing effective hematopoietic rescue with CD34-selected autografts have clearly demonstrated that this population of cells is capable of hematopoietic reconstitution following myeloablative therapy (Berenson et al., 1991; Beguin et al., 1998).

The widespread use of bone marrow as a source of CD34$^+$ cells has mostly been replaced by the use of granulocyte-colony stimulating factor (G-CSF)-mobilized peripheral blood (MPB) for both allogeneic and autologous transplantation, because of the relative ease of collection and because clinical studies have demonstrated more rapid engraftment of both granulocytes and platelets compared to bone marrow grafts (Korbling et al., 1992). It is expected that the use of additional cytokines, such as stem cell factor (SCF) or Flt-3 ligand, will further improve mobilization (Basser et al., 1998).

True HSCs are defined as capable of both self-renewal and differentiation into all hematopoietic lineages. There are numerous in vitro assays that attempt to measure stem cell content of different cell populations, but either animal or human studies are required to definitively prove stem cell function. The ability to rescue an animal from lethal radiation exposure is the best measure possible, short of performing human clinical trials, for proving that isolated cell populations contain hematopoietic stem cells. Subfractionation of CD34$^+$ progenitor cells into subsets that are further enriched for more primitive hematopoietic function has been accomplished using other surface markers, such as the histocompability locus antigen (HLA) class II antigen HLA-DR (Brandt et al., 1988, 1990; Verfaillie et al., 1990; Bernstein et al., 1991), CD38 (Terstappen et al., 1991), CD45 (Lansdorp et al., 1990), and Thy-1 (Baum et al., 1992; Craig et al., 1993), or supravital dyes such as Rhodamine 123 (Visser et al., 1984; Sutherland et al., 1989; Srour et al., 1991) and Hoechst 33342 (Wolf et al., 1993; Leemhuis et al., 1996) in combination with CD34. Combining each of these markers with CD34 allows for a greater enrichment for HSCs than CD34 alone, but it is difficult to pinpoint the true stem cell phenotype. There is significant experimental evidence available from bone marrow studies to support the conclusion that the HSC is CD34$^+$, CD38$^-$, DR$^-$, Thy-1$^+$, c-kit$^+$, and Rhodamine 123^{lo} (Watt and Visser, 1992; Spangrude, 1994). Cells defined in this way contain a greater proportion of cells capable of long-term marrow reconstitution than do unfractionated CD34$^+$ cells, but it is not clear whether this phenotype defines a pure stem cell population. In any case, the ultimate objective of defining the surface characteristics of hematopoietic stem cells is to be able to identify (and therefore isolate) a subpopulation of cells that is capable of hematopoietic rescue following myeloablative therapy but is free from disease when isolated from the marrow or mobilized peripheral blood of cancer patients. Very primitive subpopulations may be less likely to have been affected by genetic alterations making the cells malignant and therefore make better autografts than whole bone marrow, mobilized peripheral blood apheresis products, or CD34-selected apheresis products.

It is not yet clear which subpopulation of CD34$^+$ cells, if any, might be best suited to transplantation for hematopoietic malignancy. In theory, the more markers that are used to identify primitive stem cells, the more likely those cells will be disease free, but also the more difficult it becomes to isolate sufficient numbers to achieve rapid and sustained engraftment. It also appears that the ideal subset to use for hematopoietic rescue after high-dose chemotherapy may vary according to disease type. If so, then flow cytometric cell sorting has the potential to serve as a common selection platform for customized graft engineering.

The Thy-1^+ subset of CD34$^+$ cells has been tested the most extensively in a clinical setting. SyStemix has been issued patents and has pending patent applications for this definition of the hematopoietic stem cell and has now conducted phase I trials with these cells as autografts in multiple myeloma, breast cancer, and non-Hodgkin's lymphoma (NHL). The hematopoietic properties of these cells were originally described in 1992 (Baum et al., 1992). These cells are capable of engraftment and multilineage differentiation in immunodeficient mice (Uchida et al., 1994; Murray et al., 1994, 1995), and preliminary human studies have shown that CD34$^+$Thy-1^+ HSC can also rescue a cancer patient's hematopoietic system following myeloablative therapy, provided they are given in sufficient numbers (Tricot et al., 1998). Significant in vitro and animal reconstitution data exist to support the stem cell nature of these cells, but there may well be other subsets of CD34$^+$ cells identified in the future that have better engraftment kinetics or that have less residual disease when isolated from patients with different disease states.

POTENTIAL CLINICAL APPLICATIONS

Since the bone marrow and MPB from cancer patients are frequently contaminated with tumor cells, the use of either tissue as an autograft could possibly contribute to relapse. The use of either CD34-selected or highly purified HSC autografts may prevent the reintroduction of tumor cells into these patients and therefore reduce relapse rates. There are likely to be diseases for which CD34 selection alone results in a tumor-free autograft and others for which highly purified HSCs are required. In theory, the use of highly purified HSCs as autografts would be preferable to more heterogenous CD34$^+$ populations for treating hematopoietic malignancies that are thought to originate at either the progenitor or precursor states of differentiation [e.g., Multiple Myeloma (MM), NHL, acute myelocytic leukemia (AML)]. In contrast, both purified HSCs and CD34-selected autografts might prove to be equally effective as therapies for non-hematopoietic malignancies that are known to metastasize to the bone marrow (e.g., breast cancer). Preliminary studies designed to test this hypothesis in breast cancer patients have shown that CD34 selection alone may not be sufficient to completely eliminate residual tumor. Four of 58 mobilized peripheral blood samples tested contained detectable cytokeratin$^+$ tumor cells after CD34 selection but were negative after further enriching for HSCs by high-speed sorting (Hanania, et al., 1997). Furthermore, this study revealed that there is a small percentage (<5%) of breast cancers that are both cytokeratin$^+$ and CD34$^+$ and are therefore unsuitable for either selection method.

The importance of obtaining a tumor-free graft for autologous transplantation in cancer patients has been debated extensively in the last decade and is still unresolved, largely because reliable residual disease assays are not available for most cancers and because most current purging methods are incapable of completely removing all detectable tumor. Some investigators believe that relapse is more likely to originate from residual tumor cells in the graft because of recent indications that many autografts are contaminated (Gazitt and Reading, 1996) before and after purging and because of evidence from gene-marking studies that residual (marked) tumor cells in an autograft are capable of contributing to relapse (Brenner et al., 1993; Deisseroth et al., 1994). Others believe that the number of tumor cells contained in an autograft is insignificant relative to residual disease in the host (Zhang et al., 1997). Only after reliable autograft purging methods are available will it be possible to determine whether the pretransplant therapy is effective in removing tumor from the host.

For example, high dose chemotherapy with autologous marrow or mobilized peripheral blood stem cell rescue has been extensively applied for the treatment of MM, resulting in improved event-free and overall survival when compared with standard chemotherapy (Attal et al, 1996; Barlogie et al., 1998). However, relapses are common and cure is unlikely in the majority of patients. Because both bone marrow and MPB are often contaminated with myeloma cells, it is conceivable that relapse after autotransplantation originates, at least in part, from the autografted tumor cells. Positive selection of $CD34^+$ cells reduces the contamination of myeloma cells from the apheresis products up to 3 logs and provides a cell suspension capable of restoring normal hematopoiesis after a myeloablative conditioning regimen (Lemoli et al., 1998; Schiller et al, 1998). Nevertheless, there is considerable evidence that the lesion underlying the malignant B-cell proliferation observed in MM occurs in an immature precursor cell that expresses CD34. Purification of $CD34^+$ $Thy-1^+$ cells from MPB by flow cytometry has been shown to deplete myeloma cells by up to 7 logs, according to very sensitive flow cytometric analysis of CD38-bright cells and quantitative PCR analysis of patient-specific complementarity-determining region III (CDRIII) DNA sequences (Gazitt et al., 1995). Patient-specific oligonucleotide probes for the mutated CDRIII region in malignant B cells have been used to demonstrate elimination of residual tumor cells below detectable levels in the $CD34^+Thy-1^+$ subset, whereas tumor cells remain detectable in the $CD34^+Thy-1^-$ subset (Reading et al., 1996). Therefore, CD34 selection alone is likely not sufficient for producing a tumor free graft in MM. Clinical trials to test the ability of $CD34^+Thy-1^+$ cells to provide hematopoietic support after myeloablative therapy for MM and other forms of cancer are ongoing. Preliminary results are presented later in this chapter.

Similarly, selection of $CD34^+$ DR^- cells from the bone marrow of chronic-phase CML (chronic myelogenous leukemia) patients has been shown to eliminate more BCR/ABL^+ leukemic cells than merely selecting for $CD34^+$ cells (Verfaillie et al., 1992; Leemhuis et al., 1993; Fogli et al., 1998). There is also limited evidence that tumor-propagating cells in certain types of NHL express CD34 (Macintyre et al., 1995), making it theoretically necessary to purify autografts further. Detection of residual disease is more difficult in NHL; therefore it is not known whether these tumor cells segregate into the $CD34^+Thy-1^-$ population. It is known that purification of the $CD34^+Thy-1^+$ cell fraction from normal bone marrow results in the elimination of 5–6 logs of $CD19^+$ B cells. Also, the lack of expression of Thy-1 on acute myeloid leukemia cells (Blair et al., 1997) makes selecting for $CD34^+Thy-1^+$ cells attractive as a potential therapy for the small number of AML clones that do not express CD34, but there have not been any clinical trials performed to confirm this hypothesis.

Thus it seems likely that, for at least some cancers, autologous transplantation with highly purified CD34+Thy-1+ HSCs could result in the reinfusion of significantly fewer tumor-propagating cells than transplantation with CD34-selected autografts. It seems reasonable to believe that elimination of tumor cells from the graft in combination with more effective pretransplant antitumor therapy might result in clinical benefit. There may also be a clinical benefit associated with infusing highly purified HSCs after myeloablative conditioning into patients with severe autoimmune diseases because of the 4–6-log T-cell and B-cell) depletion associated with the HSC purification process. In general, the more rigorous the HSC purification process, the greater the associated T-cell depletion effect. Also, the lower the T-cell dose, the lower the probability of reinfusing autoreactive T-cells; therefore a highly purified HSC autograft should be superior to a CD34-selected autograft for treating autoimmune disorders. Typically, a purified HSC autograft containing approximately 1×10^6 HSCs/kg would contain a T-cell dose of only 1×10^3/kg. Clinical trials are required in order to test these hypotheses. Previous studies with marginally T-cell-depleted autografts have demonstrated safety, but efficacy has not been thoroughly assessed (Burt et al., 1998).

Likewise, the use of purified HSCs, essentially free of alloreactive T cells, B cells, and granulocytes, permits clinical studies aimed at eliminating graft-vs-host disease (GVHD) from occurring in HLA-mismatched allogeneic transplants. The low frequency of immunocompetent alloreactive donor T cells in a purified HSC allograft, compared to a CD34- selected allograft, is likely to reduce the risk of GVHD but increase the risk of engraftment failure. Thus, sophisticated T-cell engineering studies are required, but perhaps more patients could receive potentially curative treatment if greater genetic disparity between host and allograft were tolerable.

Purified HSCs may also be superior to CD34+ cells as targets for gene therapy due to their self-renewal potential. Genetic modification of HSCs theoretically offers the chance of a lifetime cure, whereas the benefits of transducing a heterogeneous CD34+ population may be short term. Once stable, long-term genetic modification of large numbers of HSCs is possible, diseases resulting from enzyme deficiencies, such as Factor VIII, adenosine deaminase, and glucocerebrosidase, may be cured through transduction of the appropriate normal genes into HSCs. There is also preliminary evidence to suggest that transduction of HIV resistance genes into HSCs may render progeny T cells resistant to infection (Bonyhadi et al., 1997).

Finally, the use of purified allogeneic HSCs for in utero transplantation offers an opportunity to treat and possibly cure severe combined immunodeficiency and certain genetic metabolic disorders while the fetus is still tolerant to genetic disparity. The original proof of principle was demonstrated with xenogeneic transplantation experiments that demonstrated engraftment of purified human HSCs in sheep fetuses (Srour et al., 1993; Zanjani et al., 1995). It is hoped that the disease pathology may improve after engraftment of normal stem cells that compete with defective host cells. It is especially important for this type of allograft to contain as few immunocompetent T cells as possible to avoid GVHD.

The ability to use high-speed cell sorting to isolate hematopoietic subpopulations from MPB and BM lends itself well, in theory, to stem cell transplantation therapy for hematologic malignancy. The advantages are the high purity that can be achieved and the ability to select cells based on multiple cell surface markers simultaneously. The other selection devices being used to select CD34$^+$ cells for transplantation can achieve almost as high a purity in much less time but can select for only one marker at a time and therefore do not offer the possibility of further defining the graft composition. The major disadvantages are the technical difficulty and the length of time required to purify clinically relevant doses of rare cell types. Whereas 5–10 million highly purified primitive hematopoietic stem cells are more than sufficient for most in vitro or even small-scale animal reconstitution studies, at least 20-fold higher numbers are required for testing hematopoietic reconstitution in human trials. Neither the minimum nor the optimal CD34$^+$ or purified HSC dose has been established with certainty. Most studies suggest a dose range of $1-2 \times 10^6$ CD34$^+$ cells/kg in order to ensure rapid granulocyte *and* platelet engraftment in an autologous setting and from 2 to 5×10^6 CD34$^+$ cells/kg in an allogeneic setting (Bernstein et al., 1998; Zimmerman et al., 1994). Considering that primitive HSCs comprise between 10 and 60% of the CD34$^+$ cell population, it would be reasonable to conclude that between 5×10^5 and 1×10^6 purified HSCs/kg would be sufficient for long-term engraftment. For most individuals, this translates to a target HSC dose of approximately 100 million cells. The HSC percentages can range from 0 to 2% in G-CSF mobilized peripheral blood and from 0 to 1% in bone marrow; therefore it remains a great technical challenge to use flow cytometry alone to purify these large numbers of cells. For cell types that comprise greater than 30% of the starting material, purifying 100 million cells would be feasible using flow cytometry directly, but whenever the starting percentage is low, as it is for CD34$^-$, a preenrichment step is required prior to flow cytometric sorting. Otherwise, sort times become impractical. It could be done if sorting could somehow be automated so that multiple platforms are operating simultaneously, but even with "high speed sorting," sort times would range from 150 to 300 h to isolate 100+ million CD34$^+$ cells directly from G-CSF-mobilized peripheral blood. Examples of preenrichment steps are density gradient separation (Ferrero et al., 1998), centrifugal elutriation (Noga et al., 1986, 1991; Yoder et al., 1993), chemical depletion (Gazitt et al., 1995), and immunomagnetic separation (Ishizawa et al., 1993). Each has its own associated advantages and disadvantages. If judged solely on product purity and recovery, the immunomagnetic separation devices work very well, provided a high-quality monoclonal antibody source is available.

Small-scale experiments using the various immunomagnetic CD34 selection devices can result in very high purities, with reasonable recovery, but there is often considerable variability in results. In a recent direct comparison study, CD34$^+$ cells were enriched to a median purity of 92.2% (43.5–96.1% $n = 17$) with the Isolex device, 96.5% (66.6–99.2%, $n = 17$) with the MiniMACS, and 77.9% (31.4–93.6%, $n = 15$) with the Ceprate-LC. Median recovery of CD34$^+$ cells was 30.8% (18.6–71.8%) for Isolex, 69.9% (39.1–100%) for MiniMACS, and 42.9% (23.7–100%) for Ceprate-LC (Kruger et al., 1998). A similar (unpublished) experiment at SyStemix performed with a clinically relevant (1×10^{10}) starting population gave very similar results, except that the CD34 purity was lower (50%) after using the Ceprate system, and CD34 recovery from Isolex was above 80%.

The introduction of technology that allowed for multiparameter sort decisions to be made at rates of up to 25,000 events/s, without compromising purity or recovery, has contributed to advances in our understanding of hematopoiesis by enabling various rare cell types to be isolated in sufficient numbers for animal and human transplant studies. The slower-speed, widely available commercial flow cytometers manufactured by Becton Dickinson (San Jose, CA) and Beckman-Coulter (Hialeah, FL) are best suited to the smaller research samples encountered on a daily basis in most institutions, since cell recovery, not sorting time, is the primary concern. The high-speed cell sorters currently manufactured by SyStemix (Palo Alto, CA) and by Cytomation (Fort Collins, CO) are best suited to sorting the large clinical-scale tissues, since their fluidics are designed for high throughput and they can make sort decisions about five-fold faster. The high-speed sorters do tend to have a higher setup-associated cell loss during calibration, but this is insignificant (~2%) when sorting a clinical-scale tissue. The two types of technology are complementary, since small-scale research studies allow investigators to make predictions and high-speed sorting enables them to test those predictions in animals and in humans.

Further improvement in sort rates is not likely without dramatic changes in fluidics design. Maximizing sort rates is dependent on the ability of the electronics to make sort decisions quickly and on the ability to pass cells through a laser beam, single file, at high speeds without exposing the cells to lethal pressure differentials. While there seems to be no limit to the potential for increasing the sort decision rate in the future, the physical trauma that the cells endure during the droplet generation process will likely prove to be the limiting factor for continued improvement. Not only is it hard to imagine viable cells surviving the increased stream velocity that would be required to take full advantage of faster electronics, but as the pressures and oscillation frequencies are being increased, it becomes more and more difficult to maintain a stable droplet break-off point. Future rate increases may eventually require use of a non-droplet-generating method such as that used in the optical sorter (Roslaniec et al., 1996, 1997).

SYSTEMIX PROCESS

For those diseases where rigorous and reproducible purification of HSCs is necessary to eliminate residual tumor cells or mature immunocompetent cells, flow cytometric cell sorting is the ideal method for end-stage purification, but it must follow one or more presort enrichment steps. SyStemix has developed a cell processing scheme that begins with shipping MPB apheresis products to a processing center and consists of CD34 enrichment, high-speed sorting for $CD34^+Thy-1^+$ cells, cryopreservation, shipment to the clinical site, thawing, and infusion. Since these cells are used for human transplantation studies, the manufacturing process as a whole, including the high speed sorting procedure, is compliant with current Good Manufacturing Practices (cGMPs) and with current U.S. FDA regulations governing industry-sponsored cellular therapy trials. Quality management is an integrated part of the manufacturing process.

With mobilized peripheral blood, the "process" begins with the HSC mobilization in the patient and ends after the purified HSCs are reinfused back into the patient. Mobilization regimens are changing rapidly as new cytokines are found to be beneficial, but most often mononuclear cells (and HSCs) are harvested by apheresis following the fifth and sixth day of G-CSF administration (10 µg/kg/day). Alternatively, the combination of G-CSF or GM-CSF with chemotherapy (e.g., cytoxan, 4 g/m^2) is useful for HSC mobilization in patients who have received multiple courses of prior chemotherapy. Following G-CSF mobilization, the apheresis product contains from 2 to 5×10^{10} cells and the HSC frequency ranges from 0.1 to 1%, whereas with G-CSF + cytoxan mobilization, each apheresis product typically contains between 0.1 and 1×10^{10} cells with the HSC frequency ranging from 1 to 5%. Depending on the degree of mobilization and the target number of HSCs, one to three aphereses are typically required to achieve the desired HSC dose. Before shipment to SyStemix, the individual MPB samples are diluted to a final concentration of less than 5×10^7 cells/ml with a proprietary transport medium. Once diluted, the cells can be stored (or shipped) for 24–36 h at 20–24°C without compromising HSC viability.

It is immediately apparent that it would be nearly impossible to sort such a large cell number (2×10^{10}) single file through a flow cytometer, even with recent technological developments. Such a task would require approximately 300 h of continuous sorting at 20,000 events/s; therefore, some form of CD34 enrichment must be performed in advance. The combination of counterflow centrifugal elutriation and phenylalanine methyl ester (PME) lysis was originally used as the presort process to eliminate erythrocytes and mature myeloid cells prior to sorting for CD34$^+$Thy-1$^+$ cells. This presort process reduced the sample size by approximately 1 log, but the samples still required anywhere from 12 to 36 h of continuous sorting to process. For the last few years, SyStemix has utilized an immunomagnetic CD34 selection device to enrich for CD34$^+$ cells prior to sorting for CD34$^+$Thy-1$^+$ cells. Whereas the elutriation/PME depletion method enriched CD34$^+$ cells approximately 5-fold, the positive selection method enriches CD34$^+$ cells approximately 40-fold. This reduction in presort sample size has resulted in a 4–8-h sorting process for each apheresis product and has thus greatly simplified the associated staffing issues.

The immunomagnetic CD34 selection process consists of four steps: (1) labeling cells with an anti-CD34 monoclonal antibody (MAb); (2) mixing cells with paramagnetic beads coated with sheep anti-mouse immunoglobulin G (IgG), whereby CD34$^+$ cells form rosettes with the beads; (3) attaching the rosettes to a magnet and washing CD34$^-$ cells from the system; and (4) releasing the cells from the beads using a peptide mimotope that competes with the MAb binding site of the CD34 molecule. The process is designed to accommodate a wide variety of incoming tissues arriving from distant sites for processing in a reproducible, error-proof manner and requires approximately 6 h to perform. The process for isolating HSCs from BM is identical with that for MPB except that mobilization is not necessary, there is a bone marrow harvest instead of an apheresis procedure, and a buffy coat is prepared prior to CD34 selection.

Subsequently, the cells are stained with fluoresceinated MAbs directed against a different epitope of CD34 and against the Thy-1 antigen and are sorted on the high-speed clinical cell sorter (HSCCS). The purified HSC product is tested for purity, viability, and sterility and then cryopreserved using a controlled-rate freezer. Once the target dose is achieved and all products are released, the HSC infusion products are shipped to the clinical sites in a liquid nitrogen shipper and, upon arrival, returned to the liquid phase of liquid nitrogen until the time of infusion. Upon thaw, the cells are slowly diluted with 2 volumes of saline before reinfusion via an intravenous line.

REGULATORY AND QUALITY ASSURANCE ISSUES

Regardless of the cell type used, the FDA needs to approve the "manufacturing process" before any patients can be treated. The FDA published a draft "Points to Consider" document in 1991 that addressed their regulatory strategy for cell and gene therapy products. The traditional regulatory framework for pharmaceutical products was implemented for cell therapy products. Investigational New Drugs (IND) applications were required to initiate clinical trials, and manufacturing sites and processes were required to be licensed. The 1993 Federal Register Notice reinforced the requirement for conformity to existing regulations, including cGMPs. These standards were similar to the current quality assurance/good manufacturing practices defined for blood and tissue banks by the FDA in the late 1980s. They include organizational standards to ensure appropriate medical control of patient care and laboratory standards for collecting, processing, and storing patient tissues. In 1997, the FDA published another draft proposal for the regulation of cell- and tissue-based products, whereby the agency outlined distinct areas in which no FDA submissions were required and where FDA oversight would be limited. In this proposal, the FDA did not reduce its own authority to control the use of these cellular products, but it did remove some of the need for sites to be licensed to perform such manipulations. A distinction was drawn for autologous stem cell products that, when used for homologous purposes (e.g., hematopoietic reconstitution), do not require licensing by the FDA. The term "nonhomologous" was introduced to distinguish those products being altered in some way prior to infusion (ex vivo expansion and gene modification) and those products being used for purposes other than their natural function from the "minimally manipulated" cellular products, in terms of regulatory requirements. Details of the current and proposed regulations can be obtained by writing to the agency or via the Internet at www.FDA.gov.

Whether the FDA or individual professional societies hold ultimate responsibility for the regulation of these products, the resulting quality standards will likely resemble those imposed on traditional pharmaceutical products. Substantial emphasis is placed on control of the final product by requiring lot-to-lot release based on approved specifications for purity, identity, potency, and sterility. Exceptions to the rules will have to be granted on an individual basis due to the many inherent differences involved in producing a large quantity of a defined chemical product versus producing a small quantity of a living cellular product, but the "spirit" of the regulations will be very similar. Ultimately, each individual company is responsible for assuring that its product is both safe and effective for its intended use. Proving that each and every lot manufactured meets preestablished quality standards consumes a significant percentage of the effort and cost required to produce a cellular therapy product. It is essential that a system be in place to ensure that all tissues are handled according to "good tissue practices" aimed at preventing contamination and preserving integrity and function. Most often, final product quality control testing is limited by the amount of product (i.e., the number of cells) available; therefore, a comprehensive quality assurance support system needs to be in place to supplement final product testing. For example, environmental monitoring programs and process validation experiments are crucial when there is an insufficient number of cells to test for sterility on a lot-to-lot basis.

In order to argue effectively that its products are 100% safe, a company must implement a quality assurance program that starts with proper facility and process design and includes equipment validation, raw material testing, documentation control, vendor qualification, personnel training, and internal auditing. Ideally, the manufacturing area is designed for handling potentially infectious agents in order to protect the products from cross-contamination and the personnel from the spread of infectious agents. The air-handling units within the facility should provide single-pass (100% outside) HEPA-filtered air at a slightly negative pressure relative to adjacent corridors. Separation of activities into separate processing suites with separate air-handling systems, gowning rooms, material flow patterns, cleaning stations, and waste flow patterns will ensure that cross-contamination does not occur. In order to maximize biosafety, disposable materials should be used for processing whenever possible. If reusable product-contact equipment is required, labor-intensive and expensive validation of all cleaning and sterilization procedures is necessary in order to prove that cross-contamination is impossible. It must also be kept in mind that these cellular products are only as sterile as the environments in which they are manufactured and that routine monitoring can be extremely useful in preventing accidental contamination. All open-container manipulations should occur in a class-100 biosafety cabinet that is disinfected between each use. Monitoring the air within the biosafety cabinet for microbial contamination during processing can result in data that support final product sterility claims. The document control procedures should force all manufacturing and quality control personnel to operate from the same set of current, peer-approved procedures. All raw material specifications, test methods, standard operating procedures, and manufacturing batch records should be controlled in a database capable of tracking the current status of each document. A copy of all relevant procedures is kept in the manufacturing or testing areas. All effective procedures require multidepartmental signature and significant justification for change. A history file is kept on each document so that it is possible to determine retrospectively which procedures were effective at the time a particular lot of material was manufactured.

Application of these quality assurance principles to a flow cytometric cell-sorting process is difficult, to say the least. These are not "black box" instruments designed for routine manufacturing use, especially when all fluidics, including the nozzle itself, need to be sterilized between different patients. They require talented, dedicated operators to keep them functioning reliably, and significant effort is required to verify in-process and final product quality. With these considerations in mind, SyStemix

Fig. 4.1. SyStemix high-speed clinical cell sorter. (**A**) Panoramic view of HSCCS, showing the optical table, isolator box with transfer and glove ports, electronics console, and computer interface. This is one of five instruments being used to purify human hematopoietic stem cells for clinical trials in the United States and France. (**B**) Close up of illumination chamber inside the isolator box from directly in front. Note disposable fluidics and chilled postsort sample storage capability. The camera provides a close-up view of the laser intercept and droplet break-off points. (**C**) View from the back of the instrument shows the current dual-laser configuration. Efforts are underway to replace these with less expensive air-cooled lasers. (**D**) Side view of the HSCCS showing the operator at work fine tuning the alignment during the course of a sterile sort.

chose to design and build its own HSCCS for isolating clinically relevant numbers of HSCs. Although at least three commercially available cell sorters can now available that sort at relatively high acquisition rates, viable cell recovery, instrument sterilization, and containment of potentially pathogenic aerosols remain as obstacles to using those instruments to purify cells for clinical use.

The HSCCS pictured in Figure 4.1, is capable of sorting at rates exceeding 20,000 events/s. The instrument is based on the original design of van den Engh and Stokdijk (1989) at Lawrence Livermore National Laboratories but was modified extensively at SyStemix (Sasaki et al., 1995). SyStemix has used this instrument to purify CD34$^+$Thy-1$^+$ HSCs for stem cell transplants in breast cancer, MM, and NHL. A description of the unique technical and procedural considerations required for using this instrument for production of clinical trial material follows.

SyStemix's HSCCS is a dual-laser instrument (488 nm primary and 633 nm secondary). The beams are individually focused to 40-μm spots that are spatially separated by 60 μm along the axis of the fluid jet. The detector table contains a photodiode for

FSC detection (Fig. 4.1B) and six photomultiplier tubes (Fig. 4.1C). With this configuration, up to five fluorescence (FL) signals (three primary and two secondary) may be detected simultaneously with two scatter signals. Bidirectional compensation between the FL1 and FL2 channels is built into the hardware. The instrument uses a commercially available nozzle assembly with a 70-µm tip (Becton Dickinson, San Jose, CA). A 10-l sheath tank is used to deliver phosphate-buffered saline (PBS) to the nozzle at a pressure of 44 psi for up to 30 hours of continuous, stable sorting. The nozzle assembly is oscillated at a frequency of 62 kHz, which provides a jet velocity of approximately 20 m/s and a stable droplet break-off at approximately 0.4 in. The instrument uses "off-the-shelf" mirrors and filters for ease of repair and optical customization.

Sterilization of all product-contact instrument components is essential and is one of the major unique technical challenges that must be faced when sorting cells for clinical use. Sterilization protocols must be validated and are all encompassing. Detailed, written cleaning and sterilization procedures must describe ways to sterilize all reusable components of the fluidic assembly, calibration particles, and immediate cytometer environment surrounding the sorted cells.

The fluidics can either be sterilized by chemical exposure or else be purchased sterile and replaced between patients. Some investigators use 70% ethanol followed by copious amounts of either water or PBS to flush out the ethanol. SyStemix's HSCCS was designed to utilize off-the-shelf medical tubing for easy access and replacement. The tubing is discarded and replaced between patients and is exposed to 30% hydrogen peroxide between multiple tissues from the same patient. Users of commercial cytometry equipment, such as the FACS Vantage (Becton Dickinson) or the Altra (Beckman-Coulter) are less able to routinely replace the tubing, since it is not easily accessed or available presterilized. In that situation, a chemical sterilization method could be developed, but this would require additional turn-around time between patients as well as the development and cost of tests to verify sterility and clearance of sterilant from the lines. While in-place sterilization is an option, it is certainly less convenient than the replacement system used on SyStemix's HSCCS.

Reusable components such as sheath tanks, waste tanks, vial adapters, and nozzle assemblies need to be sterilized between patients. It is best if these components are designed to be autoclavable. For the sheath tanks in particular, filters can be attached to the air-in port as well as the sheath fluid-out ports prior to autoclaving. The nozzle assembly presents a somewhat more difficult problem. Few available nozzle assemblies can withstand repeated autoclave cycles. Some, such as the one used on the HSCCS, can withstand approximately five cycles but inevitably are damaged by the high heat and pressure of the autoclave. Others, such as the BD macrosort, contain the oscillation mechanism within the nozzle assembly itself and are less likely to withstand even those five cycles. Considering the replacement cost and effort required to validate a nozzle assembly for clinical use, an alternative sterilization method is required. SyStemix's nozzle assemblies are sterilized by either radiation (e-beam) or hydrogen peroxide (H_2O_2). The latter can be labor intensive but is certainly less damaging than autoclaving. E-beam sterilization, on the other hand, is performed by an outside vendor, has a rapid turn-around time, and requires no in-house labor. The nozzle assemblies eventually wear out after repeated e-beam sterilization but have a mean time to failure of 15 cycles compared to 5 for autoclaving. SyStemix has validated that exposure to either H_2O_2 or e-beam radiation, used as circumstances require, provides sufficient sterilization for the nozzle assembly, whereas autoclaving is suitable for all the other reusable components. Once an appropriate cycle is developed, sterilization becomes a simple and well-established routine, but if any one of these components is not kept sterile, the sample, and therefore the patient, is put at risk.

Regarding reagents, it is of obvious importance to use only sterile reagents throughout the process. Some commercially available sheath fluids contain preservatives such as sodium azide that inhibit proliferation, but sterile, endotoxin-free Dulbecco's phosphate-buffered saline (DPBS) can be either made in-house or purchased from a number of vendors and aseptically pumped into presterilized tanks. All calibration particles used to align the instrument need to be sterilized prior to use, since all previous sterilization efforts are nullified if a vial of contaminated beads is used to align the instrument. Sterile calibration particles are not readily available, but most commercial particles can be sterilized by mixing with H_2O_2. Although the process is time consuming, large quantities of calibration particles can be prepared at one time and transferred to single-use vials for later use.

The final sterilization issue to be considered is the immediate environment surrounding the tissue while it is being sorted. Although commercial cytometers are equipped with a flip-up or slide-closed sample chamber, they are difficult to sterilize and not likely to maintain sterility throughout the procedure. These difficulties inspired SyStemix to develop its own proprietary cytometer isolator system (Vardanega et al., 1997). This isolator completely encloses the sample area on the cytometer and thus protects both the tissue and the operator (Fig. 4.1A). Sterilization is accomplished via an H_2O_2 vapor generator (Steris, Mentor, OH). A pressure regulator on top of the isolator is fitted with incoming and outgoing HEPA filters (Fig. 4.1D) and maintains the isolator under slight positive pressure during the sort. The transfer unit on the left of the isolator fits to a transfer port on SyStemix's customized biosafety cabinets and allows materials to be passed into the isolator under aseptic conditions. Users of commercial equipment would likely be required to either modify their cytometers for an isolator system or place the cytometer in a clean room in order to maintain the sterile environment required for clinical sorting.

IN-PROCESS AND FINAL PRODUCT TESTING

With high-speed sorting, any small error in calibration can have significant detrimental effects on the recovery of the target cell population; therefore, a battery of in-process test procedures should be implemented to minimize this possibility. Because of the high pressures involved, a leak or a cracked line could cause significant loss of the patient sample within a matter of seconds, so all fluidic lines and components are stress tested immediately prior to sample introduction. Since the signal optimization and region determination steps can be very costly in terms of cell numbers when operating at high speeds, it can be helpful to analyze a small aliquot on a desktop analyzer, such as the FACSCalibur (Becton Dickinson), prior to analyzing the sample on the HSCCS. This instrument requires very few cells to determine approximate sort percentages and it provides an immediate check of sorter alignment. If significant variance is detected, the sample can be immediately removed and the system re-aligned. Once alignment is confirmed and the sort regions have been determined, a short 50,000-event sort is performed at the expected sort rate and tested for purity on a FACSCalibur before proceeding with the clinical sort. If the purity is below 90%, the drop delay calibration is repeated. Similarly, after the HSCCS indicates that the first million cells have been collected, an aliquot is removed and a cell count performed via a hemacytometer. If the hemacytometer cell count agrees with the instrument counter (within 70%), then the full-scale sort proceeds until all the cells are sorted, but if the recovery is not acceptable, then the drop delay calibration needs to be adjusted before proceeding to sort the remainder of the tissue. During the sort, it is essential to monitor and record laser performance, signal intensities, sort percentages, flow rate, drop distance, collection volume, and center and side stream integrity. A minor drift in any of these can affect sorter performance and drastically reduce overall cell recovery. If purity or recovery has deteriorated, the sort must be stopped and a fresh collection tube used to perform another purity or recovery check.

Final product testing should focus on supporting product safety and potency claims by testing for cell number, sterility, viability, and purity. The sterility and viability assays are most important in early phase 1 and 2 clinical trials that emphasize patient safety as their endpoint. Final product dose specifications are based on the postsort purity and the total cell number; therefore, those assays must also be developed in advance of clinical trial production. A representative pre- and postsort purity test used to determine the CD34$^+$Thy-1$^+$ percentage is illustrated in Figure 4.2. Since biopotency assays take several weeks to complete and are frequently difficult to interpret, they are

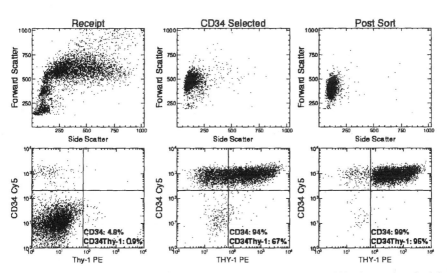

Fig. 4.2. Flow cytometric analysis of typical incoming mobilized peripheral blood at receipt, after initial CD34 selection, and after sorting for CD34$^+$Thy-1$^+$ cells. The first row shows the light scatter properties and the second row shows the CD34 and Thy-1 fluorescence of the stained cells at the various stages of purification. These data were generated on a FACSCalibur during SyStemix's routine quality control monitoring of their HSC purification process.

not often used for release testing. Nevertheless, early in the method development process it is important to develop (and ultimately validate) a relevant biopotency assay that can be used to evaluate process improvements and deviations. For example, a combination of three different biopotency assays (CFU-C, CAFC, and SCID-HU) was used early in SyStemix's development process to show that the purified HSCs were still capable of proliferating and engrafting after having been exposed to the intense laser light and high barometric pressures associated with flow cytometric cell sorting.

RESULTS

As mentioned previously, SyStemix has conducted a limited number of clinical trials designed to test the safety and efficacy of purified CD34$^+$Thy-1$^+$ HSC autografts for the treatment of MM, NHL, and breast cancer. The MM trial was performed using purified CD34$^+$Thy-1$^+$ stem cells that were isolated using the combination of elutriation and PME lysis prior to high-speed cell sorting. The breast cancer and NHL trials were performed with HSCs isolated using a combination of immunomagnetic CD34

TABLE 4.1. NHL and BrCA Clinical Trial Processing Results (%)

Presort HSC purity (Mean %CD34$^+$Thy-1$^+$)	51
Presort HSC recovery[a] (mean %)	87
Postsort viability[b] (mean %)	96
Sort rate (target event rate per second)	15,000
Sort duration (mean hours)	5
Abort rate (mean %)	23
Postsort HSC purity (Mean % CD34$^+$Thy-1$^+$)	93
Postsort HSC step recovery[a] (mean %)	48
Overall process recovery[c] (mean %)	33

[a]HSC step recovery = number of CD34+Thy-1$^+$ cells post-process divided by number of CD34$^+$Thy-1$^+$ cells preprocess.

[b]Postsort viability was determined by trypan blue dye exclusion.

[c]Overall process recovery = number of CD34+Thy-1$^+$ cells cryopreserved divided by number of CD34+Thy-1$^+$ cells received.

TABLE 4.2. SyStemix Autologous HSC Engraftment Data

HSC Dose ($\times 10^6$/kg)	Median Days to ANC, $>0.5 \times 10^9$/l	Median Days to Platelets, $>50 \times 10^9$/l
All patients	12	28
(n=46)	(9–26)	(9–NR)
Dose < 0.7	13	35
(n=31)	(11–26)	(16–NR)
Dose >0.7	10	17
(n=15)	(9–12)	(9–94)

selection and high-speed sorting. Process recovery, viability, and final product purity results from these trials are summarized in Table 4.1. SyStemix's clinical-scale high-speed sorting process now consistently results in >90% CD34 purity, most often above 95%. The CD34$^+$Thy-1$^+$ purity ranged from 85 to 95% and averaged 93%. The exceptional purity comes at the cost of a relatively low CD34 recovery. If selecting for all CD34$^+$ cells, the postsort recovery would average approximately 50% of the presort CD34$^+$ cell number and approximately 35% of the CD34$^+$ cell number in the apheresis product. Because the sort criteria intentionally select for a subset of CD34$^+$ cells (CD34$^+$Thy-1$^+$), the actual CD34 recovery is approximately 15% and the CD34$^+$Thy-1$^+$ cell recovery averages approximately 35% of the incoming product.

Short-term engraftment data (Table 4.2) clearly indicate that these cells remain functional after high-speed sorting. The HSC dose infused varied from 0.14 to 3.5 × 10^6/kg recipient body weight. This range of cell doses produced neutrophil recovery to an absolute neutrophil count (ANC) > 0.5 × 10^9/l in a median of 12 days (range 9–26 days) and a platelet count of >50 × 10^9/l at a median of 28 days, with a range of 9–NR (not reached) days.

Patients who received an HSC dose of <0.7 × 10^6/kg had slower neutrophil and platelet engraftment than those who received a larger cell dose. The hematological recovery in patients who received an HSC dose of >0.7 × 10^6/kg is comparable with that seen with CD34-selected or unselected MPB and is considered safe and clinically acceptable. Above this threshold dose, engraftment is reliably rapid, although there were outliers who showed delayed platelet recovery. This was often found to be associated with infections that began during the period of neutropenia and resulted in increased platelet consumption, despite each of the transplanted products having been proven sterile.

It is obviously too soon to know whether highly purified CD34$^+$Thy-1$^+$ HSC autografts will ultimately prove to be superior to unfractionated MPB, or even CD34-selected autografts. If so, then further development of high-speed sorting technology would be required in order to make it feasible to use for routine production of pharmaceutical-grade material. At this point in time, the technology is better suited to custom autograft production for early phase I/II clinical trials than it is for large-scale phase III trials or commercialization. Because of its ability to select cells based on multiple parameters, high-speed sorting works best as a platform for identifying and testing new cell types in a variety of disease states. Once a particular cell population has proven useful, it is currently more practical to use an immunomagnetic selection device to isolate those cells, provided that only one or maybe two selection criteria are required. Future development efforts should therefore focus on reducing the cost of the instrumentation and on increasing the capacity to purify larger quantities of cells in a shorter amount of time, perhaps by improving the ease of alignment, fluidic stability, and instrument reliability so that automation might be possible.

REFERENCES

Andrews RG, Singer JW, Bernstein ID (1990): Human hematopoietic precursors in long-term culture: Single CD34$^+$ cells that lack detectable T cell, B cell, and myeloid cell antigens produce multiple colony-forming cells when cultured with marrow stromal cells. J Exp Med 172(1):355–358.

Attal M, Harousseau JL, Stoppa AM, Sotto JJ, Fuzibet JG, Rossi JF, Casassus P, Maisonneuve H, Facon T, Ifrah N, Payen C, Bataille R (1996): A prospective, randomized trial of autologous bone marrow transplantation and chemotherapy in multiple myeloma. N Engl J Med 335:91–97.

Barclay AN, Birkeland ML, Brown MH, Beyers AD, Davis SJ, Somoza C, Williams AF (eds) (1993): *The Leukocyte Antigen Facts Book.* New York: Academic Press, p 176.

Barlogie B, Jagannath S, Tricot G, Desikan KR, Fassas A, Siegel D (1998): Advances in the treatment of multiple myeloma. Adv Intern Med 43:279–320.

Basser RL, To LB, Begley CG, Maher D, Juttner C, Cebon J, Mansfield R, Olver I, Duggan G, Szer J, Collins J, Schwartz B, Marty J, Menchaca D, Sheridan WP, Fox RM, Green MD (1998): Rapid hematopoietic recovery after multicycle high-dose chemotherapy: Enhancement of filgrastim-induced progenitor cell mobilization by recombinant human stem cell factor. J Clin Oncol 16(5):1899–1908.

Baum CM, Weissman IL, Tsukamoto AS, Buckle AM, Peault B (1992): Isolation of a candidate human hematopoietic stem cell population. Proc Natl Acad Sci USA 89(7):2804–2808.

Beguin Y, Baudoux E, Sautois B, Fraipont V, Schaaf-Lafontaine N, Pereira M, Paulus JM, Sondag D, Fillet G (1998): Hematopoietic recovery in cancer patients after transplantation of autologous peripheral blood CD34+ cells or unmanipulated peripheral blood stem and progenitor cells. Transfusion 38(2):199–208.

Berenson RJ, Bensinger WI, Hill RS, Andrews RG, Garcia-Lopez J, Kalamasz DF, Still BJ, Spitzer G, Buckner CD, Bernstein ID (1991): Engraftment after infusion of CD34$^+$ marrow cells in patients with breast cancer or neuroblastoma. Blood 77:1717–1722.

Bernstein SH, Nademanee AP, Vose JM, Tricot G, Fay JW, Negrin RS, DiPersio J, Rondon G, Champlin R, Barnett MJ, Cornetta K, Herzig GP, Vaughan W, Geils G Jr, Keating A, Messner H, Wolff SN, Miller KB, Linker C, Cairo M, Hellmann S, Ashby M, Stryker S, Nash RA (1998): A multicenter study of platelet recovery and utilization in patients after myeloablative therapy and hematopoietic stem cell transplantation. Blood 91(9):3509–3517.

Bernstein ID, Andrews RG, Zsebo KM (1991): Recombinant human stem cell factor enhances the formation of colonies by CD34$^+$ and CD34$^+$Lin$^-$ cells, and the generation of colony-forming cell progeny from CD34$^+$Lin$^-$ cells cultured with interleukin-3, granulocyte colony-stimulating factor, or granulocyte-macrophage colony-stimulating factor. Blood 77:2316–2321.

Blair A, Hogge D, Ailles L, Lansdorp P, Sutherland H (1997): Lack of expression of Thy-1 (CD90): on acute myeloid leukemia cells with long-term proliferative ability in vitro and in vivo. Blood 89:3104–3112.

Bonyhadi M, Moss K, Voytovich A, Auten J, Kalfoglou C, Plavec I, Forestell S, Boehnlein E, Kaneshima H (1997): RevM10-expressing T cells derived in vivo from transduced human hematopoietic stem-progenitor cells inhibit human immunodeficiency virus replication. J Virol 71:4704–4716.

References

Brenner MK, Rill D, Moen R, Krance R, Mirrow J, Anderson W, Ihle J (1993): Gene marking to trace origin of relapse after autologous bone marrow transplantation. Lancet 341:85–86.

Brandt J, Srour EF, vanBesien K, Briddell RA, Hoffman R (1990): Cytokine-dependent long-term culture of highly enriched precursors of hematopoietic progenitor cells from human bone marrow. J Clin Invest 86(3):932–941.

Brandt JE, Baird N, Lu L, Srour E, Hoffman R (1988): Characterization of a human hematopoietic progenitor cell capable of forming blast cell containing colonies in vitro. J Clin Invest 82:1017–1027.

Burt RK, Traynor AE, Cohen B, Karlin KH, Davis FA, Stefoski D, Terry C, Lobeck L, Russell EJ, Goolsby C, Rosen S, Gordon LI, Keever-Taylor C, Brush M, Fishman M, Burns WH (1998): T cell depleted autologous hematopoietic stem cell transplantation for multiple sclerosis: report on the first three patients. Bone Marrow Transplant 21(6):537–541.

Civin CI, Strauss LC, Brovall C, Fockler MJ, Schwartz JF, Shaper JH (1984): Antigenic analysis of hematopoiesis III: A hematopoietic progenitor cell surface antigen defined by a monoclonal antibody raised against KG-1a cells. J Immunol 133:157–161.

Craig W, Kay R, Cutler RL, Lansdorp PM (1993): Expression of Thy-1 on human hematopoietic progenitor cells. J Exp Med 177(5):1331–1342.

Deisseroth AB, Zu Z, Claxton D, Hanania EG, Fu S, Ellerson D, Goldberg L, Thomas M, Janicek K, Anderson WF, Hester J, Korbling M, Durett A, Moen R, Berenson R, Heimfeld S, Hamer J, Calvert L, Tibbits P, Talpaz M, Kantarijian H, Champlin R, Reading C (1994): Genetic marking shows that Ph$^+$ cells present in autologous transplants of chronic myelogenous leukemia (CML) contribute to relapse after autologous bone marrow transplantation in CML. Blood 83(10):3068–3076.

Ferrero D, Tarella C, Cherasco C, Bondesan P, Omede P, Ravaglia R, Caracciolo D, Castellino C, Pileri A (1998): A single step density gradient separation for large scale enrichment of mobilized peripheral blood progenitor cells collected for autotransplantation. Bone Marrow Transplant 21(4):409–413.

Fogli M, Amabile M, Martinelli G, Fortuna A, Rondelli D, Ratta M, Curti A, Tura S, Lemoli RM (1998): Selective expansion of normal haemopoietic progenitors from chronic myelogenous leukaemia marrow. Br J Haematol 101(1):119–129.

Gazitt Y, Reading C (1996): Autologous transplantation with tumor-free graft: a model for multiple myeloma patients. Leuk Lymphoma 23:203–212.

Gazitt Y, Reading C, Hoffman R, Wickrema A, Vesole DH, Jagannath S, Condino J, Lee B, Barlogie B, Tricot G (1995): Purified CD34$^+$ Lin$^-$ Thy$^+$ stem cells do not contain clonal myeloma cells. Blood 86(1):381–389.

Hanania EG, Kshirsagar B, Sriti Z, Wu J, Cataniag F, Ramanathan R, Reading C, Schnell J (1997): Detection of minimal residual disease in breast cancer using immunofluorescent microscopy and 2-color flow cytometry. Blood 90(10):352b.

Ishizawa L, Hangoc G, Van-de-Ven C, Cairo M, Burgess J, Mansour V, Gee A, Hardwick A, Traycoff C, Srour E, Hoffman R, Law P (1993): Immunomagnetic separation of CD34+ cells from human bone marrow, cord blood, and mobilized peripheral blood. J Hematotherapy 2(3):333–338.

Korbling M, Juttner C, Henon P, Kessinger A (1992): Autologous blood stem cell versus bone marrow transplantation. Bone Marrow Transplant 10(Suppl 1):144–148.

Kruger W, Gruber M, Hennings S, Fehse N, Fehse B, Gutensohn K, Kroger N, Zander AR (1998): Purging and haemopoietic progenitor cell selection by CD34$^+$ cell separation. Bone Marrow Transplant 21(7):665–671.

Lansdorp PM, Sutherland HJ, Eaves CJ (1990): Selective expression of CD45 isoforms on functional subpopulations of CD34$^+$ hemopoietic cells from human bone marrow. J Exp Med 172(1):363–366.

Leemhuis T, Yoder M, Grigsby S, Aguero B, Eder P, Srour EF (1996): Isolation of primitive human bone marrow hematopoietic progenitor cells using Hoechst 33342 and Rhodamine 123. Exp Hematol 24(10):1215–1224.

Leemhuis T, Leibowitz D, Cox G, Srour EF, Tricot G, Hoffman R (1993): Identification of BCR/ABL negative primitive hematopoietic progenitor cells within chronic myeloid leukemia marrow. Blood 81(3):801–807.

Lemoli RM, Fortuna A, Raspadori D, Ventura MA, Martinelli G, Gozzetti A, Leopardi G, Ratta M, Cavo M, Tura S (1998): Selection and transplantation of autologous hematopoietic CD34$^+$ cells for patients with multiple myeloma. Leuk Lymphoma 26(Suppl 1):1–11.

Macintyre E, Belanger C, Debert C, Canioni D, Turhan A, Azagury M, Hermine O, Varet B, Flandrin G Schmitt C (1995): Detection of clonal CD34$^+$19$^+$ progenitors in bone marrow of BCL2-IgH-positive follicular lymphoma patients. Blood 86:4691–4698.

Murray L, Chen B, Galy A, Chen S, Tushinski R, Uchida N, Negrin R, Tricot G, Jagannath S, Vesole D, Barlogie B, Hoffman R, Tsukamoto A (1995): Enrichment of human hematopoietic stem cell activity in the CD34$^+$Thy-1$^+$Lin$^-$ subpopulation from mobilized peripheral blood. Blood 85(2):368–378.

Murray L, DiGiusto D, Chen B, Chen S, Combs J, Conti A, Galy A, Negrin R, Tricot G, Tsukamoto A (1994): Analysis of human hematopoietic stem cell populations. Blood Cells 20:364–369.

Noga SJ, Davis JM, Thoburn CJ, Donnenberg AD (1991): Lymphocyte dose modification of the bone marrow allograft using elutriation. In Gee AP (ed). *Bone Marrow Transplantation.* Boca Raton, FL: CRC Press, pp 175–200.

Noga SJ, Donnenberg AD, Schwartz CL, Strauss LC, Civin CI, Santos GW (1986): Development of a simplified counterflow centrifugation elutriation procedure for depletion of lymphocytes from human bone marrow. Transplantation 41:220–229.

Reading CR, Gasitt Y, Estrov Z, Juttner C (1996): Does CD34+ cell selection enrich malignant stem cells in B-cell (and other) malignancies? J Hematother 5:97–98.

Roslaniec MC, Bell-Prince CS, Crissman HA, Fawcett JJ, Goodwin PM, Habbersett R, Jett JH, Keller RA, Martin JC, Marrone BL, Nolan JP, Park MS, Sailer BL, Sklar LA, Steinkamp JA, Cram LS (1997): New flow cytometric technologies for the 21st century. Hum Cell 10(1):3–10.

Roslaniec MC, Martin JC, Reynolds RJ, Cram LS (1996): High speed optical chromosome sorting based on light induced photoinactivation of unwanted chromosomal DNA. Cytometry 24(Suppl 8):102.

Sasaki DT, Tichenor EH, Lopez F, Combs J, Uchida N, Smith CR, Stokdijk W, Vardanega M, Buckle AM, Chen B, Tushinski R, Tsukamoto A, Hoffman R (1995): Development of a clinically applicable high-speed flow cytometer for the isolation of transplantable human hematopoietic stem cells. J Hematother 4:503–514.

Schiller G, Vescio R, Freytes C, Spitzer G, Lee M, Wu CH, Cao J, Lee JC, Lichtenstein A, Lill M, Berenson R, Berenson J (1998): Autologous CD34-selected blood progenitor cell transplants for patients with advanced multiple myeloma. Bone Marrow Transplant 21(2):141–145.

Spangrude G (1994): Biological and clinical aspects of hematopoietic stem cells. Ann Rev Med 45:93–104.

Srour EF, Zanjani ED, Cornetta K, Traycoff CM, Flake AW, Hedrick M, Brandt JE, Leemhuis T, Hoffman R (1993): Persistence of human multilineage, self-renewing lymphohematopoietic stem cells in chimeric sheep. Blood 82(11):3333–3342.

Srour EF, Leemhuis T, Brandt JE, vanBesien K, Hoffman R (1991): Simultaneous use of rhodamine 123, phycoerythrin, Texas red, and allophycocyanin for the isolation of human hematopoietic progenitor cells. Cytometry 12:179–183.

Strauss LC, Rowley SD, LaRussa VF, Sharkis SJ, Stuart RK, Civin CI (1986): Antigenic analysis of hematopoiesis. V. Characterization of My-10 antigen expression by normal lymphohematopoietic progenitor cells. Exp Hematol 14:878–886.

Sutherland HJ, Eaves CJ, Eaves AC, Dragowska W, Landsorp PM (1989): Characterization and partial purification of human marrow cells capable of initiating long-term hematopoiesis in vitro. Blood 74(5):1563–1570.

Terstappen LW, Huang S, Safford M, Landsrop PM, Loken MR (1991): Sequential generations of hematopoietic colonies derived from single nonlineage committed CD34$^+$ CD38– progenitor cells. Blood 77(6):1218–1227.

Tricot G, Gazitt Y, Leemhuis T, Jagannath S, Desikan KR, Siegel D, Fassas A, Tindle S, Nelson J, Juttner C, Tsukamoto A, Hallagan J, Atkinson K, Reading C, Hoffman R, Barlogie B (1998): Collection, tumor

References

contamination, and engraftment kinetics of highly purified hematopoietic progenitor cells to support high dose therapy in multiple myeloma. Blood 91(12):4489–4495.

Uchida N, Aguila HL, Fleming WH, Jerabek L, Weissman IL (1994): Rapid and sustained hematopoietic recovery in lethally irradiated mice transplanted with purified Thy-1.1^{lo} Lin$^-$ Sca-1$^+$ hematopoietic stem cells. Blood 83:3758–3779.

Van den Engh G, Stokdijk W (1989): Parallel processing data acquisition system for multilaser flow cytometry and cell sorting. Cytometry 10(3):282–293.

Vardanega M, Swan R, Joubran J, Medeiros D, Tichenor E, Lewis H (1997): U.S. Pat. 5,641,457 (to SyStemix, Inc, Palo Alto, CA).

Verfaillie C, Miller WJ, Boylan K, McGlave P (1992): Selection of benign hematopoietic progenitors in chronic myelogenous leukemia on the basis of HLA-DR expression. Blood 79:1003–1010.

Verfaillie C, Blakolmer K, McGlave P (1990): Purified human hematopoietic progenitor cells with long term in vitro repopulating capacity adhere selectively to irradiated bone marrow stroma. J Exp Med 172(2):509–512.

Visser JW, Bauman JG, Mulder AH, Eliason JF, DeLeeuw AM (1984): Isolation of murine pluripotent hematopoietic stem cells. J Exp Med 159:1576–1590

Watt SM, Visser JW (1992): Recent advances in the growth and isolation of primitive human hematopoietic progenitor cells. Cell Proliferation 25:263–297.

Wolf NS, Kone A, Priestley GV, Bartelmez SH (1993): In vivo and in vitro characterization of long-term repopulating primitive hematopoietic cells isolated by sequential Hoechst 33342-rhodamine 123 FACS selection. Exp Hematol 21:614–622.

Yoder MC, Du XX, Williams DA (1993): High proliferative potential colony-forming cell heterogeneity identified using counterflow centrifugal elutriation. Blood 82(2):385–391.

Zanjani ED, Srour EF, Hoffman R (1995): Retention of long-term repopulating ability of xenogeneic transplanted purified adult human bone marrow hematopoietic stem cells in sheep. J Lab Clin Med 126(1):24–28.

Zhang MJ, Baccarani M, Gale RP, McGlave PB, Atkinson K, Champlin RE, Dicke KA, Giralt S; Gluckman E, Goldman JM, Klein JP, Herzig RH, Masaoka T, O'Reilly RJ, Rozman C, Rowlings PA, Sobocinski KA, Speck-B, Zwaan FE, Horowitz MM (1997): Survival of patients with chronic myelogenous leukaemia relapsing after bone marrow transplantation: comparison with patients receiving conventional chemotherapy. Br J Haematol 99(1):23–29.

Zimmerman TM, Lee WJ, Bender JG, Mick R, Williams SF (1994): CD34 may be used to determine the adequacy of a stem cell harvest for hematologic recovery following high dose chemotherapy. Prog Clin Biol Res 389:303–308.

5

Microfabricated Fluidic Devices for Single-Cell Handling and Analysis

David J. Beebe
University of Wisconsin, Madison, Wisconsin

INTRODUCTION

While traditional flow cytometry has provided the ability to perform rapid sorting of cells, the range of modalities that can be used as a basis for sorting is limited to optical (Kruth 1982) and electrical impedance parameters (Coulter). Optical and electrical impedance detection techniques have proved invaluable for a wide range of research investigations and clinical application. The ability to study cell physiology, for example, is limited using traditional flow cytometry techniques to indirect optical intensity measurements that reflect physiological changes. As we move toward more emphasis on functional genomics, new tools will be required to measure more physiological parameters and control environmental factors at the cellular scale. The advent of new miniaturization technologies creates new opportunities for constructing devices and systems with dimensions similar to that of a single cell. In some cases, scaling the dimensions of a system to the cellular scale can offer either enhanced performance or new functionality. In this chapter, early attempts at creating cellular-scale microsystems for single-cell handing and analysis are reviewed and the potential for future development is evaluated.

Emerging Tools for Single-Cell Analysis, Edited by Gary Durack and J. Paul Robinson.
ISBN 0-471-31575-3 Copyright © 2000 Wiley-Liss, Inc.

MICROFLUIDICS AND CYTOMETRY

Flow cytometry has had an enormous impact on biological and biochemical studies since its inception. The demonstration of the hydrodynamic orientation of cells (Fulwyler, 1977), although predating recent microfabricated fluidic device technology, takes advantage of the same scaling laws and foreshadowed many recent microfluidic devices. The details of cytometric methods are covered adequately in other chapters. Thus, the focus in this chapter will be limited to microfluidics issues and microfabricated devices.

Microfluidic Basics

Before discussing microfluidic devices, it is important to understand the different issues that come into play as fluid systems are scaled down. These include the dominance of viscous forces, large surface-to-volume ratios, and the importance of Brownian motion.

Fluids can be classified into two types: (1) Newtonian fluid (e.g., air, water, honey) and (2) non-Newtonian fluid (e.g., molten polymers, mayonnaise). In most microfluidic applications we are dealing with Newtonian fluids. The equation of momentum for a Newtonian fluid is the Navier–Stokes equation

$$\rho\left[\frac{\partial u}{\partial t} + (u \cdot \nabla)u\right] = -\nabla P + \eta \nabla^2 u + b$$

where ρ is fluid density, P is applied pressure, u is the velocity vector, η is fluid viscosity, and b is the body force unit.

The left-hand side of the equation represents "inertial forces" and the right-hand side represents the forces on the fluid due to applied pressure, viscosity, and body force (such as gravity force). One of the most important scaling factors that characterizes fluid properties at the microscale is the Reynolds number (Re). The Reynolds number is the ratio of inertial forces to viscous forces and can be calculated from physical parameters using

$$\mathrm{Re} = \frac{P v \ell}{\eta}$$

where ℓ is the characteristic dimension size of a microdevice and v is inlet fluid velocity. For example, for water (of which $r = 1$ kg/m^3, $m = 855 \times 10^{-6}$ Pa s) flowing in a 100-μm-diameter channel with inlet velocity of 0.1 m/s, the value of Re is 0.01. With such a small Reynolds number, fluid flow is laminar. Purcell (1977) provides an illustrative analogy. To create an environment similar to the one that sperm experience imagine yourself swimming in molasses while not moving any part of your body faster than 1 cm/min. Clearly the viscous forces dominate, while inertial forces are negligible.

The terms *laminar* versus *turbulent* and *steady* versus *unsteady* deserve comment as they are often missunderstood. Steady flow is defined as flow in which the velocity at a specific location does not vary with time. A flow can be steady or unsteady based on the chosen frame of reference. Laminar flow can be steady or unsteady. For example, in a microchannel tapping on the driving syringe creates unsteady flow, but as long as Re is low (i.e., viscosity is high enough and velocity is low enough), it will still be laminar (i.e., not turbulent). Turbulence typically occurs at Reynolds numbers greater than 2300 and is not possible at the dimensions and flow rates typical in microfluidic systems. However, the Reynolds number is geometry dependent and should be used with caution. From another perspective laminar flow exhibits a discrete-frequency spectrum, while turbulence exhibits a continuous-frequency spectrum. As will become apparent from the description of some of the microfluidic systems below, the low Re number of microfluidic systems (and thus, laminar flow) gives rise to some interesting properties that can prove useful or troublesome depending on the desired function of the microfluidic device. For a more through treatment of fluid dynamics at the microscale as it relates to biofluids see Brody et al. (1996), Purcell (1977), and Wilding et al. (1994).

More straightforward is the issue of surface-to-volume ratio. As the channel size decreases, the surface-to-volume ratio increases as $2/r$, where r is the radius of a cylindrical channel. Thus, in small channels (e.g., for a 100-μm channel), the surface-to-volume ratio is 20,000. Again, this becomes a significant issue if biomolecules are present as protein adsorption can degrade performance in microsystems in several ways. For example, fouling due to biomolecular adsorption may interfere with pump or valve operation or disrupt electrokinetic flow.

Sheath Flow. The fluid basis of flow cytometry is a phenomenon known as sheath flow. In this chapter we are interested in several aspects of sheath flow that also apply to microfluidic systems in general. Turbulent versus laminar flow has already been discussed above. Examining Fulwyler's (1977) original paper on the hydrodynamic orientation of cells, one calculates a Reynolds number of ~15. This is close to a regime where a backward step in the flow could produce secondary flows or shedding of vortices but turbulence is not present. In modern flow cytometry typical sheath core diameters (10–15 μm), sheath stream diameters (50–100 μm), and velocities (10 m/s) give rise to much lower Reynolds numbers.

Under laminar flow conditions the combination of a sheath flow, a sample flow, and the phenomenon of hydrodynamic focusing gives rise to the basic fluid system

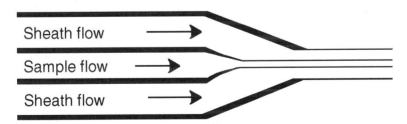

Fig. 5.1. Higher velocity of concentric sheath flow and physical constriction of nozzle act to shape the sample flow into a narrow stream.

that is the basis for modern flow cytometry. This is illustrated in Figure 5.1. The sample tube introduces the sample stream into the center of a concentric sheath flow. The sample and sheath flow are then routed through a nozzle. Since the Re number is small (i.e., laminar flow), the fluid streams are compressed into the narrower nozzle opening while maintaining their relative dimensions (i.e., the ratio of sample stream diameter to sheath flow diameter remains constant).

Diffusion (Brownian Motion). While turbulence is not an issue, diffusion between the sheath and core is an important consideration. Once the sample leaves the sample tube, it is free to diffuse with the surrounding sheath flow. This diffusion is governed by $\overline{x^2} = 2Dt$, where $\overline{x^2}$ is the mean square distance, t is time, and D is the diffusion constant.

The diffusion coefficient scales with the effective size of the particle and can vary over four orders of magnitude between small molecules (~1000 $\mu m^2/s$) and cells (<1 $\mu m^2/s$). One must always be cognizant of diffusion issues when designing and evaluating microfluidic systems. Depending on flow rates and diffusion coefficients, diffusion can play a major role in system performance. The impact of diffusion can sometimes be used in a beneficial way. For example, one can expose cells to a known sheath medium in a controlled way by careful control of sheath location/length (Pinkel and Stovel, 1985).

These and other fluid issues related to biological fluids in microscale devices are discussed by Brody et al. (1996) and Wilding et al. (1994).

Microfabricated Cytometry

Microfabrication and MicroElectroMechanical Systems. Over the last 15 years a new class of miniaturization technologies has emerged that will be loosely grouped here under the term MicroElectroMechanical Systems (MEMS). MEMS is a group of technologies that allow for the creation of mixed electrical, mechanical, chemical, and fluidic systems at the micrometer scale. While initially MEMS grew out of traditional integrated circuit techniques, it has now expanded to become a distinct and separate field encompassing a wide range of fabrication methods and materials. Two recent texts provide a good overview of these technologies (Kovacs, 1998; Madou, 1997). The focus here will be to survey work in MEMS related to flow cytometry and single-cell analysis.

Microfabricated Cytometry Devices. The phenomenon of hydrodynamic orientation of cells (discussed above) and the subsequent development of high-speed cytometry in the 1970s and early 1980s predated the current development of microfluidic systems but in fact is based on many of the same concepts. Sobek et al. (1994) recognized this and demonstrated hydrodynamic focusing and sheath flow in a fused-silica device (see Fig. 5.2). The device was fabricated from fused silica using a single mask layer and wafer-level alignment and bonding. The sample injector dimensions were 250 μm high by 280 μm wide. Hydrodynamic focusing was maintained for velocities as high as 10 m/s with a corresponding inlet-to-outlet pressure drop of 28 psi.

Microfluidics and Cytometry

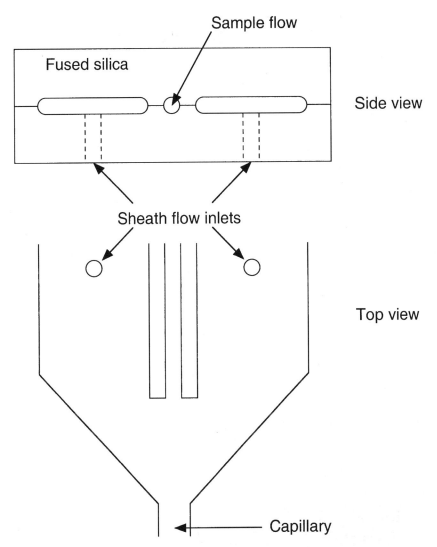

Fig. 5.2. Glass device used to demonstrate the potential of cytometry in microfabricated devices. The sample flow was introduced via a hypodermic needle coupled to the fused-silica sample flow channel. (After Sobek et al., 1994.)

Other demonstrations of simple microfabricated channels designed for cell transport and sorting soon followed. Larsen et al. (1997) presented a microchip Coulter particle counter device. The device was fabricated in silicon with a glass cover. A modified Coulter counter geometry is demonstrated. Five parallel fluid streams are brought together via five separate but parallel channels (see Fig. 5.3). The center three streams are electrolytes and create the traditional sheath flow observed in flow

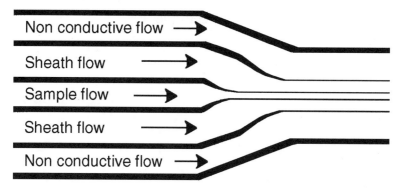

Fig. 5.3. Multiple streams can be used to provide dynamic control of the sample stream width. (After Larson et al., 1997.)

cytometry while the other two streams consist of nonconductive fluid and provide the ability to dynamically adjust the effective width of the sheath flow. Since only the middle streams are conductive, the width of those three streams determine the sensitivity of the detection. The published work only demonstrated the fluidic operation of the device; no actual detection results were presented.

Differential counting of granulocytes, lymphocytes, monocytes, red blood cells, and platelets in a silicon flow channel was demonstrated by Altendorf et al. (1997). The device consisted of a single v-groove channel (top width of 20–25 μm) (no sheath flow). Laser light scattering (both small and large angle) was the basis of differential counting. A single laser source (638 nm) with two detectors (one for small angle, one for large angle) was used. The size- and structure-dependent light scattering allows differential counting.

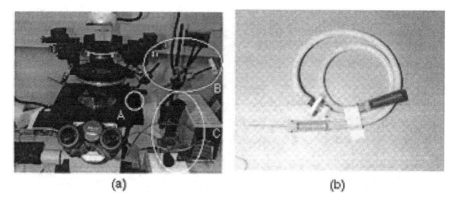

Fig. 5.4. Single-cell handling is typically accomplished with the use of a micromanipulation station and/or mouth pipette. (*a*) Typical micromanipulation station includes a microscope, pulled pipette tip (A) controlled by a micromanipulator joystick (C). Fluid transport through the pipette is controlled via a precision syringe. (*b*) Simple mouth pipette is also still commonly used to handle single cells and embryos.

SINGLE-CELL MANIPULATION AND ANALYSIS

The range of device possibilities that MEMS technologies make possible allows one to envision systems capable of more than just sorting functions. For comparison, we first briefly review the physical tools traditionally used for single-cell manipulation and analysis and then examine current efforts to construct novel microsystems for single-cell study.

Traditional Methods

Current methods for manipulating single cells and embryos have changed little in decades. Researchers still routinely use simple equipment to fabricate custom suction pipettes to hold, probe, and inject single cells and embryos (see Fig. 5.4). For example, Hogan et al. (1994) describe procedures for making holding and injection pipettes for embryo manipulation such as DNA injection and nuclear transplantation. The procedure is the equivalent of micro–glass blowing. One begins with a 1-mm-outside-diameter (OD) glass capillary. Using a microburner (e.g., a small Bunsen burner) or pipette puller (e.g., Sutter Instrument P-87) and a micro forge (e.g., Kramer Scientific), a variety of pipettes can be formed for holding and injection procedures. The inside diameter (ID) of pipettes made using these techniques can be on the order of 1 μm or as large as several hundred micrometers. Suction (mechanical or mouth) applied to a holding pipette is used to hold the cell firmly against the pipette tip, while syringes are used on injection pipettes to obtain and expel minute quantities of sample.

Microfabricated Devices and Systems

We are in the midst of rapid growth in the sophistication and variety of microdevices and systems being built to extend our ability to study life at the cellular level and below. Much of this growth is in the area of gene chips to allow rapid genomic analysis. Examples of these technologies include products by Nanogen and Affymetrix. If one thinks beyond the problem of sequencing the genome to the problem of functional genomics, then one must think beyond planar gene chip designs to systems that allow physical interaction with the cell or molecule of interest.

Mechanical Transport, Sorting, Manipulation and Characterization.
Traditionally, cell deformability has been assessed using either micropipette aspiration or filtration. When one can fabricate physical structures at the scale of the cell or molecule in a very controlled way (i.e., lithography-based microfabrication), one can begin to study the physical interaction and characteristics of single cells and molecules in more quantitative and reproducible ways than is possible with either micropipette aspiration or traditional filtration.

Ogura et al. (1991) developed a microfabricated device for the study of human red blood cell deformability. Pores down to 1 μm in diameter were created in a 0.4-μm-thick silicon nitride membrane. Cell passage times were measured for cells that had been treated in ways known to affect deformability (temperature, exposure to diamide and glutaraldehyde, and storage conditions). Results demonstrated clearly that the device was capable of differentiating cell treatments via passage time measurements. However, the use of short pore filters is largely unexplored and more study is required.

About the same time two groups began investigating the use of micromachined flow channels for the study of blood rheology and blood cell deformability to mimic the in vivo conditions of blood cells traversing capillaries. Tracey et al. (1991) used Reactive Ion Etching (RIE) to form square (4-µm) channels while Kikuchi et al. (1992) utilized anisotropic etching to realize v-groove channels with a top width of 9 µm (see Fig. 5.5). Tracey et al. (1995) measured erythrocyte length and velocity for normal erythrocytes to demonstrate their device, but in experiments with artificially hardened cells no subpopulations were found. Kikuchi et al. (1992, 1994) and Kikuchi (1995) performed a series of rheology studies using both red and white blood cells (RBC, WBC) with a variety of treatments, including formyl-L-methionyl-L-leucyl-L-phenylalanine (FMLP) and adenosine diphosphate (ADP). Direct comparisons to Nuclepore filters (Nuclepore, Pleasanton, CA) were performed by measuring the passage times. Flow rates were about 50 times higher in the Nuclepore filters (close to expected based on physical differences between filters and channels). Significant variations between channels were reported due to transient clogging and blocking at the channel entrances.

More recently Austin and co-workers have demonstrated the feasibility of using lithographic arrays for cell sorting (Brody et al., 1995; Bakajin et al., 1998; Carlson et al., 1997). Specifically, they have designed and constructed large microfabricated lattices. A typical lattice contains a variety of channel widths and lengths but in much larger numbers than those of Kikuchi (1992) and Tracy (1991), thus allowing separation of larger numbers of cells. The transport and adhesion properties of human RBCs and WBCs have been studied, and the ability to efficiently separate out the WBCs has been demonstrated (see Fig. 5.6).

Glasgow et al. (1998) have developed a system for the transport and precise positioning of single mammalian embryos. The embryos (approximately 120 µm diame-

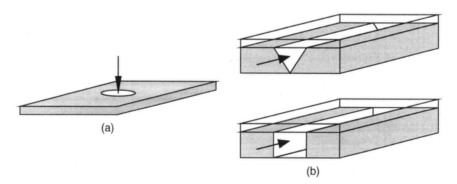

Fig. 5.5. (*a*) Simple pores can be fabricated with precise control of pore size. Single cells can then be forced through the pores enabling cell mechanics studies. (*b*) To mimic human capillaries, microchannels can be fabricated. The top illustrates the channel shape used by Kikuchi et al. (1992), while the bottom shows the shape used by Tracy et al. (1995).

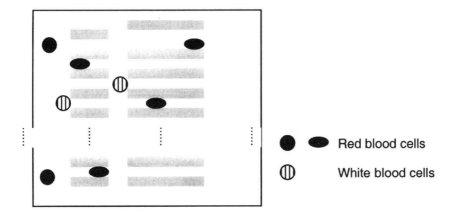

Fig. 5.6. Constrictions in the lattice are similar to those shown in Figure 5 b, bottom. The WBCs tend to get caught at the entrance to the channels rather than within the channels. (After Carlson et al., 1997.)

ter) are on the same order of size as the channels (approximately 200 μm). The similar scale of object and channel causes the embryos to roll and slip along the channel bottoms. The device also incorporates physical restrictions that allow the embryos to be positioned in precise locations (analysis stations) and then to be easily moved to another location or directed to different outlets (see Fig. 5.7). The integration of fluidic transport and positioning with other analysis techniques may lead to high-throughput analysis systems.

Electrical Transport, Sorting, and Characterization. The anatomy of single cells (i.e., phospholipid membrane, ion channels, etc.) allows for interesting interactions between cells and electric fields. Again the ability to fabricate channels with integrated electrodes provides fertile ground for new and improved manipulation and analysis at the cellular level. Several manipulations of single cells in microdevices have been demonstrated, including cell fusion, trapping, transport, and separation.

Dielectrophoresis. The principle of dielectrophoresis is central to many of these devices (see Fig. 5.8). Dielectrophoresis arises from the polarization of a particle in an electric field. If the cell is less polarizable than the surrounding media, the cell will move away from high-field regions (i.e., negative dielectrophoresis). Conversely, if the cell is more polarizable than the surrounding media, the cell will move toward areas of high field (i.e., positive dielectrophoresis, as shown in Fig. 5.8a. In highly conductive solutions (typical of cell culture solutions) the force is always repulsive (negative dielectrophoresis) (Fuhr and Shirley, 1995). Based on this concept, one can design electrode geometries and arrangements capable of trapping single cells (see Fig. 5.8b) or other cell manipulation functions. Many groups have demonstrated single-cell manipulation using dielectrophoresis. Several functions have been demonstrated, including cell trapping (Fuhr et al., 1992; Schnelle et al., 1993), cell sorting (Washizu, 1990; Kaler et al., 1996), cell transport (Masuda et al., 1987; Hagedorn et al., 1992), and cell characterization (Fuhr and Shirley, 1995; Chan et al., 1998). A good review of cell motion in time-varying fields is given by Fuhr et al. (1996).

Cell Fusion. Another cell-handling operation of importance to modern biology is cell fusion. Masuda et al. (1987) developed a microfabricated electrical cell fusion device over 10 years ago (see Fig. 5.9). The device was fabricated using patterned UV resin as a mold to form silicon rubber microchannels. More recently, Lee et al. (1995) fabricated a similar device using polyimide as the channel material. In both devices, individual cells are brought together from two separate channels and positioned at the electrodes for fusion.

Electrical Impedance. Ayliffe et al. (1998) have used microfabricated devices to directly measure the electrical impedance of single toadfish RBCs and human neutrophils. The devices are similar in geometry to the cell fusion devices described above. Electrical impedance measurements were obtained over a frequency range of 100 Hz to 2 MHz and area-specific membrane capacitances (0.98 $\mu F/cm^2$, neutrophils; 1.59 $\mu F/cm^2$, toadfish) were estimated using an RC circuit.

Magnetic Sorting. The ability to tag cells with magnetic microparticles and subsequently separate them is not new. Recently, however, several groups have demonstrated magnetic-based separations in microfabricated devices. Telleman et al. (1998) describe a device that separates magnetically labeled cells from a continuously flowing stream, as shown in Figure 5.10a. The separation efficiencies are related to field strength and flow rates Liakopoulos et al. (1997) built a device with electromagnets integrated under the flow channel, as shown in Figure 5.10b. When the magnet is turned on, magnetic particles are quickly attracted to the magnet. After rinsing, the magnet can be turned off and the cells collected.

Related Work

Several other areas of microinstrumentation research that may play an important role in future single cell analysis systems are discussed next.

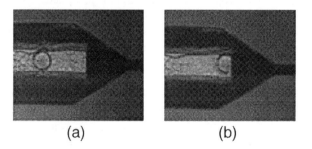

(a)   (b)

Fig. 5.7. (*a*) Mouse embryo approaching a constriction. (*b*) Mouse embryo has reached the constriction and is "parked."

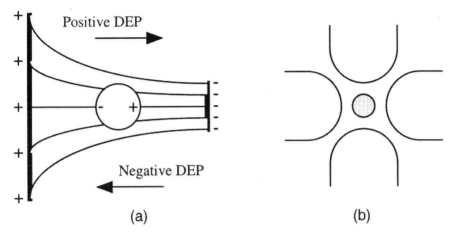

Fig. 5.8. (a) Dielectrophoresis can be used to transport cells. The direction of travel depends upon the relative polarizability of the cell and the media. (b) Through proper electrode geometry, dielectrophoretic forces can be used to trap a single cell. (After Fuhr et al., 1996.)

Micro NMR. A group at the University of Illinois at Urbana-Champaign (UIUC) has developed a family of technologies that allow nuclear magnetic resonance (NMR) techniques to be applied to sample sizes approaching the cellular scale. Webb, Sweedler, and co-workers have recently shown that high-resolution NMR spectra and images can be obtained using radiofrequency (RF) detectors with diameters as low as 150 μm. These used hand-wound solenoidal coils and were shown to give sensitivities up to two orders of magnitude greater than conventional detectors (Olson et al., 1995; Webb and Grant, 1996).

Work on planar microcoil development proceeded in parallel. The first published application of planar, lithographic microcoils, with feature sizes approaching 1 μm, was in 1994 by the UIUC microcoil research group (Peck et al., 1994). This was followed by a parallel study on glass substrates, where spectral linewidths of less than 10 Hz were reported using fused-silica capillaries loaded with 5 mM $CuSO_4$ samples placed directly on top of the coil (Peck et al., 1995). Analogous to results obtained at larger size scales (Wu et al., 1994), the spectral linewidth varied as a function of capillary wall thickness, primarily due to the localized magnetic field distortions that resulted from mismatch in the susceptibility of the coil conductor (gold) relative to the surrounding air. Later studies employing planar, lithographic microcoils provided subhertz spectral resolution using the fused-silica capillary-on-coil configuration and a susceptibility-matching fluid (Stocker et al., 1997). This same group later reported on the first integrated application of microcoils, where a monolithic connection of a GaAs-based planar microcoil, a passive impedance-matching network, and a metal semiconductor field effect transistor (MESFET) preamplifier was employed at 300 MHz for application to neutron NMR microspectroscopy at 7.05 T (Stocker et al., 1995, 1996). A separate, single-turn 1.5-cm RF coil was used for excitation. Early results provided a signal-to-noise (S/N) ratio similar to that of the

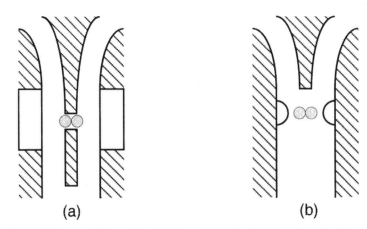

Fig. 5.9. (*a*) Cells are precisely positioned using mechanical and dielectrophoretic effects. (After Masuda et al., 1987.) (*b*) A similar device relies on dielectrophoretic effects alone to trap and orient cells for fusion. (After Lee et al., 1995.)

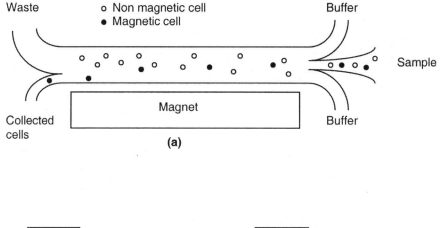

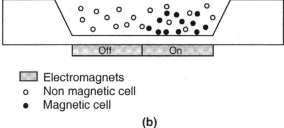

Fig. 5.10. Two approaches to magnetic separation devices. (*a*) Magnetically tagged cells are attracted toward the magnet and collected from a continuously flowing stream. (After Telleman et al., 1998.) (*b*) Magnetically tagged cells are trapped above the magnet from a flowing stream and later released for collection. (After Liakopoulos et al., 1997.)

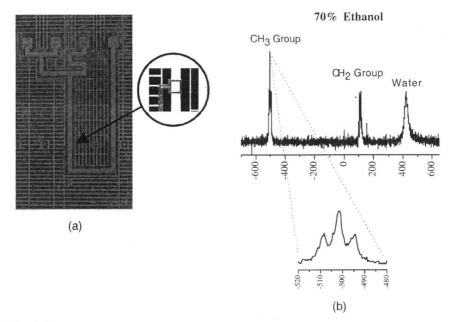

Fig. 5.11. (a) Microfluidic channel network is integrated with a small NMR receiving coil. The volume under the coil is 30 nl. Channel dimensions are 1 mm × 30 μm. (b) High-resolution spectra obtained in the device.

conventional microcoil/preamplifier configuration (in agreement with theory) but poor spectral resolution (presumably due to mismatch in susceptibility of the materials used to construct the coil and amplifier vs. the surrounding air).

The philosophy of miniature total analysis systems (μ-TASs) hinges on the integration of multiple chemical processing steps and the means of analyzing their results on the same miniaturized system. Trumbull et al. (1998) have combined microfluidics and micro-NMR in an integrated microfabricated device. Such integration and the ability to precisely control both geometric and material parameters allows for the optimization of multiple operations (e.g., separation and analysis). Figure 5.11 illustrates initial work in this area.

Cell Assays and Electrophoretic Separations. Harrison (1995), Manz (1993), and Ramsey (1994) pioneered capillary electrophoresis in microfabricated devices in the early 1990s. In the following years these groups and others demonstrated a wide range of miniaturized chemical transport, separation, and analysis techniques based on fluorescence and electrokinetic flow and electrophoretic separation principles. The extension of this work toward single-cell analysis is now being explored. Andersson et al. (1997) have demonstrated cell transport and cellular assays in an electrokinetic microchip (see Fig. 5.12). Red blood cells, yeast, and *Escherichia coli* cells were directed to desired channels within the microchannel network using electroosmotic flow. They used the system to demonstrate the ability to obtain information about single-cell reaction rates and the effect of inhibitor drugs on calcium cell influx. It should be noted that pressure-driven flow was utilized because protein adsorption on the channel walls reduced the electroosmotic flow significantly. An inhibitor was combined with the cells at the first intersection (1). Incubation occurred while the cells slowly moved down the main channel (incubation time of 4 min). An activator was mixed with the cells at the next intersection (2). A fluorescence microscope was positioned over the second intersection (2). By stopping flow appropriately, visualization of single cells was possible and the kinetics of calcium was studied.

Ewing (1998) and Mesaros et al. (1993) have coupled capillary electrophoresis to electrophoresis in narrow rectangular channels to achieve continuous analysis on small-volume samples. This approach could lead to continuous electrophoretic analysis on the effluent from single cells. The capillary is used to sample analytes and continuously distribute them along a wide, flat separation chamber, as shown in Figures 5.13a,b. The result is the ability to continuously monitor separations, as shown in Figure 5.13c. Continuous separations of dopamine and catechol have been demonstrated. The height of the separation chambers ranged from as small as 0.6 μm to as large as 21 μm. Electrochemical array detection was used to monitor the separations at the exit of the chamber (Suljak et al., 1998).

Other Areas of Interest. Another area that holds promise is the use of acoustic devices for cell transport and manipulation. Several groups have demonstrated microfabricated devices capable of transporting small particles via various acoustic methods (Kozuka et al., 1994; Luginbuhl et al., 1997; Hashimoto et al., 1997). Other techniques such as laser scissors and tweezers (Berns et al., 1991; Berns, 1998) and patch clamp techniques (Neher and Sakmann, 1992) while important, will be difficult to implement in high-throughput systems and thus are not discussed in detail here.

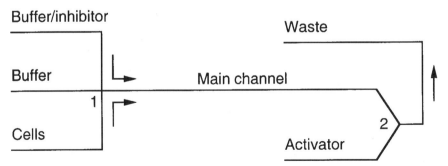

Fig. 5.12. Schematic of the fluidic network used to demonstrate cellular assays on a chip. An inhibitor is combined with the cells at 1, incubation occurs during transport down the main channel, an activator is introduced at 2, and flow is stopped and the cells are visualized just after 2. (After Andersson et al., 1997.)

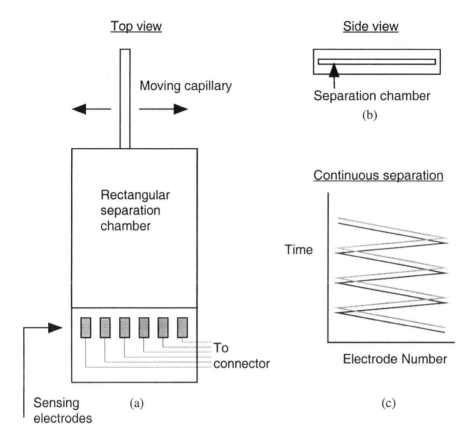

Fig. 5.13. (*a*) Top view of continuous loading scheme. A moving capillary continuously loads sample into a rectangular separation chamber. (*b*) Side view of separation chamber (heights of 0.6–21 μm have been used). (*c*) Resulting output yields a continuous separation of analytes. [After Ewing (1998) and Suljak et al. (1998).]

SUMMARY

The advent of new miniaturization technologies has the potential of providing novel tools to facilitate modern biology. Specifically, the ability to fabricate complex systems that are of the same scale as a single cell allows one to think beyond traditional methods. However, designers of new miniaturized systems for biology must be cognizant of the physics of the scale. That is, does it make sense to miniaturize? What is to be gained? The development costs of microfabricated systems is typically quite high. Thus, the two driving forces for miniaturization are greatly enhanced performance and new functionality (that cannot be achieved with other methods). The integration of these new fabrication and emerging sensing technologies may provide novel smart systems for single-cell analysis and sorting.

OTHER SOURCES OF INFORMATION

The overview of microfabricated devices for single-cell handling and analysis given in this chapter is far from exhaustive due to the space limitations. The goal of this chapter was to provide the cytometry community with an overview of potentially related work in the microfabrication community. The focus here has been on flowing systems that might potentially lead to improved or new flow cytometry devices capable of more extensive analysis of single cells. Interested readers are encouraged to examine the following conference proceedings and journals for more information on microfabricated devices and their potential application to biology.

- Proceedings of the μTAS Workshop, 1994, 1996, 1998.
- International Conference on Solid-State Sensors and Actuators, 1991, 1993, 1995, 1997.
- Proceedings of the Annual IEEE International Conference on Micro Electro Mechanical Systems, 1987–1999.
- Proceedings of the Solid-State Sensor and Actuator Workshop, Hilton Head, SC, 1990, 1992, 1994, 1996, 1998.
- *Journal of Microelectromechanical Systems, IEEE/ASME.*
- *Micromachined Transducers Sourcebook*, by G. T. A. Kovacs, CRC Press, Boca Raton, FL, 1997.
- *Fundamentals of Microfabrication*, by M. Madou, CRC Press, Boca Raton, FL, 1997.

ACKNOWLEDGEMENTS

I would like to thank the following people for advice, information, and helpful discussions: Juan Santiago, Robin Liu, Henry "Gripp" Zeringue, Ian Glasgow, Gary Durack, Tim Peck, and Andrew Webb.

REFERENCES

Altendorf E, Zebert D, Holl M, Yager P (1997): *Differential Blood Cell Counts Obtained Using a Microchannel Based Flow Cytometer.* Chicago: Transducers '97.

Andersson PE, Li PCH, Smith R, Szarka RJ, Harrison DJ (1997): Biological Cell Assays on an Electrokinetic Microchip. 1997 Int'l Conf. on Solid-State Sensors and Actuators, Chicago.

Ayliffe HE, Rabbitt RD, Frazier AB (1998): Microfabricated electric impedance chamber for the electrical characterization of single cells. SPIE.

Bakajin O, Carlson R, Chou CF, Chan S, Gabel C, Knight J, Cox T, Austin RH (1998): Sizing, Fractionation and Mixing of Biological Objects Via Microfabicated Devices. Solid-State Sensor and Actuator Workshop, Hilton Head Island, SC.

References

Berns, MW (1998): Laser scissors and tweezers. Sci Am (April): 62–67.

Berns MW, Wright WH, Wiegand-Steubing R (1991): Laser microbeam as a tool in cell biology. Int Rev Cytom 129:1–44.

Brody JP, Yager P, Goldstein RE, Austin R (1996): Biotechnology at low Reynolds numbers. Biophys J 71:3430–3441.

Brody JP, Han Y, Austin RH, Bitensky M (1995): Deformation and flow of red blood cells in a synthetic lattice: Evidence for an active cytoskeleton. Biophys J 68:2224–2232.

Carlson R, Gabel C, Chan S, Austin R, Brody J, Winkleman J (1997): Hydrodynamic activation and sorting of white blood cells in a microfabricated lattice. SPIE 2978:206215.

Chan KL, Green NG, Hughes MP, Morgen H (1998): Cellular Characterization and Separation: Dielectrophoretically Activated Cell Sorting (DACS). 20th Annual International Conference—IEEE/EMBS, Hong Kong.

Coulter WH (1956): High speed automatic blood cell counter and cell size analyzer. Proc Natl Electronics Conf 12:1034–1042.

Ewing AG (1998): Parallel Separations in Microfabricated Channels with Capillary Electrophoretic Sample Introduction. m-TAS '98, Banff, Canada.

Fuhr G, Shirley SG (1995): Cell handling and characterization using micron and submicron electrode arrays: state of the art and perspectives of semiconductor microtools. J Micromech Microeng 5:77–85.

Fuhr G, Zimmermann U, Shirley SG (1996): Cell motion in time-varying fields: Principles and potential. In *Electromanipulation of Cells*, Zimmerman U, Neu GA (eds). Boca Raton: CRC Press. Chapter 5. 259-328.

Fuhr F, Arnold WM, Hagedorn R, Muller T, Benecke W, Wagner B, Zimmerman U (1992): Levitation, holding and rotation of cells within traps made by high-frequency fields. Biochim Biophys Acta 1108: 215–223.

Fulwyler M (1977): Hydrodynamic orientation of cells. J Histochem Cytochem 25(7):781–783.

Glasgow I, Zeringue HC, Beebe DJ, Choi S-J, Lyman JT, Wheeler MB (1998): Individual Embryo Transport and Retention on a Chip. m-TAS -98, Banff, Canada.

Hagedorn R, Fuhr G, Muller T, Gimsa J (1992): Traveling-wave dielectrophoresis of microparticles. Electrophoresis 13:49–54.

Harrison DJ, Fluri K, Seiler K, Fan Z, Effenhauser CS, Manz A (1993): Electrophoresis-based chemical analysis system on a chip. Science 261:895–897.

Hashimoto K, Ikekame K, Yamaguchi M (1997): Micro actuators employing acoustic streaming caused by high-frequency ultrasonic waves. Transducers 97: 805–808

Hogan B, Beddington R, Costantini F, Lacy L (1994): *Manipulating the Mouse Embryo.* New York: Cold Spring Harbor Laboratory Press.

Jacobson SC, Hergenroder R, Koutny LB, Warmock RJ, Ramsey MJ (1994): Effects of injection schemes and column geometry on the performance of microchip electrophoresis devices. Anal Chem 66:1107–1113.

Kaler KVIS, Docoslis A, Kalogerakis N, Behie LA (1996): A Micromachined DEP Cell Filtration Device. 1996 IEEE Industry Applications Society Meeting, San Francisco, CA.

Kikuchi Y (1995): Effect of leukocytes and platelets on blood flow through a parallel array of microchannels: Micro- and macroflow relation and rheological measures of leukocyte and platelet activities. Microvas Res 50:288–300.

Kikuchi Y, Sato K, Mizuguchi Y (1994): Modified cell-flow microchannels in a single-crystal silicon substrate and flow behavior of blood cells. Microvas Res 47:126–139.

Kikuchi Y, Sato K, Ohki H, Kaneko T (1992): Optically accessible microchannels formed in a single-crystal silicon substrate for studies of blood rheology. Microvas Res 44:226–240.

Kovacs GTA (1998): *Micromachined Transducers Sourcebook.* Boston: WCB McGraw-Hill.

Kozuka T, Tuziuti T, Mitome H (1994): Acoustic Manipulation of Micro Objects Using an Ultrasonic Standing Wave. 5th International Symposium on Micromachine and Human Science Proceedings, Nagoya Congress Center.

Kruth HS (1982): Flow cytometry: Rapid biochemical analysis of single cells. Anal Biochem 125:225–242.

Larsen U, Blankenstein G, Branebjerg J (1997): Microchip Coulter Particle Counter. Transducers '97, Chicago, IL.

Lee S, Choi J, Kim Y (1995): Design of a Biological Cell Fusion Device. The 8th International Conference on Solid-state Sensors and Actuators, and Eurosensors IX, Stockholm, Sweden.

Liakopoulos TM, Choi J, Ahn CH (1997): A Bio-magnetic Bead Separator on Glass Chips Using Semi-encapsulated Spiral Electromagnets. 1997 Int'l Conference on Solid-State Sensors and Actuators, Chicago, IL.

Luginbuhl P, DeRooij N, Collins, SD, Racine G-A, Setter N, Brooks KG (1997): Acoustic wave device for the translation of microparticles. SPIE 2978:129–134.

Madou M (1997): *Fundamentals of Microfabrication*. Boca Raton, FL: CRC Press.

Manz A, Harrison DJ, Verpoorte E, Widmer HM (1993): Planar chips technology for miniaturization of separation systems: a developing perspective in chemical monitoring. Adv Chromatog 33:1–66.

Masuda S, Washizu M, Kawabata I (1988): Movements of Blood Cells in Liquid by Nonuniform Traveling Field. IEEE Trans Ind Applic 24(1):217–222.

Masuda S, Washizu M, Nanba T (1987): Novel Methods of Cell Fusion and Handling Using Fluid Integrating Circuit. Electrostatics '87, Oxford.

Mesaros JM, Luo G, Roeraade J, Ewing AG (1993): Continuous electrophoretic separations in narrow channels coupled to small-bore capillaries. Anal Chem 65:3313–3319.

Neher E, Sakmann B (1992): The patch clamp technique. Sci Am: 44–51.

Ogura E, Abatti PJ, Morizumi T (1991): Measurement of human red blood cell deformability using a single micropore on a thin Si_3N_4 film. IEEE Trans Biomed Eng 38(8):721–725.

Olson DL, Peck TL, Webb AG, Magin RL, Sweedler JV (1995): High-resolution microcoil ^{1}H-NMR for mass-limited, nanoliter-volume samples. Science 270:1967–1970.

Peck TL, Stocker JE, Chen Z, LaValle L, Magin RL (1995): Application of Planar Microcoils Fabricated on Glass Substrates to NMR Microspectroscopy. 17th Annual Int Conf of the IEEE EMBS, Montreal Canada.

Peck TL, Magin RL, Kruse J, Feng M (1994): NMR microspectroscopy using 100 μm planar RF coils fabricated on gallium arsenide substrates. IEEE Trans Biomed Engr 41(7):706–709.

Pinkel D, Stovel R (1985): Flow chambers and sample handling. In *Flow Cytometry: Instrumentation and Data Analysis*, Van Dlla MA, Dean PN, Laerum OD, Melamed MR (eds). New York: Academic Press, 77–98.

Purcell EM (1977): Life at low Reynolds number. Am J Phy 45:3–11.

Schnelle T, Hagedorn R, Fuhr G, Fiedler S, Muller T (1993): Three-dimensional electric field traps for manipulation of cells-calculation and experimental verification. Biochim Biophys Acta 1157:127–140.

Sobek D, Senturia S, Gray M (1994): Microfabricated Fused Silica Flow Chambers for Flow Cytometry. Solid-State Sensor and Actuator Workshop, Hilton Head, South Carolina.

Stocker JE, Peck TL, Webb AG, Feng M, Magin RL (1997): Nanoliter volume, high resolution NMR microspectroscopy using a 60 μm planar microcoil. IEEE Trans Biomed Eng 44(11):1122–1128.

Stocker J, Peck T, Webb A, Franke S, Feng M, Magin R (1996): Integrated Microcoil/Amplifier Detector for Microspectroscopy. 37th Experimental NMR Conference, Asilomar, CA.

Stocker JE, Peck TL, Franke SJ, Kruse J, Feng M, Magin RL (1995): Development of an Integrated Detector for NMR Microscopy. 17th Annual Int Conf of the IEEE EMBS, Montreal Canada.

Suljak SW, Thompson LA, Ewing AG (1998): Electrophoretic Separations in Ultrathin Channels Using Microelectrode Array Detection. m-TAS '98, Banff, Canada.

References

Telleman P, Larsen UD, Phillip J, Blankenstein G, Wolff A (1998): Cell Sorting in Micro Fluidic Devices. mTAS '98, Banff, Canada.

Tracey MC, Greenaway RS, Das A, Kaye PH, Barnes AJ (1995): A silicon micromachined device for use in blood cell deformability studies. IEEE Trans Biomed Engr 42(8):751–761.

Tracey MC, Kaye PH, Shepherd JN (1991): Microfabricated Microhaemorheometer. Rec Int Conf Sensors Actuators, California.

Trumbull JD, Glasgow IK, Beebe DJ, Magin RL (1998): Integrating Microfluidic Systems and NMR Spectroscopy: Preliminary Results. 1998 Solid-State Sensor and Actuator Workshop, Hilton Head Island, South Carolina.

Washizu M (1990): Handling biological cells using a fluid integrated circuit. IEEE Trans Ind Applic 26(2):352–358.

Webb AG, Grant SC (1996): Signal-to-noise and magnetic susceptibility tradeoffs in solenoidal microcoils. J Magn Reson 113:83–87.

Wilding P, Pfahler J, Bau HH, Zemel JN, Kricka LJ (1994): Manipulation and flow of biological fluids in straight channels micromachined in silicon. Clin Chem 40:43–47.

Wu N, Peck T, Webb A, Magin R, Sweedler J (1994): 1H-NMR spectroscopy on the nanoliter scale for static and on-line measurements. Anal Chem 66:38–49.

6

Single DNA Fragment Detection by Flow Cytometry

Robert C. Habbersett, James H. Jett, and Richard A. Keller
Los Alamos National Laboratory, Los Alamos, New Mexico

INTRODUCTION

Flow cytometry technology—although in this context "cytometry" is a misnomer—has been extended to enable the analysis of very small samples of individual DNA fragments, rapidly and reliably, with direct quantitation of fragment size and number. This has required a new generation of flow instrumentation, literally a molecular-level flow analyzer, which is the subject of this chapter. It is the result of extensive development work by numerous individuals at the Los Alamos National Laboratory (Ambrose et al., 1993; Petty et al., 1995; Huang et al., 1996; Goodwin et al., 1997). The initial instrument for DNA fragment size analysis, documented in 1993, evolved out of ongoing efforts to sequence DNA in flow (Goodwin et al., 1997). The simplified apparatus, to be described here, is a second-generation instrument that differs from the original in a number of significant ways. First, the range of DNA fragments that can be analyzed has been extended to both larger and smaller fragments. A different DNA-binding dye has been used in order to utilize a small, solid-state laser. A single data-acquisition card in a PC constitutes the entire data collection system. A new photon-counting detector, an avalanche photodiode (APD; Li and Davis, 1993) with high quantum efficiency, has replaced a bulky, liquid-cooled photomultiplier. Finally, DNA in solution is delivered directly into the flow cell, to within a few hundred

---

Emerging Tools for Single-Cell Analysis, Edited by Gary Durack and J. Paul Robinson.
ISBN 0-471-31575-3 Copyright © 2000 Wiley-Liss, Inc.

micrometers of the laser beam, by small-bore quartz capillary tubing. This chapter will describe and explore the nature of these changes and how they affect the performance and extend the capabilities of this technology.

There are several fundamental differences between the realm of individual molecular analysis and the familiar world of conventional flow cytometry. First, this instrument requires a photon-counting detector for maximum sensitivity, instead of a current-mode photomultiplier tube; as a result, different electronics are required (along with custom software) to record and process the photon-counting data. The second major difference is that the flow velocity is ~1000 times slower (10 mm/sec vs. 10 m/sec). Very slow flow is required in order to collect sufficient photons from the relatively low number of fluorophores bound to small DNA fragments. Also, it is essential that the detection volume—the volume within the flow cell delineated by the intersection of the laser beam energy distribution with the field of view of the fluorescence collection lens—be as small as possible to reduce the background counts.

The approach to single *fluorescent* molecule detection *in flow*—techniques and instrumentation—has been documented in detail elsewhere (Keller et al., 1996). DNA fragment sizing is the first direct application of this technology. This accomplishment required the new, highly specific DNA-binding dye TOTO-1 (Rye et al., 1992). From a family of related dimeric cyanine nucleic acid stains, TOTO-1 is essentially nonfluorescent until it binds to DNA, whereupon there is a 1400-fold enhancement of fluorescence resulting in relatively large signals, a very low background, and a low detection limit. The smallest DNA fragment clearly resolved from background in 1993 was 10,086 base pairs (bp) with about 200 photoelectrons detected from approximately 2000 dye molecules (1 : 5 dye-to-DNA bp ratio). Excellent measurement linearity was achieved for DNA fragments ranging in length from 10.1 to 48.5 kilobase pairs (kbp). Although gel electrophoresis has better resolution (at least for fragments below about 10 kbp), the flow-based measurement gives direct enumeration of the *individual* fragments analyzed. Detection of single, *small,* fluorescent molecules (<500 mol. wt.) in solution requires more elaborate equipment and complex techniques, such as time-gated detection or two-photon excitation, than those necessary to measure DNA fragment size (Keller et al., 1996; Goodwin et al., 1996; Johnson et al., 1993). The apparatus to be described in this chapter, known as the MiniSizer, can detect single phycoerythrin molecules in solution, as illustrated in Figure 6.1. However, phycoerythrin is a large macromolecule consisting of 34 individual fluorophores. Nevertheless, this is about the lower detection limit for the current configuration of the apparatus, with about 115 photons detected during the 2-ms transit time through the laser beam. Petty et al. (1995) reported detection of individual 1.5-kbp fragments in 1995 and the smallest detectable fragment continues to decrease.

An unfamiliar term needs to be introduced at this point. In this regime, with a photon-counting detector, where the fluorescence emission and background noise are resolved into single photons, a "burst" of photons denotes the passage of a single fluorescent molecule through the laser beam, as seen in the bottom panel of Figure 6.1. This term will be used synonymously with "event" or "signal," corresponding to the detection of a discrete fluorescent entity passing through the laser beam. Criteria such as burst area, burst height, and burst duration will be used to quantify these events.

Introduction

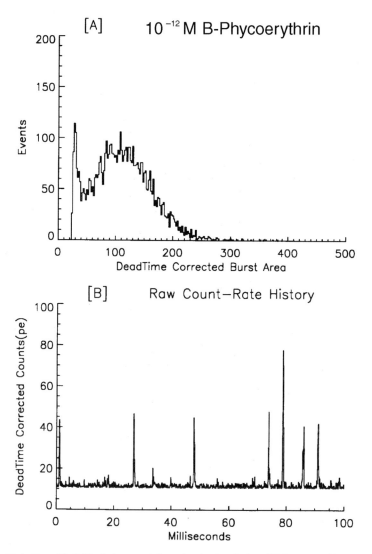

Fig. 6.1. Detection of individual phycoerythrin molecules in solution. (A) Burst area histogram of B-phycoerythrin molecules in solution at 10^{-12} M with 10 mW of laser power (532 nm). (B) Segment of the raw data, i.e., the running count rate history, showing the individual bursts of photons detected from the PE molecules with a dwell time of 100 µs. The burst duration was about 550 µs.

The first DNA fragment–sizing apparatus was fairly cumbersome, nearly filling a 5 × 8-ft optical table. Numerous modules, including a liquid-cooled photon-counting PMT, an amplifier/discriminator, a water-cooled ion laser, a multichannel scaler system with a CAMAC computer interface, and a SUN workstation, were utilized. Continual technical improvements and simplifications have resulted in the system being reported here. For the MiniSizer, the decision was made at the outset to substantially diverge from the earlier work. We decided to evaluate the related but previously unproven dye POPO-3. This dye offered the compelling advantage that we could use the new solid-state, diode-pumped miniature lasers that were becoming less expensive, and hence more available. In a proof-of-principle experiment, using a borrowed laser (532 nm, 7 mW), a state-of-the-art APD detector, and POPO-3, the 564-bp fragment of lambda DNA was observed in flow for the first time (unpublished results). Based on these results, we bought a new frequency-doubled "μGreen" laser and an actively quenched APD detector, which can count more than ten million photons per second.

The top-left graph in Figure 6.2 shows the histogram of burst areas recorded from a sample of lambda bacteriophage DNA digested with the HindIII restriction enzyme. In addition to the 564-bp fragment, well resolved at the left edge of the graph (see insert), there are several other features to point out in this figure. Only six peaks are resolved in the graph, but actually eight fragments (125, 564, 2027, 2322, 4361, 6557, 9416, 23130 bp) result from this digestion of lambda DNA. The 125-bp fragment is not visible, while two fragments (2027 and 2322 bp) that are not separately resolved produce the second peak from the left. However, the integrated area of that peak reveals that two fragments are contained therein. Interestingly, the burst duration histogram (Fig. 6.2B) has structure in it, with the bump at 1230 μs [corresponding to the peak at 3500 photoelectrons (pe) in the area histogram] representing the 23.1-kbp fragment. Another significant point is that the positions of all six peaks—derived by simultaneously fitting six Gaussians to the histogram—*appear* linearly related to the known fragment sizes. Unfortunately, in this case, this is somewhat misleading. Note that the peak in Figure 6.2A at approximately 650 pe is smaller in area than the two peaks to its right. This peak represents the 4.4-kbp fragment that contains a "cos" site where a 12-bp region is single stranded and therefore "sticky." This fragment hybridizes with its complementary sticky end on the 23.1-kbp fragment. When this happens, the 4.4-kbp fragment is underrepresented in the histogram, while the 23.1-kbp fragment becomes 27.5 kbp long. With intact lambda molecules, this sticky region is also responsible for concatamer formation. All fragments should be equally represented since eight fragments result from each and every lambda molecule completely digested. (Heating to 60°C for 10 min disassociates the 4.4- and 23.1-kbp fragments.) The interesting aspect of this is that the peak at 3500 pe is due to the 27.5-kbp aggregate instead of the individual 23.1-kbp fragments. As a result, when using the expected digest fragment sizes, the peak positions appear to be more linearly spaced (Fig. 6.2D) than they really are. The small shoulder at 3000 pe is the true 23.1-kbp peak, and it would be located further out in the histogram if the detector had a larger dynamic range. This nonlinearity is due to a fundamental limitation of the APD module caused by a fixed deadtime (about 30 ns) after each photon is detected and before the detector can again respond. Fortunately, this can easily be corrected in software, off-line, when the raw data (a running count rate history) are processed by software to extract the individual event characteristics: burst area, burst height, and width. The computer algorithm and other data processing issues will be described under Instrumentation. Another point, also to be elaborated on later, must be introduced here. In the burst-area-versus-burst-height contour plot (Fig. 6.2C), the 23.1-kbp fragment falls off the diagonal line of the smaller fragments; its area has increased but the burst height has leveled off and is no longer linearly correlated with

Introduction

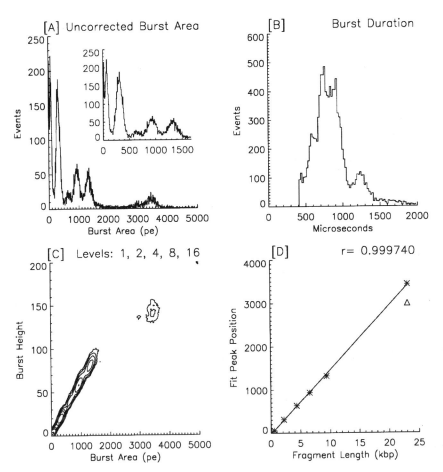

Fig. 6.2. HindIII restriction enzyme digestion of lambda DNA stained with POPO-3. In the burst area histogram (A) six peaks are well resolved; the third peak, corresponding to the 4.4-kbp fragment, is underrepresented due to aggregation with the 23-kbp fragment (see text). The small insert graph in A shows the lower end of the histogram in more detail. There is structure in the burst duration histogram (B) related to increased physical lengths of the larger fragments and thresholding effects on smaller fragments. The contour plot of burst area vs. burst height (C) reveals the largest fragment falling off the diagonal due to its extended length filling the laser beam in the vertical dimension and the burst height reaching a plateau. Linear regression analysis of fragment size in kbp vs. signal strength in photoelectrons (D) appears to have good linearity. However, "△" denotes the correct location of the peak due to the true 23.1-kbp fragment.

burst area. This is an instrumental characteristic due to the behavior of DNA molecules in flow and to the optical geometry of the MiniSizer.

The data set displayed in Figure 6.2 was deliberately chosen to provide an introduction, as well as to point out some pitfalls and complexities. Despite these issues, analyzing less than 1 pg of DNA (the equivalent of 19,000 lambda DNA molecules), this apparatus and this technique exhibit a very linear response, piecewise, over the size range of 245 bp to at least 180 kbp. Another instrument has measured concatamers of lambda DNA to nearly 400 kbp (Huang et al., 1999). The primary focus of this chapter is to provide a detailed description of the compact system as it currently exists. The system's performance will also be characterized as we strive to understand what is really going on when individual macromolecules as complex and subtle as DNA are examined very closely.

MATERIALS AND METHODS

DNA Samples and Nucleic Acid Stain

The nucleic acid stain POPO-3 is purchased from Molecular Probes (Eugene, OR; www.probes.com) at 1 mM in dimethyl sulfoxide (DMSO) and pipetted into 1-µl aliquots that are frozen at –20°C until needed for an experiment. The dye is diluted 1 : 100 (to 10^{-5} M) with 1X Tris–ethylenediaminetetraacetic acid (EDTA) [TE (10 mM Tris, 1 mM EDTA, pH 8.0)] buffer immediately before use. Samples being stained are kept in a light-tight container at room temperature, and staining procedures in general are carried out in subdued light conditions. Fluorescent lights, in particular, are to be avoided. We typically stain 450 ng of DNA in 987.7 µl TE plus 12.3 µl dilute dye solution for 1 h. These cookbook proportions yield a dye–DNA (bp) ratio of 1 : 5. Obviously, much less DNA can be prepared in correspondingly smaller volumes, but it is important to not raise the DNA concentration above about 450 pg/µl (Rye et al., 1992). However, once the samples are stained with POPO-3, they can be stored at room temperature, *in a light-tight container*, for as long as 6–8 weeks with only minimal degradation of the fluorescence distributions. For this reason, it is convenient to stain a relatively large amount of DNA and be able to use aliquots of the stock solution for several weeks. Once the stained DNA is diluted from the stock concentration (to an analysis concentration of 10^{-12}–10^{-13} M), its fluorescence is stable for no more than an hour or two. Various DNA samples have been acquired from several distributors, including Sigma, Promega, and New England Biolabs. The Kpn1-digested lambda DNA was prepared using a standard protocol for restriction enzyme digestion (Smith et al., 1976).

Instrumentation

As shown in Figure 6.3, the apparatus is straightforward and uncomplicated. It occupies about 2 ft^2 on an optical breadboard. Standard optical mounts are used to hold the various optical components, the laser, and the flow cell. Ancillary items such as two +5-V power supplies, the pressurized sample holder, a compressed nitrogen bottle and pressure regulator, and a data acquisition computer are located nearby.

Laser. The MicroGreen laser (Uniphase, San Jose, CA, www.uniphase.com), a 20-mW solid-state, diode-pumped, frequency-doubled neodymium–YAG device, requires neither cooling water nor fans. It is small, produces little heat, runs on 5 V, and has a satisfactory TEM-00 beam shape. Our unit actually produced 36 mW when new (1994), and it still has about 15 mW of output power.

Materials and Methods

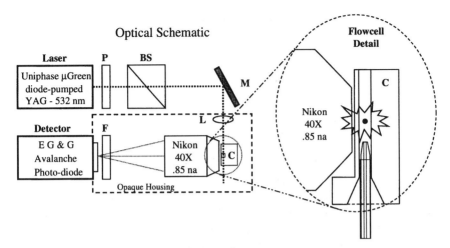

Fig. 6.3. Optical schematic of the apparatus. The laser output beam passes through a half-wave plate (P) and polarizing beam splitter (BS), which serve to attenuate the laser output power. A mirror (M) and a pair of crossed cylindrical lenses (L) direct the laser light into the flow cell. An optical band pass filter (F, 575DF30) restricts the wavelength range of the light that reaches the APD. The quartz capillary is inserted directly into the 250-μm square channel of the flow cell up to within 500 μm of the laser beam, as shown in the enlarged detail. The laser beam path is out of the page in this drawing.

Beam Shaping Optics. Light from the laser passes through a half-wave plate and a polarizing beam splitter (CVI, Albuquerque, NM; www.cvilaser.com). These two components used together permit continuous adjustment of the laser light intensity that reaches the flow cell. The birefringent element rotates the polarization of the output beam, and the beam splitter dumps the unaccepted light out through one face. The light that passes through the beam splitter is horizontally polarized. For most experiments, only 2–10 mW of laser light is used (measured prior to the mirror). The light that passes through the beam splitter is shaped into an elliptical spot, 11 μm high by 50 μm wide at the sample stream, by a pair of crossed cylindrical lenses of fused silica (Karl Lambrecht, Chicago, IL; www.klccgo.com). A 5-cm-focal-length lens defines the width and a 1-cm-focal-length lens determines the laser beam height. The lenses provide convenient and precise alignment of the laser beam on the sample stream. The laser beam spot size was calculated by the formula

$$\frac{1}{e^2} \text{ spot size} = \frac{4 \lambda f}{\pi D} \tag{6.1}$$

Where λ = wavelength, f = lens focal length, and D = input beam diameter.

Flow Cell and Fluidics. The laser light is focused on the sample stream inside a custom flow cell (NSG Precision Cells, Farmingdale, NY) with a 250-μm square cross-sectional flow channel, as shown in Figure 6.3. The flow cell is fabricated from several pieces of silica fused together, with one face ground down and polished to the thickness of a microscope cover slip (~70 μm). A modified "Ortho-type" mount supports the flow cell (with the direction of flow from bottom to top making it easier to eliminate bubbles) and provides the sheath fluid connection and sample entry access. An essential element of the system is a "drain" line at the end of the flow channel that provides back pressure and reduces the flow velocity a thousandfold compared to conventional flow cytometry (FCM). A small piece of soft silastic rubber tubing is pressed gently up to the end of the flow cell, both to create a seal and to connect a 50-cm-length of Teflon tubing (300–400 μm ID) that runs into the bottom of a partially full beaker of wastewater. High-purity water, as sheath fluid, is gravity fed from a 1-l bottle suspended on a post over the instrument. Adjusting both the height of the sheath bottle and the level of the water in the waste container controls the flow velocity. At only 3 ml/h the flow rate is stable over many hours with this simple arrangement. In 1990, Zucker used as much as 150 ft of small-bore tubing to slow the flow in an Ortho instrument (Zucker et al., 1990) and thereby increase its sensitivity. The MiniSizer also slows down the sheath flow by virtue of having the sample delivery capillary inserted directly into, and thereby largely occluding, the flow channel. The small-bore capillary provides high resistance to the sample flow that, combined with the sheath flow restriction, permits stable transit times of 10 ms or more.

Capillary and Pressurized Sample Delivery. A 30–50-cm-long piece of fused silica capillary tubing (Polymicro Technologies, Phoenix, AZ; www.polymicro.com) delivers the stained DNA molecules directly into the flow cell to within a few hundred micrometers of the laser beam. Capillary tubing of 40 μm ID is normally used, but tubing ranging from 20 to 100 μm ID has been tested. The outer diameter of the capillary, including the polyimide coating that makes it possible to have *flexible quartz* tubing, is usually about 245 μm. Sheath fluid flows around the capillary in the corners of the 250-μm square cross-sectional flow channel. Perhaps *the* single most critical item of the entire system is to have a smooth, symmetrical taper on the tip of the capillary (Fig. 6.4). In particular, the rim of the lumen (at the tip) must be as smooth as possible. A rough edge on the tapered end of the capillary invariably resulted in multiple sample streams that caused very confusing behavior of the instrument and usually produced worthless data. However, New Objective (Cambridge, MA; www.newobjective.com) fabricates capillaries for electrospray instrumentation with perfectly tapered ends that are smooth and symmetrical. The desired end result is to have stable, tight, hydrodynamic focusing of the sample stream with the end of the capillary 200–500 μm below the laser beam.

The DNA sample solution is driven through the capillary by nitrogen at a pressure ranging from roughly 0.3 to 1.0 psig. A simple pressure regulator is adequate to maintain stable sample delivery because of the high resistance to flow offered by the capillary tubing and the continuous drain line from the flow cell to the waste container. At 10^{-11} to 10^{-13} M (fragments) it is possible to regulate the sample flow rate with a conventional air pressure regulator (Fairchild Model 10, 0–2 psi; www.devicecorp.com). A sample of DNA is diluted in 1X TE buffer to an optimal concentration (about 2×10^{-12} M fragments) in a 1.5-ml Eppendorf tube. The tube fits in a plastic mount, sealed by an O-ring, which supports a rubber septum that is compressed around the capillary and the inlet nitrogen line.

Quartz Capillary Sample Delivery Tube

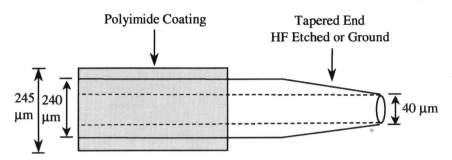

Fig. 6.4. Detail of the quartz capillary. A smoothly tapered tip is essential and the rim of the capillary lumen must also be smooth in order to prevent multiple sample streams. Examining the sample stream and tip of the capillary through a microscope while a sample of DNA is running in the instrument can reveal if there is a single sample stream and if it is tightly focused.

Detector. The detector is a photon-counting, silicon APD assembly (SPCM-AQ-121, EG & G, Canada, Vaudreuil, Quebec; www.egginc.com). This sophisticated module utilizes a proprietary circuit that actively quenches the diode and thereby enables it to count photons at rates well in excess of 10 MHz. The diode itself, made from extremely pure silicon, is mounted on a thermoelectric cooling element that eliminates the need for the cumbersome liquid cooling system typically required by a photon-counting photomultiplier tube (PMT). These features serve to keep the dark counts very low [<300 counts per second (cps)]. In this apparatus, as in conventional cytometers, the detector is not the limiting element that determines the detection limit. The dark counts of the detector are overwhelmed by background counts from the laser light impinging on the sheath/sample stream and reflected within the flow cell. An APD module with a dark count rating of < 1000 cps is more than adequate for DNA fragment sizing. Unless other considerations set the noise level, it is desirable to have the lowest possible dark counts for small single-molecule detection.

Light Collection Path. The high-numerical-aperture (NA) Nikon microscope objective (40X 0.84 NA, Melville, NY; www.nikonusa.com) is designed for a maximum working distance of 390 μm *in air*. The lens has a conventional tube length of 160 mm and a cover glass correction collar to compensate for working distance variations. More light could be collected using an immersion lens, but with the current flow cell dimensions there is not much to be gained. With one wall of the flow cell ground down to 70 μm and the sample stream in the center of the 250-μm-square flow channel, the minimum working distance is 195 μm, well within the objective's range. An opaque housing covers the flow cell, the microscope objective, the second laser-beam-focusing lens, the band pass filter (Model 575DF30, Omega Optical,

Brattleboro, VT; www.omegafilters.com) and the detector end of the APD module. This permits the system to be operated in normal room lighting.

Data Collection System. A primary raw data set consists of the history of the number of photons registered by the APD detector in small time intervals ranging from 25 to 250 μs, depending on the maximum transit time of individual fragments through the laser beam. A multichannel scaler card (MCS-2, Oxford Instruments, Oak Ridge, TN; www.oxfordinstruments.com) collects the count rate history. The card is essentially a high-speed counter with a programmable count interval and 8192 memory locations into which the count rate history is stored. It counts the transistor–transistor–logic (TTL) pulses from the APD, stores the number of detected photon pulses that occur in the dwell time interval, repeats this process until the memory is full, and writes the stored information to a disk file. The dwell time interval used depends on the experiment, that is, whether large or small fragments are being analyzed, and the sample flow velocity. A batch file controls the card, running in a DOS window under Windows 95, in an ISA-bus PC. The batch mode allows multiple scans of the card's memory to be recorded in separate data files, resulting in a discontinuous count rate history. In order to have meaningful statistics, especially if the DNA sample being analyzed has a heterogeneous mix of fragment sizes, several hundred to 1000 of these small raw data files are collected. While each scan is being written to disk, data collection is interrupted (usually for less than 0.5 s), resulting in a segmented record. Since each scan is written to a separate data file, the first data processing step is to concatenate the raw MCS files into a single large file, which then contains the entire, discontinuous count rate history. The final concatenated file is written to disk with a file header, based on the FCS2.0/3.0 file format standard, that contains all the relevant information about the data collection run, the sample ID, instrument configuration (laser power, optical detection filter, etc.), DNA source, and stain.

In the current configuration, the MCS-2 card resides in a 66-MHz 486 PC clone networked to a second PC, with a shared network drive where the raw data files are written. This second computer, a 200-MHz dual Pentium-Pro PC running WinNT 4.0, is used for all the data processing and analysis. As the raw data files arrive, a program written in the high-level language IDL (Interactive Data Language, RSI, Boulder, CO; wwwrsinc.com) performs the necessary operations on the raw count rate history. The program DNASizer concatenates all the raw data files into one large file (for subsequent storage and analysis under interactive user control). It processes the raw data to sift for the burst events, calculate the primary parameters, and create histograms and various displays. The second computer is sufficiently fast that, within a few seconds after the last raw MCS data file is acquired, the concatenated data file is saved, the entire data set has been processed, and the selected one- and two-dimensional histograms and other displays are drawn. At this point the raw MCS files are either deleted or simply overwritten.

It is possible to acquire and process the running count rate history of a photon-counting detector by other means (Agronskaia et al., 1998) but we concluded that it was best to retain the full digital data record. The importance of doing so is supported by studies on the emission dynamics of individual molecules such as phycoerythrin

Materials and Methods

which have been observed to abruptly photobleach and to rapidly turn on and off (Wu et al., 1996). These features could easily be lost during D to A and A to D conversions.

Data-Processing Software. Each raw data set contains the entire running count rate history of the APD during the data collection period. In the present photon-counting system, there is no discriminator circuitry to recognize discrete *events* and extract the pulse parameters of height, width, and area. Since the raw data are a digital record of detected photon counts in small time intervals, they must be sifted via software to identify valid events, correct for the background noise level and the detector's deadtime, and calculate the primary event parameters. When a data file is first opened, the program automatically evaluates the background noise (in regions of the data set without bursts) and then allows the user to adjust the processing threshold and offset, based on the background and the observed event rate. The processing threshold is the level that identifies the bins to be processed, and the offset is the point (above the threshold) at which a burst is defined to begin. If either of these parameters is modified, the routine sifts an entire selected scan, identifies the bursts, calculates the burst area and duration, and displays a dot plot of the area versus duration for that scan. When the value in a bin exceeds the sum of the threshold and the offset, that bin marks the start of a burst; conversely, when the counts in a bin drop below that level, the burst is terminated. A comparison is then made to determine if the burst falls within minimum and maximum burst length criteria set by the user. If the burst is acceptable, the primary parameters—area, height, and duration—are calculated and stored for subsequent use in creating one- and two-dimensional histograms. The burst area is corrected for the background count rate (the mean count rate in the absence of bursts) by subtracting the background level times the burst duration from the burst area. When the user is satisfied with the settings and the results on the selected scan, clicking on the "Process" button of the data processing panel directs the program to process the entire data set.

A small section of one raw data set is displayed in Figure 6.5A at a vertical scale that shows the full range of displayed bursts. The second line shows the same data with higher sensitivity (10X vertical gain) to reveal the low-level fluctuations in the count rate history. Many bins in this example recorded zero counts. The first stage of data processing is to smooth the raw count rate history with a fast and efficient smoothing algorithm, built into IDL, known as the Lee filter. This routine calculates local statistics and produces smoothed data that accurately preserve the overall shape and, more importantly, the area of the burst events. It also converts the raw integer data to floating-point numbers with fractional values. The next step is to remove from the data record many of the bins that contain no events and represent only the background count rate of the instrument and sample. In order to minimize coincident events, that is, more than one DNA fragment in the detection volume simultaneously, the DNA sample must be reasonably dilute (6×10^4–6×10^6 fragments/µl) and run slowly (50–250 s^{-1}, depending on the fragment size and flow velocity). Under these conditions, 50–70% of all the bins will contain only background counts. If these bins do not need to be fully processed, eliminating them early in the processing scheme saves considerable time. IDL facilitates such a selection through the use of the

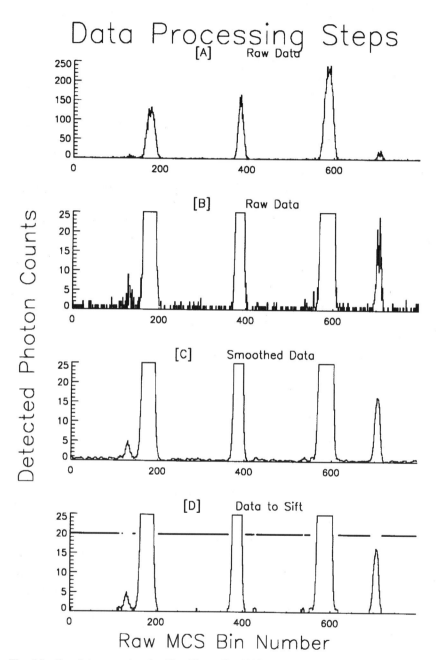

Fig. 6.5. Raw data-processing algorithm. The top line (A) is a segment of the raw data and (B) the same data at a lower y-axis scale to show small fluctuations in the count rate history. (C) Data have been smoothed and converted to floating-point values. Only a small subset of the total raw data—those bins with a count level greater than the processing threshold as drawn in (D)—actually needs to be processed to extract the real events. The horizontal line drawn at the Y value of 20 in (D) indicates the bins that are eliminated when the raw data set is truncated.

"where" function: "index = where(Raw gt threshold)." This command returns the indices of the bins (in the entire raw data set) where the bin counts are greater than the processing threshold value. One more command, "Raw = Raw(index)," truncates the entire data set, effectively bringing the events closer together than they were in the unedited count rate history. The graph in Figure 6.5D shows the remaining bins that are above the threshold; these are the bins that are sifted in the next processing step to identify and quantify the actual events. This small section of raw data makes a compelling example; of the 800 bins in the original data set, only 200 actually need to be sifted to identify the real events. For the entire data set only 2,614,155 bins out of the initial 4,915,200 required full evaluation. Instead of taking over 50 s to process all of the bins, the routine processed the truncated subset in 28 s. In the data set for Figure 6.1 only 30% of the 2.5×10^6 initial bins had to be sifted. Only the bins retained in the truncated data set receive deadtime correction and final sifting to identify the true events and extract the parametric information from each burst.

A fundamental limitation of the APD detector is that whenever the device detects a photon, it generates a 10-ns-wide TTL pulse and then is unresponsive for an additional 30 ns. However, the manufacturer carefully characterizes the deadtime for each unit (e.g., for one module, the correction factor is 1.25 at 5 mHz and 1.65 at 10 mHz), and it is simple to correct for this limitation, off-line, in software. This is accomplished by correcting the raw count rate history based on the following calculation, which takes into account the bin width (dwell time) of the raw data set and the deadtime of the APD. Because most of IDL's operators are inherently array oriented, only one line of IDL code [Eq. (6.2)], a straightforward mathematical expression, is required to correct the entire raw data set:

$$\text{True count rate} = \text{Raw} \times \left(\frac{1.0}{1.0 - (\text{Raw}/\text{b\_w}) \times \text{DeadTime}} + \text{DtFf} \right) \quad (6.2)$$

where Raw is the entire raw data array (the count rate history: 819,200 to 8,192,000 bins), b_w is the programmed dwell time [b_w = float(bin_width) × (1.0×10^{-6})], DeadTime is the specified deadtime for the individual detector, and DtFf is the detector-specific fudge factor to better fit the detector's response curve.

Figure 6.6 (HindIII-digested lambda DNA heated to 60°C for 10 min) graphically reveals the difference between the burst area histograms with and without deadtime correction. Without correction (Fig. 6.6A), the 23-kbp fragment is much closer to the 9.4-kbp peak than it should be. The lower left graph is the histogram produced after the raw count rate history has been corrected. Note that the histogram is not merely shifted to the right. The raw data are corrected on a bin-by-bin basis prior to histogram generation to adjust the measured count rate history to the level expected if the detector did not have a deadtime limitation. In Figure 6.6 the corrected count rate history was resifted and the histogram was recalculated and plotted in panel B. The peak positions plotted in Figure 6.6C are derived from simultaneously fitting six Gaussians to the histograms and then doing linear regression of the known fragment sizes versus the observed signal strength. Space limitations preclude a discussion of the many other analytical features and graphical capabilities of the program.

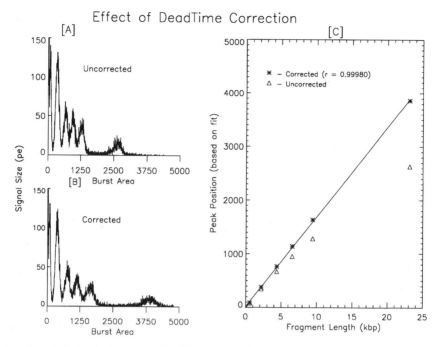

Fig. 6.6. Effect of deadtime correction. Using the manufacturer's supplied information on the detector-specific deadtime, it is possible to correct the *raw* data for the deadtime of the APD. The upper left-hand panel (A) shows the burst area histogram prior to correcting the raw data and the lower graph (B) is the result of correcting each raw data bin based on the observed count rate, the dwell time, and the deadtime of the APD. On the right is a graph showing the corrected (∗) modal channel values of Gaussians simultaneously fitted to all six peaks in the histogram. Points drawn with triangles are the peak positions prior to deadtime correction.

CRITICAL ASPECTS OF THE SYSTEM

Dye–DNA Interactions

The dye–DNA interactions, which might seem straightforward and uncomplicated, are anything but simple! For POPO-3 (and other related homodimeric, asymmetric cyanine nucleic acid stains), the DNA concentration must be tightly controlled for optimal staining to avoid either DNA precipitation or poorly resolved fragments that would otherwise be resolvable (Rye et al., 1992; Carlsson et al., 1995). Apparently, all of these dyes must be used very carefully and the range of optimal DNA concentrations must be determined experimentally. The bis-intercalating dye molecules have the potential to cross-link separate DNA fragments or to link distant regions on the same fragment, although this seems unlikely (Carlsson et al., 1995). However, the dimer has a very high binding constant compared to the monomers [10^{-11} vs 10^{-5} M (Glazer and Rye, 1992; Carlsson et al., 1995)]. As noted earlier, once optimally stained with POPO-3, a sample of DNA (at 450 ng/µl) can be kept at room temperature (in an opaque container) for several months with little degradation. An undesirable characteristic of these dyes is that they are not optimal under saturating conditions, although a dye–DNA ratio of 1 : 5–1 : 10 is optimal for flow analysis. But if the amount of DNA to be stained is not accurately known it is very easy to get suboptimal staining. This aspect alone requires that internal staining (size) standards be included along with an unknown fragment whose size is being determined. The order of reagent mixing is also important (for reasons that are not entirely clear), and the best results are produced when DNA is added to the dispersed dye solution (Glazer and Rye 1992). It is also possible to obtain good results by adding considerably diluted dye to a dispersed DNA solution (data not shown).

As with other bis-intercalating dyes, each molecule of POPO-3 intercalated unwinds the DNA helix by 60° (Spielmann et al., 1995) and each intercalator doubles the inter-base-pair spacing (Glazer and Rye 1992). This has a profound effect on the physical length of various sized DNA fragments at the dye–DNA ratio of 1 : 5 (see Table 6.1 below) which ties in with unique observations of the burst waveform made with this instrument, to be discussed under Results. We often see a gradual decline in the event rate for lambda DNA (or larger fragments), as a sample is being analyzed; at times, larger fragments seem to get stuck in the capillary. Attempts to coat the inner surface of the capillary so as to repel DNA molecules have been only marginally successful.

Sample Delivery and Flow Rate

It is necessary to set the concentration of fragments to the range where it is possible to regulate the sample event rate; in the current system, 10^{-10}–10^{-12} M is about optimal. More dilute samples can be utilized, but it takes longer to collect a reasonable amount of data. If the sample volumetric flow rate is increased too much, the coefficient of variation (CV) of the measurement increases. Increasing the sample concentration can increase the speed of data accumulation, but then it becomes difficult to avoid excessive coincidence. A delicate balance must be maintained between the long transit time (1–10 ms) through the laser beam and the sample concentration or event rate.

In operation, only a fraction of a microliter is actually analyzed. This is another advantage of using such small-bore tubing to deliver the sample to the point of analysis. The volume of a 30-cm length of 40-μm-ID tubing is 0.38 μl. If it were possible to conveniently handle 1 μl samples, this instrument could analyze 10^4–10^5 DNA frag-

TABLE 6.1. Physical Length of Stained and Unstained DNA Fragments

Fragment	Size (kbp)	Physical length (μm)	Stained length (μm)
Lambda/Kpn1	1.5	0.51	0.71
Lambda/Kpn1	17.0	5.8	8.1
Lambda/Kpn1	30.0	10.2	14.3
Lambda	48.5	16.5	23.1
T5	110.0	37.4	52.4

ments and still have more than 90% of the microliter left; if 10^4 fragments are analyzed (at 10^{-12} M), only about 30 nl of the DNA sample will be consumed.

Resolving the fragment populations present in a sample of HindIII-digested lambda DNA is relatively easy—all the child fragments are equally represented. Each initial lambda molecule yields one, and only one, member of each of the fragment classes (125, 564, 2027, 2322, 4361, 6557, 9416, 23130 bp) if the digestion is complete. (The 125 bp is not visible and the 2027- and 2333-bp fragments are not separately resolved.) However, it can be very difficult to cleanly resolve a relatively rare, large fragment within an abundance of small fragments such as from the host bacterium in a plasmid preparation with a high background of genomic DNA fragments. These situations demand careful and deliberate sample preparation.

Detector

The EG&G APD detector has a significant advantage over a photon-counting PMT simply because it has a much higher quantum efficiency, >60% (at 633 nm) as opposed to 10–20% for the PMT. However, due to the longer deadtime (40 ns vs. <1 ns), the APD has a lower *maximum* count rate than a PMT and hence a smaller useful dynamic range. The APD may also be more sensitive to damage by high light levels; in the specification sheet EG&G states that if more than 10^6 photons impinge on the diode in 1 µs, it probably will be destroyed! In routine operation we prefer to keep the average count rate below 500 kHz, but it is the *instantaneous* count rate that must be constrained, as far as the APD is concerned. Also, the APD cannot differentiate between one photon and multiple photons which might arrive simultaneously. Overall, the APD-based system is more sensitive but a PMT can inherently handle higher fluorescence intensities. However, features of the fluidics and optical geometry of the MiniSizer, coupled with the way DNA molecules behave in flow, somewhat negate the smaller dynamic range of the APD. This will be fully discussed further on. The APD module is expensive (~$4000) but it is a very small, self-contained unit that requires only +5 V at 3 A. It counts individual photons of light at significantly greater than 10 MHz, with better than 60% detection efficiency, enabling the system to detect single phycoerythrin molecules in solution or pieces of DNA molecules as small as 245 bp up to at least as large as 177 kbp (data not shown).

Detection Volume

An essential aspect of this type of instrument is to reduce the detection volume—the portion of the sample stream that is actually interrogated by the optical system. A convolution of the laser beam dimensions and the field of view of the collection lens (including any spatial filters) define this. A picoliter occupies a volume equal to 1000 µm³ (10 × 10 × 10 µm). In the current system, the sample stream is about 1 µm in diameter. Therefore, the volume of fluid from which fluorescence is optimally collected is approximately 0.01 pl. As a frame of reference, with free dye molecules at 1 µM, this volume would still contain $\cong 10^4$ molecules.

The efficiency of the dye, combined with the high NA of the fluorescence collection lens and the sensitivity of the detector, results in relatively strong signals. In an early configuration of this instrument, an adjustable slit (Oriel, Stratford, CT; www.oriel.com) was placed at the focal point of the objective in order to constrain the light reaching the detector to that originating from the sample stream. In the MiniSizer, the 200-µm-diameter active region of the APD itself, placed at the focal point of the objective, serves as the system aperture. This results in a simpler system with two fewer parts, eliminating the adjustable slit and a second microscope objective (to refocus the diverging light passing through the slit onto the active area of the APD). This approach appears to be at least as sensitive as the original configuration, and the dynamic range and linearity of the system are unaffected (data not shown).

RESULTS AND DISCUSSION

Using the best available nucleic acid stains, it is possible to stain a small sample of DNA in about an hour and thereafter, within minutes, to have a histogram of analyzed fragment sizes. In a series of experiments, the flow-based measurement has very good linearity over a large size range (245 bp to nearly 400 kbp) and there is direct enumeration of discrete populations of fragments. Unlike gel electrophoresis, the resolution of fragment size in the molecular flow analyzer improves with increasing size.

Small Fragment Sizing

The application of flow "cytometry" to measure the size of individual DNA fragments in solution is unique in that there are precisely known standards that can be used to calibrate the machine and its performance. A molecule of lambda DNA has *exactly* 48,502 bp (strain c1857 *ind* 1 *Sam* 7, Promega). Also, the sizes of the fragments that result from restriction enzyme digestion of lambda DNA are known exactly. Therefore, we know what a properly prepared sample of that DNA should look like, and, if such a sample produces the correct distribution, we can have confidence in the instrument when we use it to analyze an unknown sample of DNA. Unfortunately, POPO-3 fluorescence is optimal within a narrow range of dye–DNA ratios, and this can be elusive when the mass of DNA in the sample being analyzed is not known within a factor of 10. This limitation can be overcome by the use of internal staining standards, just as known size standards are run in adjacent lanes of electrophoresis gels. In Figure 6.7, four small DNA fragments are analyzed along with a 4.1-kbp standard. This graph illustrates the performance of the system at the instrument's detection limit. The 245-bp fragment has only about 50 POPO-3 molecules intercalated. If the four fragments were mixed together prior to analysis, they would not be individually resolved. The minimum resolvable size difference (in one batch of fragments) is about 1.5 kbp as can be seen in Figure 6.6 with the HindIII digest of lambda DNA. However, with internal staining standards, the sizing accuracy of the system is affirmed and its response is clearly linear.

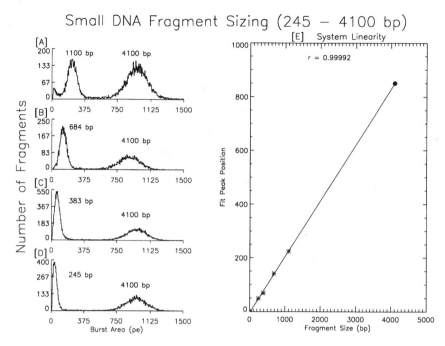

Fig. 6.7. Small DNA fragment sizing. Four different fragments were individually mixed with a 4100-bp standard (A–D) and analyzed. The 245-bp fragment (D) is at about the detection limit of the current system. With a dye–DNA (bp) ratio of 1 : 5 there are approximately 50 dye molecules bound to the smallest fragment. Nevertheless the linearity of the system is excellent all the way down to the detection limit. The position of the 4100-bp standard and 0 defined the limits of the fit. The small fragments were purified from electrophoretic gels following restriction enzyme digestion.

Burst Outline Analysis

Figure 6.8 illustrates two interesting points condensed from analyzing the same sample of Kpn1-digested lambda DNA at three different dwell time settings, 16, 25, and 50 µs. First, it can be seen that there is enhanced resolution of the three fragments (produced by this digestion: 1.5, 17.0, 30.0 kbp) in the burst *duration* graph at 25 and 16 µs dwell times. There is a threshold effect, particularly on the smaller fragments, but there is a real difference in the burst duration as the fragment size increases. Second, up to a certain size (≅22 kbp, Figs. 6.2 and 6.8) as the burst height increases, so does the burst area. Beyond that point, the burst height remains constant but the burst duration increases so that the integrated burst area continues to increase (Fig. 6.8B, middle left-hand graph). This is consistent with the dye molecules being intercalated relatively uniformly along the length of the fragment and the physical length of the 30-kbp fragment being stretched out in flow larger than the laser beam's vertical dimension (11 µm high). Therefore, until the laser beam is entirely filled by a fragment of sufficient length, the burst height increases as the fragment size increases. When the length of a fragment oriented in flow is greater than

Results and Discussion

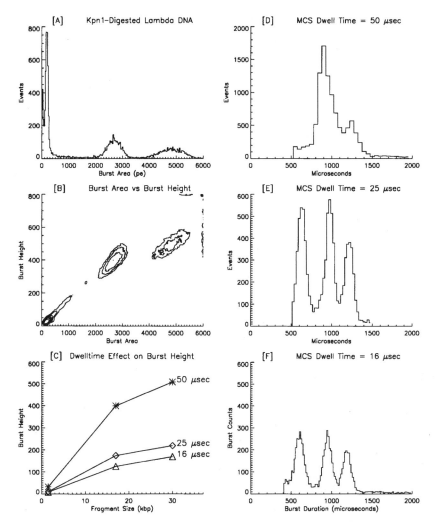

Fig. 6.8. Dwell time effect on burst height and duration histograms. This figure illustrates the effect of the MCS dwell time on the burst height and duration using a Kpn1-digested sample of lambda DNA containing three fragments: 1.5, 17, and 30 kbp. Decreasing the dwell time from 50 to 25 μs results in a dramatic change in the burst duration histogram, whereas a dwell time of 16 μs does not make much more difference. Changes in the burst height at different dwell times are due simply to the fact that fewer photons are detected in shorter time intervals. The integrated burst area histogram is unaffected by these dwell time changes.

the height of the laser beam, the burst height plateaus at the maximum value (because the entire beam height is filled) while the burst duration extends in proportion to the total fragment length. The burst height increases beyond this point only if the fragment is not stretched out, kinked, looped, or oriented perpendicular to the flow or if more than one piece of double-stranded DNA is present, so that more dye molecules are in the laser beam simultaneously. If a portion of the fragment is not double stranded, that region will have fewer dye molecules along its length and the burst height will be proportionately lower. Although both the fragment size (up to about 22 kbp) and the dwell time of the MCS card (fewer photons are counted in smaller time intervals) affect the burst height, the burst area distribution is unaffected; hence only one histogram is drawn in the upper left graph in Figure 6.8A.

DNA Extension in Flow

Intact unstained lambda DNA can be physically stretched to more than 15 μm in length (Kornberg and Baker, 1992). Furthermore, DNA molecules are easily stretched out by fluid forces in flow (Perkins et al., 1994). A reasonable approximation of the physical length of a piece of DNA can be calculated by the equation

$$\text{Physical length (μm)} = (0.34 \text{ μm/kbp}) \times (\text{fragment size in kbp}) \quad (6.3)$$

When DNA is stained with an intercalating dye—POPO-3 is a bis-intercalating *dimer*—the inter-base-pair spacing doubles [from 0.34 to 0.68 nm (Glazer and Rye, 1992)] where each dye molecule is inserted. With the dye–DNA ratio of 1 : 5 there is one dimer for every 5 bp and, therefore, a 40% maximum occupancy of available slots (it is possible to have both legs intercalated with one dye molecule for every 4 bp):

40% occupancy (1 dye dimer for 5 bp) $0.4 \times 0.68 = 0.272$
Therefore 60% of the spaces are 0.34 nm $0.6 \times 0.34 = 0.204$
$0.272 + 0.204 = 0.476$

Here, 0.476 × size (in kbp) gives the stained length (in micrometers), as indicated in Table 6.1.

From Table 6.1 it can be seen that the physical length of the stained 30-kbp Kpn1 fragment is longer than the vertical dimension of the laser beam spot size (11 μm high), while the 17-kbp fragment is not. Therefore, the fragment length that would give the maximum burst height (just long enough to entirely fill the laser beam) is about 22 kbp. The effect of the molecules stretching out in flow means that the APD detector has to contend with a smaller dynamic range of signal intensities than it would if the DNA remained tightly condensed in a globular form. As a result, the APD's enhanced sensitivity, compared to a PMT, is a real bonus, and its smaller dynamic range is not as detrimental as it might otherwise be. Also, this is why the dead-time correction is based on count rate, not on burst area. Figure 6.9 clearly illustrates this point with the measurement of a mixture of lambda and T5 DNA. At about 110 kbp, the T5 coliphage DNA is also a double-stranded linear molecule like lambda. It appears that there is only one distinct population in the burst height distribution (panel B), whereas the burst area histogram (panel A) shows both DNA molecules being equally represented. In panel D, the contour plot of burst area versus burst height, the burst height distribution levels off at about 20,000 pe (the lambda peak),

Results and Discussion

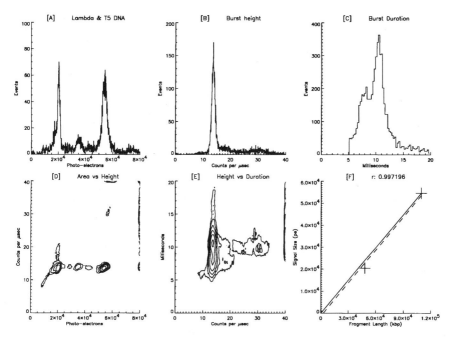

Fig. 6.9. Large DNA fragment analysis. Various graphs of the analysis of a mixed sample of lambda (48.5 kbp) and T5 bacteriophage DNA (110 kbp) using 2.5 mW of laser power and 250 μs dwell time, diluted 1:20 in 1X TE just prior to analysis. The plateau in burst height is very apparent in graph (D). Because the burst height is the same for both fragments (B), the burst duration must increase in order for T5 DNA to register as larger than lambda DNA. The increase in burst duration with constant burst height is clearly seen in (E). Some aggregation and coincident events appear in the burst height histogram. The source of the small peak at 3.5×10^4 pe in (A) is unknown.

but the burst area continues increasing out to 55,000 pe for the T5 peak. Two populations are visible in the burst duration histogram with the lambda peak at 7 ms and the T5 at almost 11 ms. It is worth postulating that circular, double-stranded DNA (with two double strands in the laser beam simultaneously) should give twice the burst height, compared to linear double-stranded molecules. This has been observed in preliminary experiments.

Future Directions

With the DNA molecule extended in flow, resulting in signals stretched out in time, there is effectively some scanning resolution along the length of the molecule. The current configuration of the instrument is designed to accommodate a second APD detector on the thick side of the flow cell, although a longer-working-distance objective will be required. Just as in conventional cytometry, two-parameter analysis is much more useful and powerful than single-parameter analysis. For example, the second APD will permit detection of a sequence-specific hybridization probe using a second fluorescent dye. Temporal location of the second signal, with respect to the size measurement signal, will give low-resolution mapping of the specific sequence along the length of the fragment. A new data collection system that can process the correlated count rate history from two photon-counting detectors has recently been completed. This system will process the raw data in real time, monitor the background level, and produce the primary data directly: burst height, width, and area. As a result, the photon-counting system will behave like a conventional cytometer that produces event-based data and a standard listmode data file. Furthermore, this system has a conventional light-scatter channel for use in experiments (on particles large enough to scatter light) where the highest possible fluorescence sensitivity is required.

This type of flow-based instrument, a multiparameter *molecular analyzer* with photon-counting sensitivity, promises to open up entirely new areas of individual, submicroscopic "particle" analysis. The intense interest in every aspect of the structure and function of DNA has driven the development of exquisitely selective and sensitive DNA binding dyes. This has been facilitated by the polymer nature of DNA and intercalating dye molecules that yield fluorescence signals proportional to the length of the DNA fragment.

As new and better fluorescent probes are developed for other biologically important molecules, and as this type of instrumentation becomes more widely available, it will be possible to study a vast range of molecular properties and interactions. This technology should be extensible to the study of protein assembly and folding and of protein–protein interactions using fluorescence energy transfer probes or other high-sensitivity probes directly at the level of the individual molecule. It should be possible to study individual receptor–ligand interactions and individual messenger RNA molecules. Furthermore, the very high sensitivity of the system will facilitate the analysis in flow of subcellular components and organelles, such as lysosomes or Golgi vesicles, and mitochondria. DNA fragment sizing is merely the first clearly demonstrated application of this new instrumentation. This technology opens the gateway to a whole new universe of flow-based analytical possibilities.

ACKNOWLEDGMENT

This work was supported by the Department of Energy (NN20) and by the National Institutes of Health–funded National Flow Cytometry Resource (RR-01315).

REFERENCES

Agronskaia A, Florians A, van der Werf K, Schins J, de Grooth B, Greve J (1998): Photon-counting device compatible with conventional flow cytometric data acquisition electronics. Cytometry 32:255–259.

Ambrose WP, Goodwin PM, Jett JH, Johnson ME, Martin JC, Marrone BL, Schecker JA, Wilkerson CW, Keller RA (1993): Application of single molecule detection to DNA sequencing and sizing. Ber Bunsenges Phys Chem 97:1535–1542.

Carlsson C, Jonsson J, Akerman B (1995): Double bonds in DNA gel electrophoresis caused by bis-intercalating-dyes. Nucl Acids Res 23(13):2413–2420.

Glazer AN, Rye HS (1992): Stable dye-DNA intercalation complex as reagents for high sensitivity fluorescence detection. Nature 359:859–861.

Goodwin PM, Cai H, Jett JH, Ishaug-Riley SL, Machara NP, Semin DJ, Van Orden A, Keller RA (1997): Application of single molecule detection to DNA sequencing. Nucleosides & Nucleotides 16(5 &6):543–550.

Goodwin PM, Ambrose WP, Keller RA (1996): Single-molecule detection in liquids by laser-induced fluorescence. Acc Chem Res 29:607–613.

Goodwin PM, Johnson ME, Martin JC, Ambrose WP, Marrone BL, Jett JH, Keller RA (1993): Rapid sizing of individual fluorescently stained DNA fragments by flow cytometry. Nucl Acids Res 21(4):803–806.

Huang Z, Jett JH, Keller RA (1999): Bacteria genome fingerprinting by flow cytometry. Cytometry 35:169–175.

Huang Z, Petty JT, O'Quinn B, Longmire JT, Brown NC, Jett JH, Keller RA (1996): Large DNA fragment sizing by flow cytometry: application to the characterization of P1 artificial chromosome (PAC) clones. Nucl Acids Res 24:4202–4209.

Johnson ME, Goodwin PM, Ambrose WP, Martin JC, Marrone BL, Jett JH, Keller RA (1993): Sizing of DNA fragments by flow cytometry. Proc SPIE 1895:69.

Keller RA, Ambrose PA, Goodwin PM, Jett JH, Martin JC, Wu M (1996): Single molecule fluorescence analysis in solution. Appl Spectros 50:12A–32A.

Kornberg A, Baker T (1992): *DNA Replication,* 2nd ed. New York: W.H. Freeman.

Li LQ, Davis LM (1993): Single photon avalanche diode for single molecule detection. Rev Sci Instrum 64(6):1524–1529.

Perkins TT, Quake SR, Smith DE, Chu S (1994): Relaxation of a single DNA molecule observed by optical microscopy. Science 264:822–826.

Petty JT, Johnson ME, Goodwin PM, Martin JC, Jett JH, Keller RA (1995): Characterization of DNA size determination of small fragments by flow cytometry. Anal Chem 67(10):1755–1761.

Rye HS, Yue S, Wemmer DE, Quesada MA, Haugland RP, Mathies RA, Glazer AN (1992): Stable fluorescent complexes of double-stranded DNA with bis-intercalating asymmetric cyanine dyes: Properties and applications. Nucl Acids Res 20(11):2803–2812.

Smith DI, Blattner FR, Davies J (1976): The isolation and partial characterization of a new restriction endonuclease from Providencia stuartii. Nucl Acids Res 3:343–353.

Spielmann HP, Wemmer DE, Jacobsen JP (1995): Solution structure of a DNA complex with the fluorescent bis-intercalator TOTO determined by NMR spectroscopy. Biochemistry 34:8542–8553.

Wu M, Goodwin PM, Ambrose WP, Keller RA (1996): Photochemistry and fluorescence emission dynamics of single molecules in solution: b-phycoerythrin. J Phys Chem 100(43):17406–17409.

Zucker RM, Elstein KH, Gershey EL, Massaro EJ (1990): Increasing sensitivity of the Ortho analytical cytafluorograph by modifying the fluid system. Cytometry 11:848–851.

7

Fluorescence Lifetime Imaging: New Microscopy Technologies

Weiming Yu, William W. Mantulin, and Enrico Gratton
University of Illinois at Urbana-Champaign, Urbana, Illinois

INTRODUCTION

In recent years there has been remarkable progress in the field of fluorescence microscopy and time-resolved imaging both in the technical aspects and in the variety of applications. Fluorescence spectroscopy is now a mature technique, and one finds its applications in many areas throughout science. During the last decade, there has been growing interest, especially in the biological sciences, in the determination of fluorescence lifetime information directly on living biological specimens. To this end, one naturally needs to expand from classical single-point measurement to simultaneous multiple-point measurement for the formation of an image. This technical expansion raises a new challenge, namely, the ability to acquire images with fluorescence lifetime information within a reasonable time frame and with similar accuracy and precision comparable to that of single-point spectroscopy measurement. This goal is the main scope of this chapter, which will provide an overview of the different techniques used in fluorescence lifetime imaging microscopy (FLIM).

After the atomic age, the space age, and now the information age, it is easy to forget how long ago fluorescence lifetimes had been first experimentally measured, the molecular aspects of fluorescence studied, and the theoretical background formulated. To realize where the future of the field lies and in which direction we would

Emerging Tools for Single-Cell Analysis, Edited by Gary Durack and J. Paul Robinson.
ISBN 0-471-31575-3   Copyright © 2000 Wiley-Liss, Inc.

like to move, it is important and useful to know where we are now by considering the historical perspective. In the field of fluorescence, many important experiments were done before the first appearance of the photomultiplier tube in 1948. After being excited to an electronic level, a molecule may follow fluorescence decay. For a single molecular species with one excited electronic state, the decay is an exponential: $I(t)=I_0\exp(-t/\tau)$, where $I(t)$ is the fluorescence intensity at time t, I_0 is the fluorescence intensity at $t = 0$, and τ is the fluorescence lifetime when the intensity reduces to $1/e$ of the original time-zero intensity. The first measured lifetime value of a fluorescent molecule was determined for fluorescein as 4.5 ± 0.5 ns in 1927 by Gaviola using his first phase-sensitive fluorometer, a sophisticated instrument at the time (Gaviola, 1927). Earlier than that measurement, fluorescence resonance energy transfer (FRET), in which the electronic energy of the excited state of a donor molecule transfers nonradiatively to a receptor, was discovered and demonstrated in 1922 by Cario and Franck (1922). It was later found that energy transfer between molecules depolarized the fluorescence and shortened its lifetime (Gaviola and Pringsheim, 1927). Now FRET is an important tool and has found many applications from studying protein and DNA molecular structure, conformation and kinetics to the organization of membrane structures and drug discoveries (Miki et al., 1998; Visser, 1997; Silverman et al., 1998; Subramaniam et al., 1998; Toth et al., 1998; Clegg, 1996, 1992).

Generally, two methods are utilized to measure the fluorescence decay time, one being the time domain, which includes the time-gating approach, and the other being the frequency domain, or phase modulation method. In the time domain, the fluorescence intensity is recorded as a function of time after a very short pulsed excitation. In the frequency domain, one uses a sinusoidally modulated excitation light source; the phase shift and demodulation of the fluorescence emission are used to calculate the fluorescence decay time. Although in the original fluorescence lifetime measurement the phase modulation method was used, a clever time-domain lifetime measurement without the luxury of fast electronics and detectors was done very early by implementing two electro-optical modulators (Pringsheim, 1949). In this apparatus, the sample was placed between the two Pockels cells in an L-shaped configuration and the detector was placed after the second Pockels cell at the emission side. By varying the distance between the sample and the second Pockels cell for each excitation, the fast fluorescence intensity decay curve can be reconstructed. This type of "time-shifting" method is still used extensively today in both time-domain and frequency-domain lifetime imaging systems. After 1950, driven by the needs and interests of biological science, time-resolved spectroscopy developed rapidly using both the frequency- and time-domain method. In the 1980s, the frequency-domain method was significantly refined and automated to be able to measure decays of multiple fluorescent components simultaneously by using multifrequency measurements (Gratton and Limkeman, 1983; Jameson et al, 1984; Gratton et al., 1984).

Overview of Recent Developments in Lifetime Imaging

Currently, fluorescence microscopy and optical microscopy are enjoying a rapid and exciting development period that creates many new possibilities for biomedical research. Since a picture is worth a thousand words, it is the biologist's dream to be able to observe the spatial distribution of relevant signals. Fluorescence time-resolved microscopy imaging is still a relatively new technology. In principle, one may use a normal lifetime spectroscopy instrument by replacing the cuvette sample holder with a microscope and reconstructing a lifetime image point after point (Dix and Verkman, 1990; Keating and Wensel, 1991). This kind of single-point measurement allowed researchers to study specific regions of their samples using a microscope. Certainly, one may choose to spend a large amount of time to obtain a lifetime image this way. To increase the imaging speed, the next development in time-resolved imaging is to use parallel data acquisition with devices such as a charge-coupled device (CCD) camera (Marriott et al., 1991; Wang et al., 1991) or a multianode photomultiplier tube (Morgan et al., 1992). The implementation of laser scanning confocal microscopy, which offers enhanced spatial resolution and enhanced contrast by eliminating out-of-focus emission, increases the sensitivity of lifetime imaging (Piston et al., 1992). Time-resolved fluorescence microscopy using two-photon excitation gives similar confocal spatial resolution with greater background rejection and reduced sample photobleaching (So et al., 1994). In addition, there are many other advantages, such as the reduced effect of photodamage and increased penetration depth of the sample, in using two-photon excitation for imaging living cells (Yu et al., 1996; French et al., 1997) and other vital samples (Masters et al., 1997). Time-resolved microscopy using the time-gated technique (Wang et al., 1991) and phase modulation methods (Marriott et al., 1991; Lakowicz et al., 1992; Piston et al., 1992) dates back to the early 1990s. There are variations of detection techniques used in the frequency-domain fluorescence lifetime imaging. In the case of homodyning detection, where the excitation light source and the detector are modulated at the same frequency, the phase shift and modulation of the fluorescence signal are measured (Clegg et al., 1996; Lakowicz et al., 1992). In the case of heterodyning detection, the detector is gain modulated at a cross-correlation frequency in addition to the modulation frequency for the excitation source. The phase shift and modulation of the signal at the cross-correlation frequency are acquired for calculation of the lifetime value (Mantulin et al., 1993; So et al., 1994). Both homodyning and heterodyning detection methods require a detector with a high-frequency bandwidth in order to measure short-lifetime components. A very interesting variation of the conventional approach is the use of time-correlated optical mixing (Dong et al., 1995, 1997; Buist et al., 1997; Mueller et al., 1995) where two lasers are tuned to different wavelengths, one laser being at the absorption band and the other at the emission band of the fluorescent probe, overlapped in space. The resulting fluorescence signal can be acquired without using fast electronics and detectors. The time-gated method typically uses single-photon counting techniques for the detection and data acquisition, and the phase modulation method mainly uses analog detection and data acquisition schemes. Both techniques have their own advantages and disadvantages that largely depend on the applications and experimental conditions. We will discuss the technical details of both methods later.

Unique Properties of Fluorescence Lifetime

Since the early single-point lifetime microscopy measurements, there have been great technical advances in microscopy and areas of applications using the time-resolved imaging technique. Again, biology plays an important role in enlarging the field of FLIM. Normal biological systems are organized in structurally and functionally unique complexes, which are not homogeneous. Everyone working with fluorescence knows that intensities may vary significantly within the same sample. Often we do not know all the details of the photochemical and photophysical process of the fluorescent probe. It is difficult to quantify what one is interested in measuring with the intensity signal alone. Unlike fluorescence intensity, the lifetime is a fundamental property of a fluorescent molecule, which directly relates to its structure and dynamics. Fluorescence lifetime values typically fall within the range of a few nanoseconds to a few hundred nanoseconds, and the lifetime values of different fluorescent molecules are often different. In some cases the same fluorescent molecule can have very different lifetime values as the probe's environment changes. For example, ethidium bromide, a nucleotide probe, will have about a 24-ns lifetime when bound to DNA as compared to 1.8 ns when the probe is free in solution (Atherton and Beaumont, 1984). Unlike the intensity measurement, the fluorescent lifetime is not sensitive to the local probe concentration. The lifetime measurement is also much less affected by sample photobleaching. Furthermore, the lifetime measurement provides a different level of discrimination against noise that makes it more immune to scattering and other background noise. An experiment based upon lifetime detection can be very specific as we can choose a probe with distinct and characteristic lifetimes. It can be very sensitive for the signal we are interested in, as we may use the differences in the lifetime to discriminate against autofluorescence, often a significant factor in biological sample. There are also fluorescence probes that are designed specifically for sensing environmental variables such as pH, concentrations of positive and negative ions, viscosity, and oxygen concentration. The inherently high temporal resolution of fluorescence lifetime measurement also allows us to probe rapid dynamics of biological molecules in the time window from microseconds to picoseconds. Integration of the lifetime technique into the conventional confocal microscope and multiphoton microscope will also allow us to have three-dimensional resolution and all the benefits of confocal and multiphoton microscopy. As the lifetime τ of a fluorescent molecule is directly related to its quantum yield ε ($\tau = \varepsilon/\kappa$, κ is the rate constant of the fluorescence decay), lifetime microscopy imaging provides a true quantitative method for conventional microscopy.

FLIM IN THE TIME DOMAIN

Background

It is popular to use time-gated detection in conventional fluorescence lifetime spectroscopy. Several laboratories have developed the lifetime imaging technique using the time-domain approach (Wang et al., 1991; Ni and Melton, 1996; Buurman et al., 1992; Cubeddu et al., 1997; Dowling et al., 1997). There is also a recent report implementing the two-photon scanning technique into the time-domain laser scanning FLIM (Sytsma et al., 1997). The general idea of time-gated detection is to use a short

FLIM in the Time Domain 143

pulse excitation, divide the fluorescence decay curve into several windows of equal time width, record the total fluorescence intensity of each window, and calculate the lifetime. Assuming there is only a single exponential fluorescence decay, the fluorescence signal after a pulse excitation has the form

$$F(t) = A \exp\left(-\frac{t}{\tau}\right) \qquad (7.1)$$

where τ is the lifetime, $F(t)$ is the fluorescence intensity at time t after excitation, and A, the preexponential factor, is the zero-time intensity of the fluorescence. The conventional method for cuvette measurement is to acquire many data points along the fluorescence decay curve and to extract the lifetime value τ and the preexponential factor by fitting the data analytically. This approach has been reported in the recent literature using methods similar to the conventional methods used for multiexponential lifetime imaging (Scully et al., 1997; Dowing et al., 1998). This method is difficult to implement in practical real-time imaging with thousands of data points to fit. However, with the help of fast computers and electronics, it is possible to acquire many images along the decay curve and analytically solve multiexponential decays within minutes (Dowing et al., 1998). For FLIM applications, one typically uses the configuration with two time windows (Fig. 7.1), which allows numerical approaches to rapidly determine the lifetime (Woods et al., 1984), and we will discuss this method in more

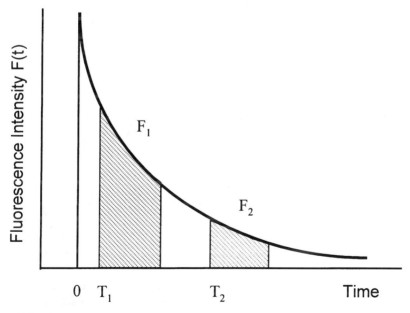

Fig. 7.1. Time-gating approach for the fluorescence lifetime imaging. T_1 and T_2 are the beginning time of each time window, F_1 and F_2 are the total fluorescence intensity collected within the gate width.

detail. If δt is the width of the time window, the integrated fluorescence signal within each time window for each pixel (i) within the image follows the equations

$$F_{1,i} = \int_{T_1}^{T_1 + \delta t} A_i \exp\left(-\frac{t}{\tau_i}\right) dt \tag{7.2}$$

$$F_{2,i} = \int_{T_2}^{T_2 + \delta t} A_i \exp\left(-\frac{t}{\tau_i}\right) dt \tag{7.3}$$

Here T_2 and T_1 are the times at the beginning of each window and $F_{1,i}$ and $F_{2,i}$ are the total intensities within each time window at the i th pixel. The averaged fluorescent decay time τ can be determined from the ratio of the two integrated intensities (Woods et al., 1984; Ballew and Demas, 1989):

$$\tau_i = \frac{T_2 - T_1}{\ln(F_{1,i}/F_{2,i})} \tag{7.4}$$

The preexponential factor can also be calculated as follows:

$$A_i = \frac{F_{1,i}}{\tau_i \exp(-T_1/\tau_i)[1 - \exp(-\delta t/\tau_i)]} \tag{7.5}$$

The mathematical calculation for the above method is straightforward. Previous work has shown that in the case $\delta t = T_2 - T_1$, when the fluorescence signal is optimally used, this rapid lifetime determination method is numerically stable for $0.2\delta t < \tau < 5\,\delta t$. To achieve the best sensitivity for a given value of lifetime τ, one should select the gate width or time window to be $\delta t = 2.5\tau$ (Ballew and Demas, 1989).

The Instrument

Typical components of a time-gated FLIM apparatus include a microscope, a pulsed laser source, gated image intensifier, CCD camera, image digitizer, and delay pulse generator. A simplified version of a FLIM instrument using the time-domain method is depicted in Figure 7.2. The fast photodiode picks up the oscillation frequency of the laser and divides it into adequate frequency ranges for synchronizing the master pulse generator and the delay pulse generator. The delayed pulse can be controlled through an interface with the computer and is used to gate the image intensifier. The image formed on the phosphor screen is focused onto a CCD camera that can be thermal-electronically cooled to reduce noise. Finally, the gated image, transferred to a computer through an image digitizer, is used for calculating the lifetime.

Pulsed Light Source. A short laser pulse is desirable for the time-gated FLIM method. For the laser pulse repetition rate, it is clear that the higher it is, the faster one may collect sufficient photons. However, higher repetition rates are not always better. For example, the pulse repetition rate of an 80- or 76-MHz Ti–sapphire laser is normally too high for most of the fluorescent probes with nanosecond decay times; such high rates can result in incomplete probe recovery for the next excitation pulse and thus complicate data analysis. Pulse pickers, cavity dumpers, and other electro-

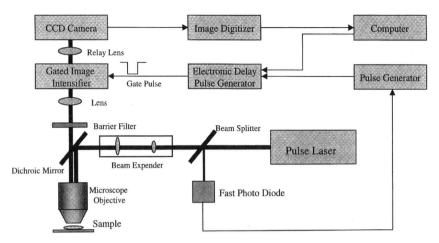

Fig. 7.2. Block diagram of a time-gated FLIM instrument.

optical switches are typically implemented to select slower pulse repetition rates in the time-gated detection method. In reality, one finds that the limiting step for selecting the repetition rate is the gating speed of the microchannel plate image intensifier, which is around several tens of kilohertz. In order to efficiently use the photons, one should match the maximum gating speed of the image intensifier with the laser pulse rate. This matching can be done through simple custom-made frequency divider circuitry (Periasamy et al., 1996)

Time Gating of the Detector. The gating mechanism of the microchannel plate image intensifier is achieved by controlling the potential of the photocathode (see Fig. 7.10 later for a drawing of a similar image intensifier). The default potential of the photocathode is set to be positive, which stops the photoelectrons from reaching the microchannel plate, and consequently the image intensifier is in the off state. When a gate pulse arrives at the photocathode with a negative voltage, the image intensifier is switched on and the photoelectrons hit the microchannel plate surface and are amplified, eventually forming an optical image on the phosphor screen. Typically, the gated pulse is controlled through TTL and the gate pulse width is allowed to vary from a few nanoseconds to a few hundred microseconds. Gate widths down to about 100 ps are also commercially available and have been used in time-domain FLIM (Dowling et al., 1998).

FLIM USING FREQUENCY-DOMAIN HOMODYNING METHOD

Background of Frequency-Domain Method

The theoretical background of phase-resolved fluorometry had been given in detail early in the 1930s by Dushinsky (1933), and one can also find details in recent reviews (Clegg and Schneider, 1996). In the frequency-domain method, the fluorescence sample is excited by an intensity-modulated excitation light source, and typically the modulation is sinusoidal. Under this condition, the fluorescence emission is forced to be sinusoidal, but with a phase delay with respect to the excitation source due to the finite fluorescence decay time. If we assume that the excitation source is modulated at an angular frequency ω, we can describe the fluorescence emission as

$$F(t) = DC_0 + AC_0 \cos(\omega t - \phi) \quad (7.6)$$

where $F(t)$ is the fluorescence intensity, ϕ is the phase delay, and DC_0 and AC_0 are the amplitude of the dc and ac components, respectively. Mathematically, Equation (7.6) is equivalent with Equation (7.1) after Fourier transformation into the frequency space. The ac amplitude of the emission is typically reduced compared to the excitation, a process called amplitude demodulation. Both the phase shift ϕ and the modulation factor m are functions of the modulation frequency ω. One may use both phase and modulation information to calculate the fluorescence decay time. This scenario of the frequency-domain method is illustrated in Figure 7.3. The lifetime τ, modulation frequency ω, phase shift ϕ, and modulation factor m preserve the following relationships in a single-exponential case:

$$\phi = \arctan(\omega\phi) \quad (7.7)$$

$$m = \frac{1}{\sqrt{1 + (\omega\phi)^2}} \quad (7.8)$$

The Fourier spectrum (Fig. 7.4) of a single-exponential decay graphically illustrates the relationship among the phase shift ϕ, the modulation factor m, and the frequency ω. When the modulation frequency $\omega = 1/\tau$, the fluorescence has a 45° phase shift, and this phase shift progressively approaches 90° with complete demodulation of the fluorescence signal as the modulation frequency is increased.

The above equations are derived under the assumption that the fluorescence decay is from a homogeneous sample with monoexponential decay constants. However, one can still use them to evaluate multiexponential and nonexponential fluorescence decay. If τ_{phase} is the lifetime calculated from the phase delay and τ_{mod} is from the modulation, in the case of a simple single-exponential decay, $\tau_{phase} = \tau_{mod}$. Indeed, if we measure both τ_{phase} and τ_{mod} in our experiment, we should be able to tell whether or not we have a single-exponential decay. If the fluorescence is either from a heterogeneous sample with multiple fluorescence sources each having a different decay constant or from a single fluorescent molecule with multiple decay constants, the calculated lifetimes τ_{phase} and τ_{mod} using Equations (7.7) and (7.8) are no longer the same, and each of them represents the apparent lifetime of the system. If we are in a situation where the decay is multiexponential, the following relationship holds: $\tau_{phase} < \tau_{mod}$. In the case of a nonexponential decay with a positive preexponential factor, the situation is the same, where the modulation lifetime τ_{mod} weights the long-lifetime components and τ_{phase} weights the short lifetime components. The averaged lifetime $\tau_{average}$ of a multicomponent system, as weighted by the intensity fraction f, will not be smaller than

FLIM Using Frequency-Domain Homodyning Method

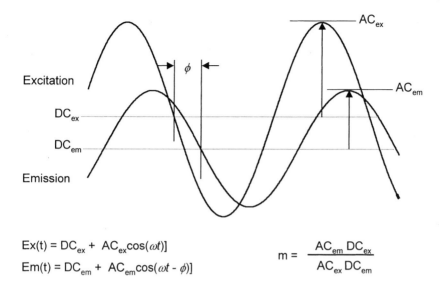

Fig. 7.3. Principle of the frequency-domain approach. Ex(t) is the sinusoidally modulated excitation function with its DC_{ex} and AC_{ex} components, Em(t) is the corresponding emission function with its own DC_{em} and AC_{em} amplitudies plus a phase shift ϕ in respect to the excitation. The m is the modulation factor of the fluorescence signal.

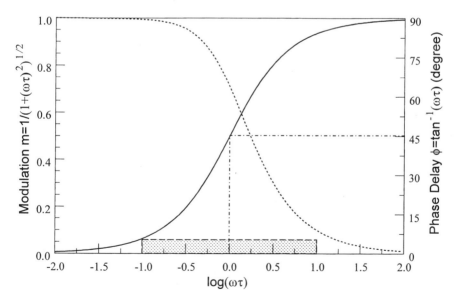

Fig. 7.4. Frequency response of the modulation and phase lag. When the angular modulation frequency $\omega = 1/\tau$, then $\phi = 45°$. For a given value of τ, increasing the modulation frequency increases the phase lag and decreases the modulation. To measure a lifetime, one may use modulation frequency ω so that $\log(\omega\tau)$ is within -1 to 1 (shaded area). The best frequency range is when $\log(\omega\tau)$ is between -0.5 to 0.5.

τ_{phase} in any case. However, the relationship between τ_{mod} and $\tau_{average}$ changes following the change of the modulation frequency. Using the equations provided in previous reviews (Clegg and Schneider, 1996; Jameson et al., 1984), the relationship among τ_{phase}, τ_{mod}, $\tau_{average}$, f, and ω for a system with two-exponential decay, 2 and 20 ns, is graphically represented in Figures 7.5 and 7.6. With the given values of the fractional contribution of the two individual components, it is seen that both the apparent τ_{phase} and τ_{mod} progressively decrease with τ_{phase} approaching the value of the lower lifetime component as the modulation frequency increases (Fig. 7.5), while $\tau_{average}$ remains stationary. With a given modulation frequency, both τ_{phase} and τ_{mod} are a function of the fractional contribution. However, at high modulation frequency, τ_{phase} is less dependent on the fraction and stays in the vicinity of the lower lifetime value of the two components up to a sharp turning point.

In the frequency-domain homodyning case, the excitation light source and the fluorescence detector are modulated at the same frequency. In order to measure fluorescence lifetimes in the nanosecond range, the modulation frequency for the light source should be on the order of 10–100 MHz. To directly measure the phase shift and demodulation factor at these frequencies, one requires a digitizer or similar data acquisition device sampling at high speed. Although there are commercially available digitizers working at megahertz and even gigahertz speed, in practice one chooses a different approach for lifetime imaging. Most of the research groups working with homodyning FLIM use the gain-modulated microchannel plate image intensifier and a variable-delay line that controls the phase shift of the modulation

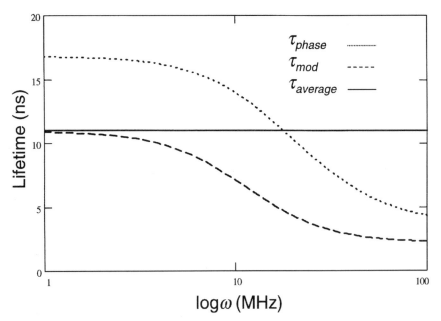

Fig. 7.5. Plot of the apparent lifetime of a system with two equally populated exponential decay (2 ns and 20 ns) components as functions of the modulation frequency ω.

FLIM Using Frequency-Domain Homodyning Method

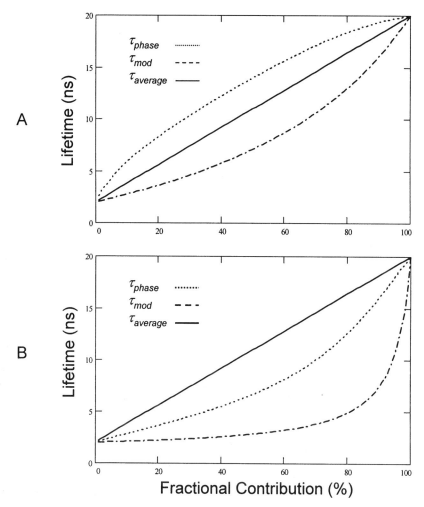

Fig. 7.6. Plot of the apparent lifetimes as functions of the fractional contribution of a system with two exponential decays (2 ns and 20 ns) at a given modulation frequency of A: 10 MHz; B: 40 MHz.

applied on the image intensifier (Hartmann et al., 1997; Mizeret et al., 1997; Clegg et al., 1996; Lakowicz et al., 1994a; Morgan et al., 1992). Using the phase shift of the image intensifier is the key that allows the fluorescence signal at the high modulation frequency to be recorded as steady-state images. Therefore, a slow-scan CCD camera or any digital camera can be used to record the phase-resolved images. One may achieve high accuracy of measurement and phase sensitivity by integrating each image for a length of time without sacrificing any temporal resolution for lifetime imaging.

By using this phase shift method, in principle one may acquire any arbitrary number of images at any different phase values between the detector and the excitation light source. To recover the lifetime of every pixel from the whole image, the most computationally efficient method is the fast Fourier transform algorithm:

$$F_{\sin,i} = \sum_{j=1}^{j=N} \sin(\Delta\theta_{j,i}) I_{j,i} \qquad (7.9)$$

$$F_{\cos,i} = \sum_{j=1}^{j=N} \cos(\Delta\theta_{j,i}) I_{j,i} \qquad (7.10)$$

$$DC_i = \left(\frac{1}{N}\right) \sum_{j=1}^{j=N} I_{j,i} \qquad (7.11)$$

$$AC_i = \sqrt{F_{\sin,i}^2 + F_{\cos,i}^2} \qquad (7.12)$$

where $I_{j,i}$ is the total fluorescence intensity for the ith pixel at the jth phase shift which has the phase difference $\Delta\theta_{j,i}$ with respect to the zero phase and $F_{\sin,i}$ and $F_{\cos,i}$ are the sine and cosine Fourier components. Respectively, DC_i and AC_i in Equations (7.11) and (7.12) are the dc and ac components. From both the modulation factor,

$$m = \frac{AC_i}{DC_i} \qquad (7.13)$$

and the absolute phase shift of the fluorescence signal,

$$\phi = \arctan\left(\frac{F_{\sin,i}}{F_{\cos,i}}\right) \qquad (7.14)$$

one may retrieve the lifetime value. The zero-phase delay is normally obtained by measuring a scattering sample or samples with known lifetime.

The Instrument

A typical homodyning FLIM setup, which in some ways resembles the time-domain FLIM, is shown in Figure 7.7. The FLIM configuration using the homodyning method can also be adapted for endoscopic imaging with small modifications (Mizeret et al., 1997; Schneider and Clegg, 1997). Due to the nature of the endoscopic measurement, images of laser reflection from the tissue may be used as the reference for the zero-phase correction.

Modulation of the Excitation Light Source. A microscopy imaging system using the homodyning method requires an intensity-modulated excitation source. There are several configurations of modulated source one may choose from, either using a pulsed laser (Lakowicz et al., 1994a; So et al., 1995) or using a continuous-wave laser in combination with acoustic-optical or electro-optical modulators

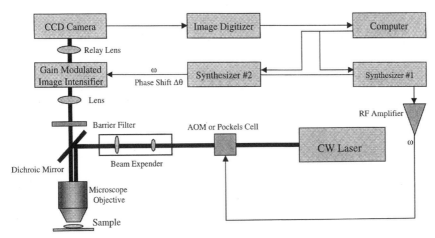

Fig. 7.7. Block diagram of a frequency-domain homodyning FLIM.

(Morgan et al., 1992; So et al., 1994; Clegg et al., 1996). Both configurations allow modulation of the excitation light source to hundreds of megahertz. These are not the components that limit the bandwidth of the whole system. Early systems using direct modulation of the deuterium lamp offered a cheap alternative, but with limited frequency response (Morgan et al., 1992, 1995).

Modulation of the Microchannel Plate Image Intensifier. In the frequency domain, the microchannel plate image intensifier is gain modulated at the appropriate frequency ranges. The gain modulation of the image intensifier can be done through either variation of the acceleration voltage at the photocathode or variation of the actual high voltage across the microchannel plate where the electrons are multiplied. One great advantage of modulation through the photocathode is the possibility of achieving a large dynamic range of the modulation frequency from a few hundred kilohertz to hundreds of megahertz. However, there are several drawbacks of modulating the photocathode. The common problems are iris effect and electronic defocusing. The iris effect is related to the characteristic of the photocathode requiring a certain amount of time to equilibrate after any changes of applied electronic potential. Electronic defocusing occurs when the modulation voltage is going through zero and the electrons at the photocathode diffuse to the nearby channels, causing blurring of the image at the end of the microchannel plate. The iris effect of the cathode will limit the bandwidth of the modulation and will also lead to nonuniform phase shift across the photocathode. To solve these problems caused by the iris effect, it is possible to reduce the impedance of the photocathode by thicker coating of the cathode surface to achieve higher frequency response. The direct modulation of the gain of the microchannel plate eliminates both the iris effect and the defocusing effect and result in greater phase stability. However, it is normally difficult using this method to reach high-frequency modulation and it is typically performed below tens of megahertz, which is still adequate for the measurement of common fluorescence probes with lifetime of a few nanoseconds.

Signal Detection. The image formed at the phosphor screen can be refocused onto a CCD camera through a lens or fiber optic coupling and digitized by the image digitizer via a computer for data processing. Typically, for the CCD camera, many research groups preferred the cooled slow-scan system as compared to the normal video-rate camera, for its reduced read-out noise and integration capability (Clegg et al., 1996; Morgan et al., 1992; Marriott et al., 1991).

FLIM USING THE FREQUENCY-DOMAIN HETERODYNING METHOD

Background

Frequency-domain heterodyning is achieved by modulating the excitation light source and the fluorescence detector at different frequencies. The idea of heterodyning is to convert the fluorescence phase shift and demodulation information from the very high frequency to the low cross-correlation frequency $\Delta\omega$, which is the frequency difference between the modulation of the excitation and the detector, thereby facilitating data acquisition (Spencer and Weber, 1969). Previously, our group has developed FLIM systems using the heterodyning method (French et al., 1992; So et al., 1995). Other research groups have also developed similar FLIM systems using wide-field illumination (Clegg et al., 1996), as well as confocal laser scanning (Piston et al., 1992).

Using the sinusoidal modulation of the excitation source, the fluorescence has the form $F(t)$ as in Equation (7.6). The fluorescence signal $C(t)$ with cross-correlation modulation at the detector is given by the equation

$$C(t) = DC_C + AC_C \cos(\omega_c t - \phi_C)] \qquad (7.15)$$

where ϕ_c is a constant phase shift for the instrument, and DC_c and AC_c are the dc and ac components, respectively, when the detector is modulated at angular frequency ω_c. The low-frequency component at the cross-correlation frequency is a new function $CC(t)$ from the product of $F(t)C(t)$ given by

$$CC(t) = DC_0 DC_C + \frac{AC_0 AC_C}{2} \cos(\Delta\omega\, t - \Delta\theta) \qquad (7.16)$$

where $\Delta\phi = \phi_c - \phi$. Practically, the cross-correlation can be achieved by modulating the gain of the photomultiplier tube or the photocathode of the image intensifier at $\omega + \Delta\omega$. By using a preamplifier with a low-pass filter, we can readily obtain the cross-correlation signals electronically from the high-frequency signals.

For spectroscopy measurements, one can in principle choose any cross-correlation frequency compatible with the bandwidth of the digitizer. For lifetime imaging, the cross-correlation frequency is determined differently for the scanning configuration and the wide-field illumination configuration. For an intensified CCD camera, used for the wide-field illumination configuration, frequency is determined by the frame rate of the camera. In the case of the laser scanning system, the cross-correlation frequency should be chosen as high as possible for the fastest scanning rate under the condition that there are enough photons. For normal scanning microscopy measurement, we typically use a cross-correlation frequency between 10 and 20 kHz.

To obtain the phase shift and modulation information, in both scanning and camera systems, we acquire four data points per cross-correlation waveform (Fig. 7.8) and use the fast Fourier transform to analyze the data for its computational efficiency. The mathematical calculation for the ac, dc, phase, and modulation in this case is the same as Equations (7.11) and (7.14) with $\Delta\theta_{j,i}=0, \ldots, (\pi/2) \cdot j, j = 1, 2, 3, 4$. An approach of three different digital filters is applied in analyzing the raw data to remove any random noise as well as noise at higher harmonics. The first digital filter is the waveform folding, which averages contiguous waveforms at the cross-correlation frequency. The folding of the waveform removes any random noise not at the specific cross-correlation frequency and its harmonics. Using the folding method, the signal-to-noise ratio typically increases linearly with the number of folds (Feddersen et al., 1989). The high-frequency noise is normally reduced electronically through preamplification of the signal. The fast Fourier transform of the in-phase averaged data is effectively a second filter, which separates the signal at the cross-correlation frequency from the noise at higher harmonics. The frame averaging at every pixel is a third filter, which also removes random noise, such as that due to dark counts and photon statistics. The last frame-averaging method is less efficient, compared with the folding, as the increase of the signal-to-noise ratio is only proportional to the square root of the number of averages. In practice, it is possible to perform waveform

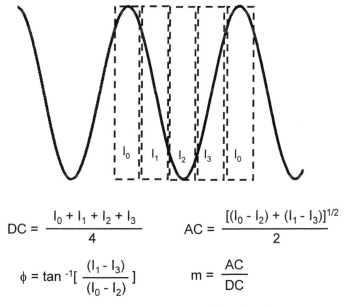

$$DC = \frac{I_0 + I_1 + I_2 + I_3}{4} \qquad AC = \frac{[(I_0 - I_2) + (I_1 - I_3)]^{1/2}}{2}$$

$$\phi = \tan^{-1}\left[\frac{(I_1 - I_3)}{(I_0 - I_2)}\right] \qquad m = \frac{AC}{DC}$$

Fig. 7.8. Calculation of the DC, AC, ϕ, and m values by sampling 4 data points per cross-correlation waveform.

folding and frame averaging to achieve high lifetime sensitivity, but at the price of slowing down the final data collection rate.

With a fast computer and a fast data acquisition card, it is possible to digitize increased number of points for each waveform. It is conceivable that analyzing more data points will allow us to better discriminate the signal at the cross-correlation frequency against noise, especially the high-frequency noise, and to achieve better phase sensitivity and lifetime measurement accuracy.

Instrumentation

Intensified Camera-Based Microscopy System. Many time-resolved microscopy systems, in both the time domain and frequency domain, adapted the intensified parallel data acquisition using either a CCD camera or a multianode photomultiplier tube (PMT) (Dowling et al., 1998; Hartmann et al., 1997; Mizeret et al., 1997; Clegg et al., 1996; McLoskey et al., 1996; Ni and Melton, 1996; Periasamy et al., 1996; So et al., 1994; Lakowicz et al., 1994a; Oida et al., 1993; Morgan et al., 1992). The image intensifier is a microchannel plate–based system, which can be either gated for time-domain measurements or gain modulated for frequency-domain experiments. The instrument diagram of our time-resolved camera is illustrated in Figure 7.9.

Modulation of the Laser. The cw beam of the argon-ion laser (model 2025, Spectra Physics, Mountain View, CA) is sinusoidally modulated by either an acoustic-optical modulator (AOM) or a Pockels cell (ISS, Champaign, IL). The choice

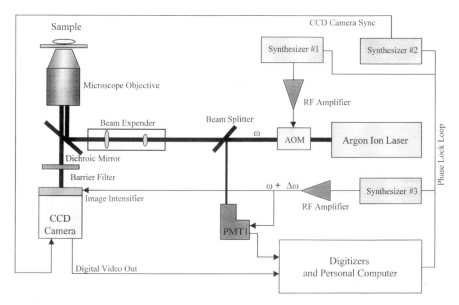

Fig. 7.9. Instrument diagram of a time-resolved fluorescence microscope using gain-modulated image intensifier in the frequency-domain heterodyning mode.

between an AOM and a Pockels cell depends on two factors; one is the characteristic of the frequency response of the device and the other is the light throughput. To obtain the same depth of light modulation, the Pockels cell attenuates more light compared to the AOM. On the other hand, the Pockels cell may have a linear frequency response over the full range between 0 and 500 MHz as compared to the AOM, which typically falls in the ranges between 20 and 120 MHz and can be modulated only at certain frequencies determined by the physical dimension of the crystal. The easiest way to achieve high-frequency modulation for the excitation source is to use a short pulsed laser beam, which can have a harmonic content exceeding a gigahertz. We may choose the alternative by using laser systems such as the Nd–YAG laser (Antares, Coherent, Santa Clara, CA)-pumped dye laser (model 700, Coherent) combined with a cavity dumper (model 7220, Coherent) or an argon-ion laser (Innova 300, Coherent)–pumped Ti–sapphire laser (Mira 900, Coherent) combined with a pulse picker (model 9200, Coherent) and a second-harmonic generator (model 5-050, Coherent). The trade-off in such a system is the high cost.

No matter which laser system one uses there will always be intensity jitter and phase instability from either the laser or the modulator. In the case of using the AOM, the phase stability of the laser modulation is strongly dependent upon the thermal stability of the AOM crystal. A water bath capable of maintaining a highly stable temperature is normally required. In order to correct the phase jitter and instability during the experiment, we choose to use a second detector, a gain-modulated photomultiplier tube (R928, Hamamatsu, Bridgewater, NJ), as a reference.

Signal Detection and Data Acquisition System. The central component of the whole system is the gain modulated microchannel plate (Fig. 7.10). The microchannel plate (model V6390U) is custom specified, made by Hamamatsu (Bridgewater, NJ) for its high-frequency response. We can modulate the image intensifier up to 500 MHz, as compared to the normal unit, which dies off at about 200 MHz. The modulation of the image intensifier is achieved via modulating the voltage of the photocathode.

The laser beam after the modulator is then coupled into the microscope (Axiovert 35, Carl Zeiss, Thornwood, NY) using an epi-illumination configuration. The fluorescence passes through a long-pass dichroic mirror and a barrier filter before hitting the photocathode of the image intensifier. After amplification in the microchannel plate, the fluorescence from the sample forms an image at the phosphor screen and can be captured by the CCD camera (model CA-D1-256, Waterloo, Ontario). This CCD camera is capable of frame rates of about 200 Hz that allow up to 50 Hz for cross-correlation. The slow refresh rate of the phosphor screen is effectively a low-pass filter for removing the high-frequency noise.

The digital image at the CCD camera is captured by an image digitizer (MATROX Electronic Systems, Dorval, Quebec) via an 8-bit digital output port on the camera. The fast Fourier transform is performed at every pixel of the whole image based on four data points per cross-correlation waveform. Before transferring the time-resolved image to the main computer, we perform on-board frame averaging to reduce noise. The reference signal from the PMT is acquired via a separate digitizer

(A2D-160, DRA Laboratories, Sterling, VA). The whole imaging system is synchronized through either the external clock or the external trigger.

Sequential Laser Scanning-Based Microscopy System. Our time-resolved laser scanning microscope system is based on multiphoton excitation and frequency-domain heterodyning techniques (So et al., 1995). Fluorescence microscopy using two-photon excitation has become popular since the introduction of this technique into microscopy in the early 1990s (Denk et al., 1990) due to its superior background noise rejection and reduction of overall sample photodamage and photobleaching. Two-photon excitation is based on the theory that the chromophore molecule may simultaneously absorb two red photons, each with half the energy required to excite the molecule. The number of photon pairs, n, absorbed by the molecule for each laser pulse has the form

$$n \approx \frac{\delta}{t}\left(E\frac{\pi A^2}{hc\lambda}\right)^2 \qquad (7.17)$$

where δ is the two-photon cross section of the chromophore, t is the pulse width of the laser, E is the energy of each laser pulse, A is the numerical aperture (NA) of the objective, λ is the wavelength of the laser, c is the speed of light, and h is Planck's constant. We can see that the number n is proportional to the square of the laser energy and to the fourth power of the numerical aperture A. This highly nonlinear relationship means that two-photon excitation appreciably occurs only at the focal

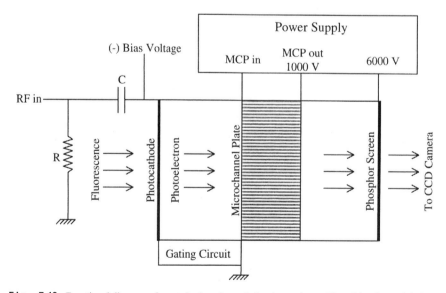

Fig. 7.10. Functional diagram of a gated microchannel plate image intensifier with gain modulation at the photocathode.

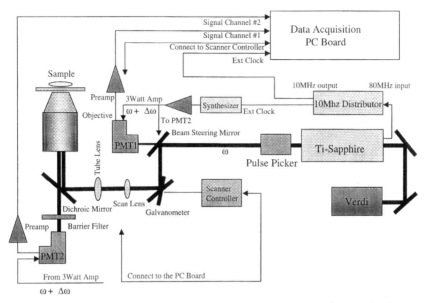

Fig. 7.11. Time-resolved laser scanning fluorescence microscope instrument diagram using frequency-domain heterodyning.

point of the microscope objective, where the excitation photon density is high. Due to this fact, microscopy using two-photon excitation has an inherent three-dimensional sectioning effect; moreover, photodamage of the sample and photobleaching of the chromophore are also localized to the focal point. It is also clear from Equation (7.17) that high efficiency of two-photon excitation can be achieved by using a very short pulse combined with a microscope objective of high numerical aperture. It has been demonstrated that there are significant advantages using two-photon excitation for imaging photosensitive dye–labeled samples (Yu et al., 1996) as well as living cells (French et al., 1997; Yu et al., 1996; Koenig et al., 1996), live embryos (Centonze and White, 1998), and other vital tissue samples (Master et al., 1997; Piston et al., 1995).

Excitation Source and Optics. The configuration of our time-resolved laser scanning microscope is illustrated in Figure 7.11. The Ti–sapphire laser (Mira 900, Coherent) pumped by a 5-W solid-state laser (Verdi, Coherent) produces an 80-MHz pulse train with about 100-fs pulse width. A 4-MHz laser pulse train, obtained by passing the original 80-MHz pulse through a pulse picker (model 9200, Coherent), forms the fundamental laser modulation frequency. It is steered onto the scanner mirrors (Cambridge Technology, Watertown, MA) and then beam expanded before being reflected by the dichroic mirror into an inverted microscope (Axiovert 35, Carl Zeiss). For the objective, we typically use a high-NA oil-immersion objective such as the F-Flura 40x/1.3 or Plan-Apochromats 63x/1.4 from Zeiss. After passing the dichroic mirror and barrier filters, the fluorescence signal is collected by a PMT (R928, Hamamatsu), which is gain modulated at the desired frequencies.

Synchronization of the Instrument and Data Acquisition. On the other hand, the 80-MHz signal output from the photodiode monitor of the Ti–sapphire laser is amplified and divided by a custom made divider box, which generates a 10-MHz transistor–transistor logic (TTL) pulse used as the main clock for the whole instrument. A high-bandwidth frequency synthesizer (HP8341B, Hewlett Packard, Palo Alto, CA) externally referenced to the 10-MHz main clock, produces a signal at the harmonic frequency of the laser fundamental plus a cross-correlation frequency and is then amplified by a high-frequency amplifier (403L ENI, Rochester, NY) for the gain modulation of the PMT. The digitizer (A2D-160, DRA Laboratories) located inside a PC is externally clocked, using the main clock line. The on-board counter uses the external clock to produce the pixel clock, the line clock, and the frame clock that is sent to the scanner controller unit for synchronous laser scanning. An example of the relation among the laser, main clock, digitization rate, cross-correlation frequency, and pixel clock is graphically shown in Figure 7.12. The solid-state laser-pumped Ti–sapphire laser produces a very stable laser pulse and provides a stable clock. The frequency drift of the laser during a typical working day is on the order of 20–40Hz when the laser system is properly tuned. In principle, it is not necessary to use the second detector in this case. The raw data recorded by the digitizer are transferred to the main computer for analysis and retrieval of the lifetime information. The time-resolved images are then displayed on the computer screen. The frame rate of a 256 × 256 pixel image depends on the

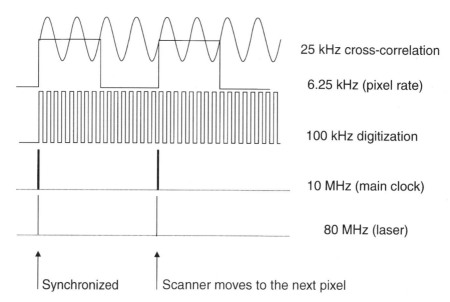

Fig. 7.12. Clock map for a 100-kHz single-channel digitizer sampling a 25-kHz signal with 4 times folding. The resulting pixel rate is 160 μs. The 80-MHz laser frequency and the 10-MHz main clock are shown only symbolically.

cross-correlation frequency and how many folds of the waveform are performed. Under our default condition of 12.5-kHz cross-correlation and two times folding, the imaging frame rate is about 13 s.

FLIM USING OPTICAL MIXING METHODS

Background

Pump–probe is a general terminology for using two laser beams overlapped in space and interacting with the sample molecules. The stimulated emission technique refers to the use of a second laser beam at an emission wavelength to stimulate the emission of a fluorescent molecule from the excited state induced by the first laser beam. The pump–probe technique is well established in laser spectroscopy for studying ultrafast molecular processes and dynamics (Diffey et al., 1998; Vanbrederode et al., 1998; Kumble et al., 1998). The use of stimulated emission for spectroscopy study has also been demonstrated with a simple system (Lakowicz et al., 1994b; Kusha et al., 1994). The capability of using stimulated emission for time-resolved fluorescence imaging has been reported in the literature using both time-domain (Mueller et al., 1995) and frequency-domain (Dong et al., 1995) configurations. We have adapted the frequency-domain heterodyning pump–probe technique (Elzinga et al., 1987a,b) for time-resolved microscopy imaging.

The basic idea of the stimulated emission technique is graphically illustrated in Figure 7.13. Two laser beams are required for this technique. The fluorescence molecule is excited by the first laser beam, the pump beam, which is intensity modulated at a given frequency ω. The second laser beam, the probe beam, which is intensity modulated at frequency $\omega + \Delta\omega$, is tuned within the wavelength of the fluorescence emission to stimulate the decay transition of the molecule from the excited states. One may choose to detect either the stimulated emission or the remaining fluorescence emission. In the epi-illumination configuration we used, we chose to detect the isotropic emission of the fluorescence, rather than the stimulated emission. The reasons are threefold. First, the stimulated emission travels collinearly in the direction of the pumping or the probing beam (Sargent et al., 1974) that is not seen directly by the microscope objective. Second, there is strong high-frequency noise and laser interference competing with the stimulated emission signal. Third, equivalent information is contained within either the fluorescence or the stimulated emission, since the system conserves the total number of the excited-state population. The important function of the second probe beam is to modulate the excited state population of the fluorescent molecules. The consequence of beating the two laser beams at different frequencies, asynchronous pump and probe, is the modulation of the intensity of the fluorescence emission at the cross-correlation frequency $\Delta\omega$ and its harmonics. The modulation depth of the fluorescence signal depends upon the degree of ground-state saturation. It is not necessary to deplete the ground-state population for time-resolved measurements. Therefore, relatively low laser power can be used as compared with spectroscopic studies using stimulated emission (Dong et al., 1995). To obtain the lifetime information, the algorithm of phase delay and modulation factor can be implemented with a simple detection and data acquisition scheme.

Fast detectors and associated fast electronics are eliminated using this pump–probe approach. Since short laser pulses contain high harmonics up to hundreds of gigahertz, this method has an extended frequency bandwidth that is not limited by the speed of the detector. The only limiting factor for achieving high temporal resolution using this method is the pulse width of the laser.

Using this pump-probe method, the observable fluorescence signal depends on how well the pump and probe lasers are overlapped in space as well as in time, since the signal can come only from the fluorescent molecules that interact with both lasers. In fact, the nonlinear relationship between the fluorescence intensity and the overlap integral of the intensity of both pump and probe lasers provides high spatial resolution, similar to one-photon confocal and two-photon microscopy in both radial and axial directions (Dong et al., 1997).

The Instrument

Our instrument configuration, shown in Fig 7.14, requires two lasers synchronized by a common clock. The detection part of the instrument is simpler than in the normal heterodyning configuration. In particular, the detector does not need to be gain modulated, since the heterodyning occurs in the sample.

Pump and Probe Laser Sources. We use a 10-MHz TTL pulse generated from the master synthesizer (HP8341B, Hewlett Packard) to phase lock both the pump

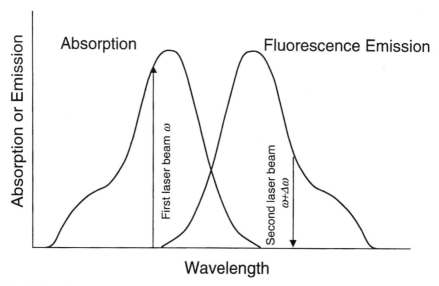

Fig. 7.13. Principle of the frequency-domain pump-probe method. The first laser beam, oscillating at frequency ω, excites the molecule to the excited states. The second laser beam, oscillating at frequency $\omega+\Delta\omega$, stimulates the molecules back to the ground state. The resulting fluorescence is modulated at frequency $\Delta\omega$ and its harmonics.

FLIM Using Optical Mixing Methods

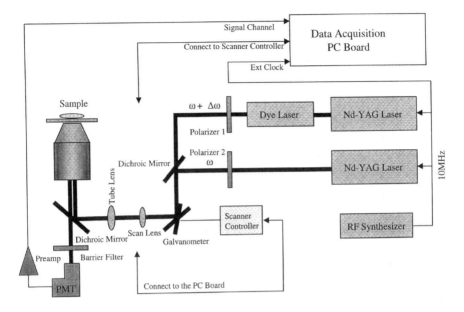

Fig. 7.14. Instrument diagram of a pump-probe fluorescence microscope for time-resolved imaging in the frequency-domain, deterodyning mode.

laser, an Nd–YAG laser (Antares, Coherent), and the probe laser, a second Nd–YAG laser (Antares, Coherent)–pumped DCM dye laser (Model 700, Coherent). The 532-nm laser line of the Nd–YAG laser passing through a second-harmonic generator is used to excite the molecule. The wavelength of the DCM dye laser can be tuned to a wavelength appropriate for the fluorescence emission spectrum of the molecule under investigation. For rhodamine dyes, the DCM laser is typically tuned to around 640 nm. Both Nd–YAG lasers typically operate around 76.2 MHz with a frequency difference of 5 kHz. A combination of Glan-Thompson polarizers can be used to control the laser power and to set the relative polarization between the pump and probe beams. The magic angle condition, which eliminates the polarization effect on the lifetime measurement, can be satisfied conveniently by setting the probe beam 54.7° with respect to the pump beam so that the parallel and perpendicular fluorescence components have a 1:2 ratio going into the detector. Using this setup, the upper limit of the laser power for the pump laser is on the order of microwatts and for the probe beam, it is on the order of milliwatts. The linearity of the signal response to the laser power has been characterized for rhodamine-B in water (Dong et al., 1995). The upper power limit for the pump beam was established at 10 μW and 7 mW for the probe beam for this dye before deviation of the signal.

In addition to the laser arrangement of Figure 7.14, we can also use an alternative configuration in which two Ti–sapphire lasers (Mira 900, Coherent) both pumped by argon-ion lasers (Innova 300, Coherent) are phase locked via a Synchrolock system (model 900, Coherent). In this case, the 80-MHz self-oscillation frequency of one Ti–sapphire laser is used as the clock for locking the second Ti–sapphire laser. The advantage of using the Ti–sapphire lasers is mainly twofold. First, the short pulse width of the Ti–sapphire laser of about 100 fs provides harmonics at hundreds of gigahertz before significant demodulation. Second, the intensity stability of the laser facilitates the measurement and subsequent data analysis. The potential problem with this configuration is the intolerance to any jitter in the phase-locked loop between the two lasers, since the duty cycle of the laser is about 12.5 ns, which is considerably longer than the 100-fs pulse width.

Optics, Signal Detection and Data Acquisition. Both the pump and probe laser beams are collinearly combined on a dichroic mirror before being directed onto the scanning mirrors of the galvanoscanner (model 603X, Cambridge Technology, Watertown, MA). The mirror movement of the scanner is synchronized with the main 10 MHz clock and controlled by the computer. As shown in Figure 7.14, a second dichroic mirror located inside the microscope reflects both laser beams into a high-NA microscope objective. The fluorescence signal is collected using the same objective, passed through the same dichroic mirror and 600±20-nm bandpass barrier filters before detection by a PMT (R928 or R1104, Hamamatsu). The signal at the cross-correlation frequency is first electronically filtered using a preamplifier (Stanford Research, Sunnyvale, CA) and then captured by the data acquisition card for lifetime analysis. As in the case for normal heterodyning using laser scanning, the frame rate depends on the cross-correlation frequency and the number of in-phase averaging. In a typical pump–probe lifetime measurement, the default pixel dwell time is 400 μs, which corresponds to a frame rate of 32 s for a 256 × 256 pixel image.

EXAMPLES OF LIFETIME IMAGING

FLIM is still a relatively new research method not commercially available for many biology laboratories. FLIM research has demonstrated that this new microscopy method is useful for many applications in biological and other systems, including the determination of cellular Ca^{2+} (Periasamy et al., 1996; So et al., 1995; Piston et al., 1994; Lakowicz et al., 1994a); studying DNA-RNA distribution and local concentrations (Sytsma et al., 1998; Piston et al., 1992); pH variations in different cellular environments (Sanders et al., 1995); membrane fluidity (Yu et al., 1996; Dix and Verkman, 1990); chlorophyll structure and function (Sanders et al., 1996); collagen and elastin autofluorescence (Dowling et al., 1998); tumor detection (Cubcddu et al., 1997; Mizeret et al., 1997; Itoh et al., 1997); oxygen concentration (Hartmann et al., 1997); and thermodynamic studies of combustion (Ni and Melton, 1996). Previously, we have used the capability of our FLIM systems for studying hapten presentation processes in the macrophage (French et al., 1997), membrane fluidity of vital cells (Yu et al., 1996), and cell stress under UVA and near-IR irradiation (Koenig et al., 1996). Important observations have been made in those individual studies because of the lifetime images. Using energy transfer, FLIM has also proven valuable for studying receptor-mediated cellular activities and intracellular vesicle fusion processes (Gadella and Jovin, 1995; Oida et al., 1993).

Examples of Lifetime Imaging

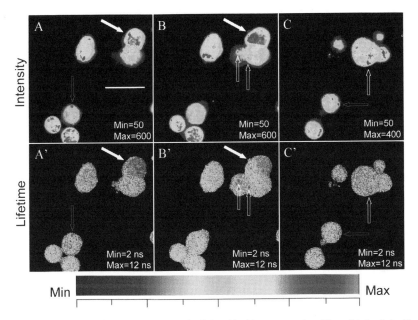

Fig. 7.15. Intensity and lifetime imaging of cellular chloride concentration. The cell is loaded with dih-MEQ* and imaged under conditions with 80 MHz and below 4mW laser power at 747 nm using 40x/1.3NA F-Fluar Zeiss objective. The lifetime images are from the modulation measurements. The scale bar is 20 micron. *See color plates.*

We will now show examples of how using lifetime microscopy imaging is useful in the determination of chloride concentrations within living cells. As a co-factor, chloride regulates cell membrane potential, which is fundamental for cell functions, especially for the cerebral activities (Wagner et al., 1997). Traditionally, chloride concentration is determined by electrophysiological measurement (Krnjevic and Schwartz, 1967). The dynamic quenching by chloride of fluorescence probes such as MEQ, a quinoline derivative, also measures chloride concentrations (Inglefield and Schwartz-Bloom, 1997). However, quenching measurements based solely upon the intensity are not adequate for heterogeneous systems like cells. We will illustrate this point more clearly through the example images.

As a model system, we used the pheochromocytoma cell from rat adrenal gland [PC12 cell, American Type Culture Collection (ATCC), Manassas, VA] cultured in Ham's F12K medium with 15% horse serum and 2.5% fetal bovine serum. The cell was labeled with the membrane-permeable chloride probe dih-MEQ synthesized from MEQ* (Molecular Probes, Eugene, OR) following the procedures recommended by Molecular Probes. The labeled pc12 cell was washed and suspended in phosphate-buffered saline (PBS) and mounted on a hanging drop microscope slide with a #1.5 glass cover slip.

*MEQ: 6-methoxy-N-ethylquinolinium iodide

Examples of the intensity and lifetime image using two-photon scanning FLIM are displayed in Figure 7.15. The intensity images of the cells in panels A, B, and C show high heterogeneity. The areas shown in red color are about 10 times more intense than the areas in dark blue and 3 times more intense than the areas shown in dark green. The lifetime images, on the other hand, show relatively uniform color (panels A′, B′, and C′), which indicates a relative uniform lifetime distributions, partially because the lifetime is fluorophore concentration independent. However, there is still spatial lifetime heterogeneity within these cells. Intensity and lifetime do not have a monotonic relationship. Examples of high intensity correlated with high values of lifetime are indicated by the red arrows in images A, A′ and C, C′. The lifetime value for these red spots is about 12–14 ns, which corresponds to a very low chloride concentration by referencing to the calibration curve (Fig. 7.16) measured in the microscope at the same modulation frequency as that of the cell measurement. This result would imply that the cell accumulates dih-MEQ molecules in a very localized area where the chloride concentration is very low. Examples of high intensity correlated with low lifetime are indicated with white block arrows in images B, B′ and C, C′. The average lifetime value in these regions is about 4.5 ns, which corresponds to 50 mM chloride or more. If we rely only on the intensity data and use intensity quenching as a reference, it would be impossible to correctly interpret the image data in this case. Interestingly, the edge of the cells in the lifetime images show larger than 10 ns lifetime values. This result would be consistent with the exclusion of chloride ion from the plasma membrane. Images A and B are taken in the same field of view but with a few

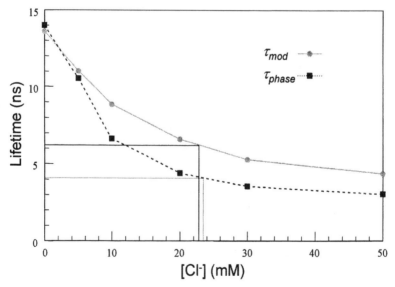

Fig. 7.16. Calibration curve of chloride concentration in deionized water using τ_{mod} and τ_{phase}. The measurement was performed in the microscope using the 80 MHz modulation frequency at 747nm with a 100-fs Ti-sapphire laser. The apparent lifetime value is low under this condition and is valid only for calibration of measurements at this modulation frequency.

Examples of Lifetime Imaging

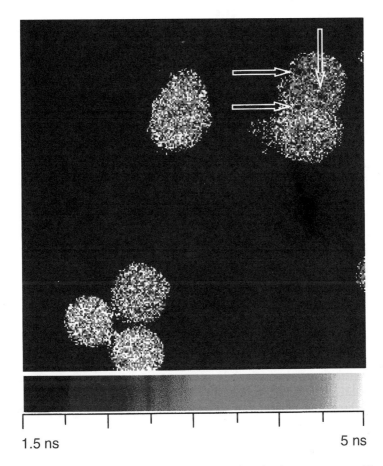

Fig. 7.17. Lifetime image of cellular chloride concentration from the phase measurement. The above image displays τ_{phase} value of the same measurement shown in Fig. 7.16B′. Very low lifetime features are indicated by white arrows. The lifetime value for these spots is between 2.0–2.5 ns, which is likely due to NADH (NAD⁺) autofluorescence in the mitochondria. The fact that τ_{phase} measures the low lifetime components is illustrated in Fig. 7.6B, where the phase lifetime stays low at high modulation even with less than 30% of total intensity for the component with lower lifetime value. *See color plates.*

micrometers differences in the focal plane. The cell indicated by the solid white arrows in images A, A′ and B, B′ has a lifetime gradient. This effect is most clearly shown in image B′, where the right side of the cell is blue with average lifetime about 4.5 ns and the left side is more yellow-green with a 5.7-ns average lifetime. In image A′ the same cell is "blue" with about 4.5 ns on average. This result suggests the existence of a chloride gradient from about 30 to 50 m*M* within a single cell under our experimental conditions.

We have discussed the properties of τ_{mod} and τ_{phase} in a multicomponent system when using the frequency-domain method. Indeed, the phase lifetime image (Fig. 7.17) shows features of a low lifetime value that are not clearly visible from the modulation lifetime (Fig. 7.15A') and certainly are not visible from the intensity image (Fig. 7.15A). The averaged lifetime value for the cellular chloride measurement over five to six images gives $\tau_{mod} = 6.25$ ns and $\tau_{phase} = 4.0$ ns, both falling into an average of about 23 mM of chloride according to our calibration. The fact that $\tau_{mod} \neq \tau_{phase}$, indicates a heterogeneous lifetime distribution within individual pixels.

COMPARISON OF DIFFERENT FLIM TECHNIQUES

In some sense, all the FLIM techniques introduced above are unique. The characteristics of the method, such as the sensitivity, photon economy, data acquisition speed, signal-to-noise ratio, and time required for one measurement, very much depend on how the instrument performs and how one does the measurement. There are many variables. It is rather a complicated mathematical task to make a comparison of all the differences. We will point out only some of the major differences, advantages, and shortcomings. To a large extent, some of the shortcomings of a particular method can be overcome by new technological development.

Lifetime Resolution

In the time domain, using the normal two-gated-window method, the width of the excitation pulse and the minimum gate width of the time-gating windows determine the lifetime resolution of the method. This is easy to understand from Figure 7.1, as one can measure the fluorescence only after the decay of the excitation pulse and within the decay time of the fluorescence. If one uses the fastest laser available, the shortest lifetime one may achieve is then limited by the speed of the electronics. It is still possible to measure picosecond lifetimes by using nanosecond gated windows. However, in this case the second gated window will not see much light, and one expects extensive integration will be necessary to collect enough photons. In the case of the frequency domain, using the single-frequency method, the applied modulation frequency to either the excitation source or the detector determines the sensitivity of this method. When a short pulse laser, such as a picosecond or femtosecond laser, is used, the modulation frequency bandwidth of the detector determines the lifetime resolution. Using a typical PMT (R928, Hamamatsu), which can be gain modulated up to 500 MHz–1 GHz (ISS, Champaign, IL), the measurable fluorescence lifetime limit is in the hundred-picosecond range. Certainly, a faster detector such as a modified microchannel plate, which can be modulated up to 10 GHZ, will facilitate measurement of shorter lifetimes.

Photon and Time Efficiency

In the time domain, when the signal from two time-gated windows is collected sequentially after each excitation pulse, the photon and time efficiency is high. It is possible to reach close to 100% efficiency using the emitted fluorescent light (Sytsma et al., 1998) after each pulse. However, in the sequential collection mode, the signal from the two gates cannot be imaged with a single CCD camera. If the signal from the two gates is recorded in two separate passes, the photon efficiency will decrease to one half of that in the previous case. One expects even lower photon efficiency when more than two gates are used. In the frequency domain, when the excitation source and the detectors are sinusoidally modulated, half the fluorescence emission is suppressed. When using the frequency-domain homodyning or heterodyning method, the best photon efficiency is about 50%. However, the amount of fluorescence photons generated strongly depends on the nature of excitation. In the case of pulsed excitation, the number of generated fluorescence photons depends on the frequency of the excitation. In the case of sinusoidally modulated excitation, the photon number depends upon the fluorescence lifetime of the molecule. Indeed, using sinusoidally modulated excitation, more photons are available in the same period of time than with many other means of excitation. Additionally, in the time domain there is a restriction of the highest pulse repetition rate one can use so that there is no second excitation before completion of the previous fluorescence decay.

To decide which method, time domain or frequency domain, is appropriate for a specific application, one should also consider the different detection methods, that is, the single-photon counting or the analog methods. Photon-counting detection has higher photon sensitivity but is susceptible to saturation as compared to the analog detection method. In general, the choice between the time domain and frequency domain depends largely on the fluorescence intensity level of the specimen under investigation. For bright samples the choice is the analog detection and for dim samples the choice is the photon-counting method.

Laser Scanning Versus Wide-Field Illumination

Wide-field illumination is the most common method used in both time-domain and frequency-domain FLIM. With the wide-field illumination one needs an intensified camera either time gated or gain modulated. The cost of such a camera is high. Laser scanning provides an alternative illumination method requiring only single-point detection. The system using a scanner is considerably more cost effective and can be used in both the time-domain FLIM (Sytsma et al., 1998) and the frequency-domain FLIM (Dong et al., 1995; So et al., 1995). The performance and characteristics of the instrument depend on which illumination source is used. Laser scanning using either a two-photon or confocal microscope directly provides three-dimensional sectioned images. The major difference between the two illumination methods is at the detector. Both methods (laser scanning and wide-field illumination) have their advantages and shortcomings and we compare the two cases in the following:

1. Due to the single-point nature of laser scanning, the method is usually associated with single-point detectors such as a PMT. For wide-field illumination, two-dimensional array detectors are typically used. An important advantage of laser scanning is that it allows for imaging regions of interest with arbitrary shapes and for single-point measurements as a function of time. These imaging modes are very useful in biological studies, especially in studies of physiolog-

ical functions of live samples. On the other hand, two-dimensional array detectors typically allow imaging of one particular shape depending on the physical arrangement of the detector arrays. The drawback of using scanning is the slower frame rate as compared to that of the array detectors. For fluorescence lifetime imaging, one may reach pixel rates of about 5–10 µs at best using a scanner. With this scanning rate, for a 256 × 256-pixel frame, a frame rate about 0.3–0.6 s is expected, significantly slower than for using normal two-dimensional array detectors.

2. For a particular measurement, the laser power per pixel one needs to use is related to the photon sensitivity of the detector and how long one has to integrate for good photon statistics. A camera-based detector made for FLIM applications has an image intensifier that has the same photon sensitivity as a photomultiplier. For a typical image, there are about 10^5–10^6 pixels. Using the same laser, the average laser power is about 10^5–10^6 times less for wide-field illumination than for laser scanning. To detect the same amount of fluorescence per pixel, the camera has to integrate longer in the case of wide-field illumination as compared to the sequential single-point scanning. However, the total amount of laser energy used for a measurement in the imaging area is about the same.

3. One can use the FLIM method to measure slow decay fluorescence and phosphorescence. In this situation, when using laser scanning, the pixel rate is determined by the decay rate of the fluorophore, allowing sufficient time for the excited molecule to relax to the ground state, as well as collecting enough photons for an accurate lifetime calculation. For instance, if a probe's lifetime is 1 ms, the required minimum pixel dwell time is approximately 0.01 s and the resulting frame rate for a 256 × 256 pixel image will be on the order of 10 min. Certainly, that would be impractical. For measuring long-lifetime probes, one still needs to integrate over a sufficiently long period for good photon statistics; the practical approach is to use wide-field illumination and two-dimensional array detectors.

Background Rejection and Noise Immunity

For many FLIM applications, we cannot avoid background noise. This noise may arise from autofluorescence or other fluorescence contaminants as well as from stray and scattered light. The lifetime sensitivity of any FLIM method depends on how well the instrument can reject the background noise. To discuss the noise problem, we need to separate the noise that is fluorescent in nature from nonfluorescence noise. For nonfluorescence noise, the frequency-domain heterodyning method provides several noise-suppressing filters, especially the in-phase folding averaging filter, which we believe is superior to other methods. The homodyning method provides a fast Fourier transform and averaging filters for noise suppression. However, by shifting the relative phase between detector and excitation, homodyning measures the dc signal that is subject to interference from very-low-frequency noise. The only filter used in the time-domain method is time-correlated averaging, which is also susceptible to low-frequency noise.

For noise having a decay time, the background varies when using different methods, either the time domain or the frequency domain. In the time domain, the noise is an additional value to the overall intensity. For time-resolved spectroscopy, in the time domain, simple subtraction is sufficient to remove the noise. The scattering noise ($\tau=0$) does not interfere with the measurement. In the frequency domain, the noise, whether the source is fluorescence or scattering, appears as an additional component with its own phase value and modulation amplitude. It is still possible, though complicated, to make background corrections (Reinhart et al., 1991; Swift and Mitchell, 1991; Lakowicz et al., 1987). In principle, the same background correction approach can be used for the FLIM measurement with knowledge of the fractional contribution of the noise for every pixel.

Another issue related to the background correction is to use FLIM to enhance or suppress the contrast of certain lifetime components while not obtaining the lifetime value per se. In this respect, the homodyning FLIM offers a unique mechanism for changing the intensity contrast of a particular lifetime component. This is easy to understand by using Equation (7.16). In the case of homodyning $\Delta\omega = 0$,

$$CC(\Delta\theta) = DC_0 DC_c + \frac{AC_0 AC_c}{2} \cos(\Delta\theta - \phi) \qquad (7.18)$$

and $CC(\Delta\theta)$ becomes a dc value. When the phase difference between the detector and the excitation is set at $\Delta\theta = \phi$, the cosine has its maximum value and one obtains the maximum intensity contrast for the lifetime component that has phase delay ϕ at the particular modulation frequency. Similarly, when $\Delta\theta = \phi - \pi/2$, the intensity of the lifetime component with delay ϕ will be suppressed.

Effects of Photobleaching

For FLIM applications, photobleaching normally does not affect the measurement. This is especially true for the time-domain method since the gated channel records only a correlated signal following the fluorescence decay curve. In the frequency domain (homodyning and heterodyning), because the signal is recorded at a lower frequency with respect to the modulation frequency, a steadily decreasing signal could bias the fast Fourier transform, resulting in distortion of phase and modulation values. Photobleaching affects the lifetime measurement at a different time scale for the homodyning and the heterodyning method. In principle, for heterodyning, the measured phase and modulation of the signal will be biased if significant photobleaching happened within the applied cross-correlation frequency, typically on the order of kilohertz for a scanning system. For homodyning, the effect of photobleaching depends on the time factor to acquire the series of images for a single waveform, typically on the order of a second. This photobleaching problem can be addressed by acquiring more data points per waveform not in sequence for homodyning and/or performing a complete Fourier transform, which recovers the correct waveform, as well as the bleaching rate.

Single- or Multiple-Exponential Decay

For real-time imaging, the time-domain FLIM provides results for a single apparent lifetime assuming a single-exponential decay. To solve the multiexponential problem, one needs data-fitting routines and more than two gates are necessary. For frequency-domain FLIM, using a single modulation frequency, one obtains two apparent lifetime values, τ_{phase} and τ_{mod}, which can be used to evaluate whether or not one has a single-exponential case. For systems with multiexponential decays, τ_{phase} and τ_{mod} are not equal and each value can be used to evaluate different lifetime components, as we discussed in great length previously under Background of Frequency-Domain Method. With two apparent lifetimes obtained in the frequency domain, one has more information about the systems under investigation than with just one lifetime value in the case of the time domain. This point is also demonstrated in our lifetime measurement of cellular chloride concentrations. To resolve multiexponential decay, that is, individual lifetimes as well as the fractional contribution of each component, one needs to use more than one modulation frequency. However, for biological applications, we find that interest is often focused on the differences in the apparent lifetime at different spatial locations, and not so much on solving the fractional components with different lifetime values.

SUMMARY

We have described the theories and their applications of the major techniques used for fluorescence lifetime imaging. FLIM is one step forward toward quantitative microscopy, which we believe is the future of microscopy studies. The availability of the instrumentation to general laboratories depends on the cost, which is still high as most of the techniques require high-speed electronic or optical components in addition to the laser. In this respect, the pump–probe technique is attractive since it eliminates the cost of the high-speed detectors. To practically realize the FLIM methodologies, it is important to make these techniques user friendly. We have compared some of the major differences between the different FLIM methods that are the basis for choosing one technique over the other in practical applications. We have demonstrated, by measuring cellular chloride concentrations, that the capability of lifetime imaging is superior to simple intensity imaging. Valuable information can be derived from these quantitative measurements.

ACKNOWLEDGMENTS

We wish to acknowledge the support of the National Institutes of Health (RR03315) for the research on two-photon time-resolved microscopy. We thank Qiaoqiao Ruan for culturing the PC12 cell and Osman Akcakir for the chloride calibration measurement. Our thanks also go to Robert Clegg, who recently joined the Laboratory for Fluorescence Dynamics, for the valuable discussions we had on FLIM.

REFERENCES

Atherton SJ, Beaumont PC (1984): Photobiochem Photobiophys 8:103–113.

Ballew RM, Demas JN (1989): Anal Chem 61:30–33.

Buist AH, Mueller M, Gijsbers EJ, Brakenhoff GJ, Sosnowski TS, Norris TB, Squier J (1997): J Microscopy 186(3):212–220.

Buurman EP, Sanders R, Draaijer A, Gerritsen HC, van Veen JJF, Houpt PM, Levine YK (1992): Scanning 14:155–159.

Cario G, Franck J (1922): Z Phys 11:161.

Centonze VE, White JG (1998): Biophys J 75(4):2015–2024.

Clegg RM (1996): In *Fluorescence Imaging Spectroscopy and Microscopy,* Vol. 137. Wang XF, Herman B (eds). Chemical Analysis Series, New York: Wiley, pp 179–252.

Clegg RM (1992): Methods Enzymol 211:353–88.

Clegg RM, Schneider PC, Jovin TM (1996): In *Biomedical Optical Instrumentation and Laser-Assisted Biotechnology.* Dordrecht, The Netherlands: Kluwer Academic, pp 143–156.

Clegg RM, Schneider PC (1996): In *Fluorescence Microscopy and Fluorescent Probes.* New York: Plenum Press.

Cubeddu R, Canti G, Pifferi A, Taroni P, Valentini G (1997): Photochem Photobiol 66(2):229–236.

Diffey WM, Homoelle BJ, Edington MD, Beck WF (1998): J Phys Chem B 102(15):2776–2786.

Dix JA, Verkman AS (1990): Biochem 29:1949–1954.

Denk W, Strickler JH, Webb WW (1990): Science 248:73–76.

Dong CY, So PTC, French T, Gratton E (1995): Biophys J 69:2234–2242.

Dong CY, So PTC, Buchler C, Gratton E (1997): Optik 106(1):7–14.

Dowling K, Hyde SCW, Dainty JC, French PMW, Hares JD (1997): Opt Commun 135:27–31.

Dowling K, Dayel MJ, Hyde SCW, Dainty JC, French PMW, Vourdas P, Lever MJ, Dymoke-Bradshaw AKL, Hares JD, Kellett PA (1998): IEEE J of Selected Topics in quantum Electronics 4(2):370–375.

Dushinsky FZ (1933): Physik 81:7.

Elzinga PA, Lytle FE, Jian Y, King GB, Laurendeau NM (1987a): Appl Spectroscopy 41:2–4.

Elzinga PA, Kneisler RJ, Lytle FE, Jian Y, King GB, Laurendeau NM (1987b): Appl Opt 26:4303–4309.

Feddersen BM, Piston DW, Gratton E (1989): Rev Sci Inst 60(9):2929–2936.

French T, Gratton E, Maier J (1992): Time-resolved laser spectroscopy in biochemistry III. SPIE Proc 1640:254–261.

French T, So PTC, Weaver Jr DJ, Coelho-Sampaio T, Gratton E, Voss Jr EW, Carrero J (1997): J Microscopy, 185:339–353.

Gadella Jr TWJ, Jovin TM J (1995): Cell Biol 129:1543–1548.

Gaviola E (1927):Z Phys 42:853.

Gaviola E, Pringsheim P (1927): Z Phys 43:384.

Gratton E, Limkeman M (1983): Biophys J 44:315–324.

Gratton E, Jameson DM, Hall R (1984): Ann Rev Biophys Bioeng 13:105–124.

Hartmann P, Ziegler W, Holst G, Luebbers DW (1997): Sensors Actuators B 38–39:110–115.

Inglefield JR, Schwartz-Bloom RD (1997): J Neurosci Methods, 75:127–135.

Itoh H, Evenzahav A, Kinoshita K, Inagaki Y, Mizushima H, Takahashi A, Hayakawa T, Kinosita K Jr (1997): SPIE Proc 2979:733–740.

Jameson DM, Gratton E, Hall R (1984): App Spec Rev 20:55–106.

Keating SM, Wensel TG (1991): Biophys J 59:186–202.

Koenig K, So PTC, Mantulin WW, Tromberg BJ, Gratton E (1996): J Microsc 183:197–204.

Krnjevic K, Schwartz S (1967): Exp Brain Res 3:320–336.

Kumble R, Palese S, Lin VSY, Therien MJ, Hochstrasser RM (1998): J Amer Chem Soc 120(44):11489–11498.

Kusba J, Bogdanov V, Gryczynski I, Lakowicz JR (1994): Biophys J 67:2024–2040.

Lakowicz JR, Jayaweera R, Joshi N, Gryczynski I (1987): Anal Biochem 160:471–479.

Lakowicz JR, Szmacinski H, Nowaczyk K, Berndt KW, Johnson M (1992): Anal Chem 202:316–330.

Lakowicz JR, Szmacinski H, Nowaczyk K, Lederer WJ, Kirby MS, Johnson ML (1994a): Cell Calcium 15:7–27.

Lakowicz JR, Gryczynski I, Bogdanov V, Kusba J (1994b): J Phys Chem 98:334–342.

Mantulin WW, French T, Gratton E (1993): Medical Lasers and Systems II SPIE Proc 1892:158–166.

Marriott G, Clegg RM, Arndt-Jovin DJ, Jovin TM (1991): Biophys J 60:1374–1387.

Masters BR, So PTC, Gratton E (1997): Biophys J 72:2405–2412.

McLoskey D, Birch DJ, Sanderson A, Suhling K (1996): Rev Sci Instrum 67(6):2228–2237.

Miki M, Miura T, Sano K, Kimura H, Kondo H, Ishida H, Maeda Y (1998): J Biochem 123(6):1104–1111.

Mizeret J, Wagnieres G, Stepinac T, van den Bergh H (1997): Lasers in Med Sci 12:209–217.

Morgan CG, Mitchell AC, Murray JG (1992): J Microsc 165(1):49–60.

Morgan CG, Mitchell AC, Peacock N, Murray JG (1995): Rev Sci Instrum 66(1):48–51.

Mueller M, Ghauharali R, Visscher K, Brakenhoff G (1995): J Microsc 177(2):171–179.

Ni T, Melton LA (1996): Appl Spectrosc 50(9):1112–1116.

Oida T, Sako Y, Kusumi A (1993): Biophys J 64:676–685.

Periasamy A, Wodnicki P, Wang XF, Kwon S, Gordon GW, Herman B (1996): Rev Sci Instrum 67(10):3722–3731.

Piston DW, Sandison DR, Webb WW (1992): SPIE Proc 1640:379–389.

Piston DW, Kirby MS, Cheng H, Lederer WJ, Webb WW (1994): Appl Optics 33:662–669.

Piston DW, Masters BR, Webb WW (1995): J Microsc 178(1):20–27.

Pringsheim P (1949): *Fluorescence and Phosphorescence.* New York: Interscience.

Reinhart G, Marzola P, Jameson D, Gratton E (1991): J Fluorescence 1(3):153–161.

Sanders R, Draaijer A, Gerritsen HC, Houpt PM, Levine YK (1995): Anal Biochem 227:302–308.

Sanders R, van Zandvoort MAMJ, Draaijer A, Levine YK, Gerritsen HC (1996): Photochem Photobiol 64(5):817–820.

Sargent III M, Scully MO, Lamb Jr WE (1974): *Laser Physics.* London: Addison-Wesley.

Schneider P, Clegg RN (1997): Rev Sci Instrum 68(11):4107–4119.

Scully AD, Ostler RB, Phillips D, O'Neil PO, Townsend KMS, Parker AW, MacRobert AJ (1997): Bioimaging 5:9–18.

Silverman L, Campbell R, Broach JR (1998): Curr Opin Chem Biol 2(3):397–403.

So PTC, French T, Gratton E (1994): Time resolved Laser Spectroscopy in Biochemistry IV SPIE Proc 2137:83–92.

So PTC, French T, Yu WM, Berland KM, Dong CY, Gratton E (1995): Bioimaging 3:49–63.

Subramaniam V, Kirsch AK, Jovin TM (1998): Cell Mol Biol 44(5):689–700.

Swift KM, Mitchell GW (1991): SPIE Proc 1431:171–178.

Sytsma J, Vroom JM, De Grauw CJ, Gerritsen HC (1998): J Microsc 191(1):39–51.

Toth K, Sauermann V, Langowski J (1998): Biochemistry 37(22):8173–8179.

Vanbrederode ME, Ridge JP, Vanstokkum IHM, Vanmourik F, Jones MR, Vangrondelle R (1998): Photosyn Res 55(2–3):141–146.

Visser AJWG (1997): Curr Opin Colloid & Interface Sci 2(1):27–36.

Wagner S, Castel M, Gainer H, Yarom Y (1997): Nature 387:598–603.

Wang XF, Uchida T, Coleman DM, Minami S (1991): Appl Spectrosc 45(3):360–366.

Woods RJ, Scypinski S, Love LJC, Ashworth HA (1984): Anal Chem 56:1395.

Yu WM, So PTC, French T, Gratton E (1996): Biophys J 70:626–636.

8

Fluorescence Lifetime Flow Cytometry

John A. Steinkamp
Los Alamos National Laboratory, Los Alamos, New Mexico

INTRODUCTION

Conventional flow cytometry (FCM) instruments rapidly measure biochemical, functional, and physical properties of individual cells and subcellular components. They are used in clinical diagnostic medicine [e.g., acquired immunodeficiency syndrome (AIDS) diagnosis/treatment monitoring] and in biological and biomedical research applications. Routine clinical tests and biological measurements are primarily based on labeling cells with multiple-color fluorochromes for correlated analysis of biomolecules, such as DNA, RNA, proteins, enzymes, lipids, and cell surface receptors (human lymphocyte subsets); for determining specific DNA sequences [fluorescence *in situ* hybrid-hybridization (FISH)]; and for measuring subcellular components, such as mitochondria and chromosomes. Multicolor detection methods employing color-separating dichroic and bandpass optical filters have been routinely used for measuring multiple fluorochrome emissions when spectra are sufficiently separated and the fluorochromes require only one excitation source. Electronic compensation is presently used to separate signals from partially overlapping fluorescence emission spectra to eliminate spectral cross-talk between adjacent detection channels (Loken et al., 1977). Overlapping spectral emissions also have been resolved by recording and analyzing emission spectra using fast Fourier transform (FFT) computational methods (Buican, 1990). If fluorochromes cannot be resolved by these means but

Emerging Tools for Single-Cell Analysis, Edited by Gary Durack and J. Paul Robinson.
ISBN 0-471-31575-3 Copyright © 2000 Wiley-Liss, Inc.

have separated excitation spectra, multiple excitation sources can be employed to sequentially excite labeled cells and spatially resolve spectrally overlapping emission signals (Beavis and Pennline, 1994; Roederer et al., 1997; Steinkamp et al., 1991). In addition to utilizing the spectral emission properties of fluorescent probes (color/intensity) to measure cellular features, excited-state lifetimes (decay times) provide a means to discriminate among fluorescent markers and serve as spectroscopic probes to study the interaction of fluorescent markers with their cellular targets, each other, and the surrounding microenviroment.

The direct measurement of excited-state lifetimes by flow cytometry is important because it provides information about fluorophore/cell interactions at the molecular level. An advantage of lifetime measurements is that lifetimes can be considered in some instances as absolute quantities. However, the lifetimes of fluorophores bound to cellular macromolecules can be influenced by physical and chemical factors near the binding site, such as solvent polarity, cations, pH, energy transfer, and excited-state reactions. Often such changes are accompanied by a change in the temporal nature of the fluorescence decay (e.g., single exponential, multiexponential, or nonexponential). Therefore, it is expected that lifetime measurements can be used to probe cellular complexes and subcompartments, such as (1) the chemical and structural changes that occur in DNA and chromatin during the cell cycle, in differentiating cells, and in apoptotic cells with damaged chromatin (Sailer et al., 1996, 1997a, 1998b); (2) DNA and double-stranded RNA (Sailer, 1998a); (3) the cell surface by quenching of fluoroscein isothiocyanate (FITC) -conjugated antibodies (Deka et al., 1996); and (4) the binding of cytotoxic chemicals to DNA and other molecular targets in cells (Sailer et al., 1997b). Table 8.1 lists the lifetimes of fluorescent markers used to measure cellular DNA, RNA and protein content, mitochondria, antibody labeling of cellular antigens, and the lifetime of cellular [line Chinese hamster ovary (CHO)] autofluorescence.

Static spectroscopic fluorescence measurements of excited-state lifetimes are made in the time domain by time-correlated, single-photon counting (Demas, 1983; Lakowicz, 1983) or in the frequency domain by determining the frequency response of the fluorescence emission to a continuous intensity-modulated excitation. Time-domain methods have been employed in fluorescence microscopy to measure the lifetime of fluorophores bound to cells (Keating and Wensel, 1991; Vigo et al., 1987) and of single molecules in flow (Wilkerson et al., 1993). In the frequency-domain method, the fluorescence emission signal has the same frequency content as the excitation but is shifted in phase, and the depth of modulation is decreased due to the finite lifetimes of the excited states. There has been remarkable progress in frequency-domain spectrofluorometric developments during the past several years (Berndt et al., 1990; Gratton and Limkeman, 1983; Gratton et al., 1984; Lakowicz et al., 1985, 1986; Mitchell and Swift, 1989; Thompson et al., 1992). These developments have been applied to microscope-based cellular imaging systems (Gadella et al., 1997, 1993; Lakowicz et al., 1992; So et al., 1994, 1995) and to flow cytometers for measuring lifetimes of fluorophores bound to cells using real-time analog (Pinsky et al., 1993; Steinkamp et al., 1993b) and digital signal (Beisker and Klocke, 1997; Deka et al., 1994; Durack et al., 1998) processing methods. The technology for combined fluorescence lifetime measurements on fluorophore-labeled particles and the

TABLE 8.1. Examples of Fluorescence Lifetimes and Corresponding Phase Shifts at Various Excitation Frequencies for Fluorochromes Used to Label Cellular Complexes and Cells

Fluorescent dye/compound	Excitation wavelength (nm)	Fluorescence lifetime (ns)	Phase shift[a]		
			At 10 MHz (deg)	At 30 MHz (deg)	At 50 Mhz (deg)
Hoechst 33342 (DNA)[b]	360	2.6	9.2	26.1	39.2
DAPI (DNA)[b]	360	3.5	12.3	33.4	47.7
Mithramycin (cells)[c]	420	3.0	10.7	29.5	43.3
Propidium iodide[c]	515	1.2	4.3	12.7	20.6
Propidium iodide (cells)[c]	515	13.0	39.2	67.8	76.2
Ethidium bromide (cells)[c]	515	19.0	50.0	74.4	80.5
Ethidium bromide (DNA)[d]	515	22.5	54.6	76.7	81.9
7-AAD (DNA)[e]	515	0.8	2.9	8.6	14.1
Acridine Orange (cells)[c]	480	3(Gn), 13(Rd)	10.7, 39.2	29.5, 67.7	43.3, 76.2
Pyronin Y (cells)[c]	530	0.6, 2.3	2.2, 8.2	6.5, 23.4	10.7, 35.8
FITC[f]	480	3.6	12.5	34.0	48.5
Fluorescein[c]	480	4.7	16.4	41.5	55.9
Rhodamine 123 (cells)[g]	511	2.0, 4.0	7.2, 14.1	20.6, 37.0	32.1, 51.5
Phycoerythrin-Avidin[c]	530	3.5	12.6	33.4	47.7
Texas Red-Avidin[c]	530	4.6	16.1	40.9	55.3
Fluorescamine[h]	uv	7.0	23.8	52.8	65.5
Dansyl chloride[h]	uv	14.0	41.2	69.4	77.2
CHO Cells (autofluores.)[i]	365	1.8	6.4	18.7	29.5

[a] Phase shift equals arctan $\omega\tau$, where $\omega = 2\pi f$ is the angular frequency and τ is the fluorescence lifetime.
[b] Biophys J 72:567, 1997.
[c] Unpublished data (J. Martin).
[d] Biochemistry 16:3647, 1977.
[e] Proc SPIE 2137:462, 1994.
[f] Anal Let. 18(A4):393, 1985.
[g] J Fluoresc 6:209, 1996.
[h] Clin Biochem 21:139, 1988.
[i] J Microsc 183:197, 1996.

surrounding fluorophore solution by flow cytometry also has been described (Steinkamp and Keij, 1999). Although frequency-domain, analog signal-processing methods are more suitable than time-domain methods for making lifetime measurements on a cell-by-cell basis in flow in real time, time-resolved lifetime measurements on single cells have been demonstrated (Deka and Steinkamp, 1996).

Fluorescence lifetimes also can be used to electronically separate the overlapping emissions from fluorescent probes based on differences in their lifetimes expressed as phase shifts using phase-sensitive detection in both static (Lakowicz and Cherek, 1981) and flow systems (Steinkamp and Crissman, 1993). Because signals from fluorophores are resolved electronically, rather than spectroscopically, the number of fluorescent probes that can be used on the same sample is increased; because different markers bind to different targets, increasing the number of potential fluorochrome probes increases the number of biological parameters that can be studied concurrently and correlated, and light losses from optical filtering are eliminated. In addition, background interferences, for example, autofluorescence, unbound fluorophores, nonspecific fluorophore labeling, and Rayleigh scatter, may be reduced or eliminated in the measurement process (Steinkamp et al., 1997, 1999b).

The resolution of signals from fluorescence emissions by phase-sensitive detection using flow cytometry was first demonstrated on cells stained with propidium iodide (PI) and FITC for total cellular DNA and protein content, respectively (Steinkamp and Crissman, 1993). Although the PI and FITC fluorescence emission signals are readily separable by conventional FCM methods (Crissman and Steinkamp, 1982), they were separated electronically using a single photomultiplier tube (PMT), a long-pass (barrier) filter to block scattered laser excitation light, and two phase-sensitive detection channels, one for PI and the other for FITC. In addition, phycoerythrin (PE)–antiThy 1.1 antibody–labeled rat thymus cells in suspension have been stained with PI (overlapping emission spectra) for locating "dead cells" and analyzed by phase-resolved methods to electronically separate cells that were PI positive only, PE–antiThy 1.1 labeled and PI positive, and PE–antiThy 1.1 labeled only (Steinkamp et al., 1999a). These studies also used murine thymus cells labeled with PE/Texas Red–antiThy 1.2 antibody and stained in suspension with PI (highly overlapping emission spectra). The application of phase-sensitive detection to eliminate autofluorescence from lung fibroblasts labeled with a cell surface FITC antibody has been described (Steinkamp et al., 1999b), and viable cells labeled with Hoechst 33342 and monobromobimane (overlapping fluorescence emissions) have been analyzed using phase-sensitive flow cytometry to determine relative DNA and glutathione content, respectively (Keij et al., 1999).

MATERIALS AND METHODS

Theory of Operation

Fluorochrome-labeled cells are analyzed as they intersect an optically focused continuous wave (cw) laser beam that is sinusoidally modulated in intensity at a high frequency (see Fig. 8.1). The time-dependent fluorescence emission signal $v(t)$ is a Gaussian-shaped, modulated pulse that results from the passage of the cell across the laser beam and can be expressed in an approximate form as

$$v(t) = V[1 + m\sin(\omega t - \phi)]e^{-a^2(t-t_0)^2} \qquad (8.1)$$

where V is the signal intensity, ω is the angular excitation frequency (modulation), ϕ and m are the respective signal phase shift and demodulation terms associated with a single fluorescence decay time (τ), t is time, and a is a term related to the laser beam height and the velocity of a cell crossing the laser beam at time t_0 (Zarrin et al., 1987). An exact derivation of this relationship has been given by Deka et al. (1994). The cw-excited (dc), low-frequency signal component is extracted using a low-pass filter to give conventional FCM fluorescence intensity information (see Fig. 8.4A below). The high-frequency modulated sine wave signal component, which is shifted in phase (ϕ) by an amount

$$\phi = \arctan \omega\tau \tag{8.2}$$

relative to the excitation frequency and demodulated by a factor m (described below), is processed by phase-sensitive detectors (PSDs) (Blair and Sydenham, 1975; Meade, 1982) to resolve fluorescence emission signals based on differences in lifetimes (expressed as phase shifts) and to quantify lifetime directly as a parameter (see Figs. 8.4B and 8.4C below). Each PSD consists of a multiplier, a sine wave reference signal with phase shifter to shift the phase of the reference (ϕ_R) with respect to the modulated fluorescence signal input to the multiplier, and a low-pass filter. The PSD outputs are Gaussian-shaped signals that are proportional to the fluorescence intensity, the demodulation factor, and the $\cos(\phi - \phi_R)$, expressed as

$$v_o(t) = \frac{1}{2}mV\cos(\phi - \phi_R)\, e^{-a^2(t-t_0)^2} \tag{8.3}$$

Measurement of Fluorescence Lifetime.
Fluorescence lifetime as measured by phase shift is quantified by the two-phase method (Meade, 1982) using two PSDs (see Fig. 8.4C below). A quadrature phase hybrid circuit is used to form two reference sine wave signals that are 90° out of phase with each other. These signals are input as references along with the modulated fluorescence signal to two PSDs, the outputs are expressed as

$$v_{\phi-90}(t) = \frac{1}{2}mV\sin\phi\, e^{-a^2(t-t_0)^2} \quad \text{and} \quad v_\phi(t) = \frac{1}{2}mV\cos\phi\, e^{-a^2(t-t_0)^2} \tag{8.4}$$

where ϕ is the signal phase shift [Eq. (2)] and ϕ_R has been set to zero. The $v_{\phi-90}(t)/v_\phi(t)$ ratio expression results in $\tan\phi$, which is directly proportional to the fluorescence decay time, expressed as

$$\tau = \frac{1}{\omega}\tan\phi = \frac{1}{\omega}\left(\frac{V(\phi-90)}{V(\phi)}\right) \tag{8.5}$$

Where $V(\phi - 90)$ and $V(\phi)$ are the peak values in Equation (4).

Fluorescence lifetime may also be determined by measuring the relative depth of amplitude modulation (m) of the emission signal (m_{em}) with respect to the excitation signal (m_{ex}). The relative modulation or demodulation factor m is determined from the ratio

$$m = \frac{m_{em}}{m_{ex}} = \frac{\text{modulation of fluorescence}}{\text{modulation of excitation}} = \cos\phi = \frac{1}{[1+(\omega\tau)^2]^{1/2}} \tag{8.6}$$

In the steady-state system it is only necessary to measure the ac and dc fluorescence emission and excitation signal components and determine the relative modulation by ratio calculations (Spencer and Weber, 1969). In our system, the amplitude demodulation factor may be determined by measuring the maximum and minimum signal components at the peak height of the Gaussian-shaped fluorescence detector output signal (V_{em}) and the ac and dc components of the steady-state laser excitation monitor signal (V_{ex}) (see Fig. 8.1), either by digital (Deka et al., 1994) or analog methods (Steinkamp et al., 1998). The relative modulation is then expressed as

$$m = \left[\frac{(V_{max} - V_{min})}{(V_{max} + V_{min})}\right]_{em} / \left[\frac{(V_{max} - V_{min})}{(V_{max} + V_{min})}\right]_{ex} \quad (8.7)$$

The real-time ratio results in cos φ which is proportional to the fluorescence decay lifetime, expressed as

$$\tau = \frac{\tan(\cos^{-1} m)}{\omega} \quad (8.8)$$

The above equations are derived on the assumption of a single-component exponential decay of fluorescence from a homogeneous emitting fluorophore population. This is often cited as the major shortcoming of the single-frequency method, because the existence of a unique single-component decay is presupposed, but not demonstrated by the measurement. This is indeed true if only one of the two quantities, that is, lifetime by phase shift or amplitude demodulation, is measured. However, if both are measured, the existence of an exponential can be demonstrated. In the heterogeneous fluorophore population, the lifetime measured by the degree of amplitude demodulation will almost

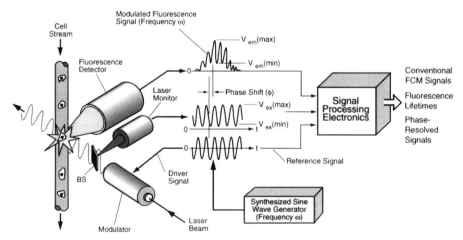

Fig. 8.1. Conceptual diagram of single-frequency, fluorescence lifetime flow cytometer illustrating the laser excitation beam, modulator, modulated laser beam, beam splitter (BS), laser monitor, cell sample stream, cell-stream laser-beam intersection point in the flow chamber, fluorescence detector, modulated fluorescence and reference signals, and synthesized sine-wave signal generator. The modulated fluorescence signals are processed electronically to give conventional FCM signals, fluorescence lifetimes, and phase-resolved fluorescence signals.

Materials and Methods

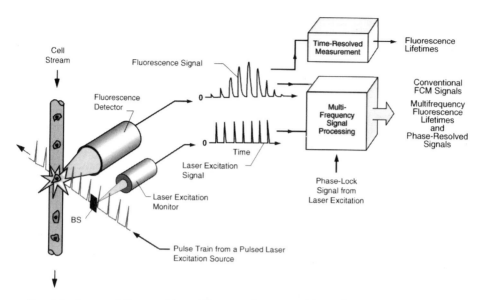

Fig. 8.2. Conceptual diagram of the multifrequency, fluorescence lifetime flow cytometer illustrating the laser excitation pulse train, beam splitter (BS), laser monitor, cell sample stream, cell-stream laser-beam intersection point in the flow chamber, fluorescence detector, fluorescence and laser monitor signals, and time and multifrequency measurement signal-processing electronics for making lifetime-based measurements.

always be larger than the weighted average of the individual component lifetime values, whereas the lifetime determined by phase shift will always be shorter than the weighted average (Spencer and Weber, 1969). It is only when there is a single exponential decay that both methods give the same result. Also, phase shift measurements at two or more frequencies can be used to detect heterogeneous fluorescence decays (Deka et al., 1995). If the decay is multiexponential, measurement by time-resolved methods (Deka and Steinkamp, 1996; Demas, 1983; Lakowicz, 1983) or by multifrequency methods over a wide range of frequencies (Jameson et al., 1984; Lakowicz et al., 1984) is required, as illustrated in Figure 8.2, using the multiharmonic frequency content from the pulse train of a mode-locked laser (Alcala et al., 1985; Lakowicz et al., 1986).

Phase-Resolved Separation of Fluorescence Emission Signals. The principle of phase suppression, as applied to flow cytometry, for separating two fluorescence emission signals having different lifetimes (i.e., phase shifts ϕ_1 and ϕ_2) by phase-sensitive detection, is illustrated below and in Figure 8.4B. The output of the PSD (by superposition of the two signals) is

$$v_0(t) = \frac{1}{2}m_1 V_1 \cos(\phi_1 - \phi_R)e^{-a^2(t-t_0)^2} + \frac{1}{2}m_2 V_2 \cos(\phi_2 - \phi_R)e^{-a^2(t-t_0)^2} \quad (8.9)$$

where V_1 and V_2 are the signal intensities, m_1 and m_2 are the demodulation factors, and ϕ_1 and ϕ_2 are the phase shifts that result when a cell stained with two fluorochromes, each having a different lifetime τ_1 and τ_2, is excited by a single modulated source. To resolve either of the two signals, the reference phase is shifted by an amount $\pi/2 + \phi_1$ or $-\pi/2 + \phi_2$ degrees (Veselova et al., 1970). This results in one signal being passed and the other being nulled. For example, if the reference phase is adjusted to equal $-\pi/2 + \phi_2$ degrees, the detector output is expressed as

$$v_o(t) = \frac{1}{2}m_1 V_1 \sin(\phi_2 - \phi_1)e^{-a^2(t-t_0)^2} \tag{8.10}$$

Similarly, if the reference phase is adjusted to equal $\pi/2 + \phi_1$ degrees, the output is expressed as

$$v_o(t) = \frac{1}{2}m_2 V_2 \sin(\phi_2 - \phi_1)e^{-a^2(t-t_0)^2} \tag{8.11}$$

Both signals are resolved, but with a loss in amplitude [$\sin(\phi_2 - \phi_1)$]. When fluorescence signals are processed by two PSDs operating in parallel, the contributions to the total fluorescence signal are resolved by setting one detector reference to $\pi/2 + \phi_1$ degrees and the other detector reference to $-\pi/2 + \phi_2$ degrees (Steinkamp and Crissman, 1993).

Signal Processing. The present signal detection/processing electronics (Steinkamp et al., 1993) are based on the homodyne signal-processing scheme. Signal homodyning (Blair and Sydenham, 1975) relies on direct measurement of the phase shift by multiplying the signal with a suitable reference at the same frequency using any number of electronic devices, such as a double-balanced mixer. This is conceptually the simplest form of signal processing. The main disadvantages are susceptibility to noise interference; requirements for high-frequency precision electronics; loss of phase resolution at high frequencies due to limited resolution of variable time delays; and processing that can be implemented only in the analog mode for high-frequency measurements.

To avoid the need for high-frequency signal-processing electronics and to better isolate the signal from noise interference, a frequency-heterodyning technique has been developed for static spectrofluorometric frequency-domain lifetime measurement (Spencer and Weber, 1969). This technique works by mixing the fluorescence signal, at the detector PMT base or an external mixer, with a second signal of different frequency (i.e., frequency heterodyning). The resulting difference frequency contains the same information as the original high-frequency signal, but the difference can be set to any suitable lower value to suit the measurement conditions, for example, signal processing (digital) speed. In addition, the frequency-heterodyning signal-processing method can be used with either single- or multifrequency excitation schemes, such as a mode-locked laser pulse train (Alcala et al., 1985; Lakowicz et al., 1986). Modern multifrequency lifetime spectrofluorometers utilize pulsed excitation sources, coupled with frequency heterodyne methods, to measure excited-state lifetimes in bulk solutions (Mitchell and Swift, 1989).

The analog electronics commonly found in static and flow frequency-domain measurement instruments are limited to collecting only one frequency at a time. As the

Color Plates

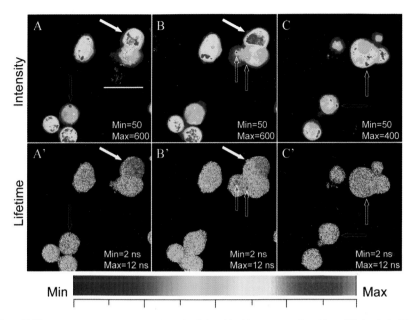

Fig. 7.15. Intensity and lifetime imaging of cellular chloride concentration. The cell is loaded with dih-MEQ* and imaged under conditions with 80 MHz and below 4mW laser power at 747 nm using 40x/1.3NA F-Fluar Zeiss objective. The lifetime images are from the modulation measurements. The scale bar is 20 micron.

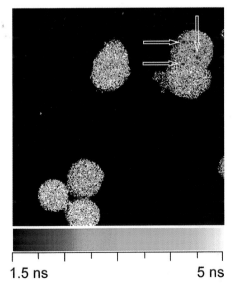

Fig. 7.17. Lifetime image of cellular chloride concentration from the phase measurement. The above image displays τ_{phase} value of the same measurement shown in Fig. 7.16B'. Very low lifetime features are indicated by white arrows. The lifetime value for these spots is between 2.0–2.5 ns, which is likely due to NADH (NAD$^+$) autofluorescence in the mitochondria. The fact that τ_{phase} measures the low lifetime components is illustrated in Fig. 7.6B, where the phase lifetime stays low at high modulation even with less than 30% of total intensity for the component with lower lifetime value.

Color Plates

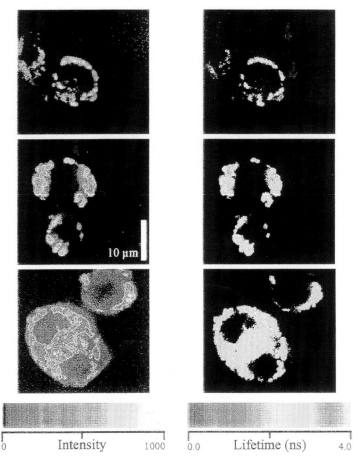

Fig. 9.4. Two-photon imaging of macrophage cell line J774 incubated with FITC–BSA for various periods of time. Macrophage cell line J774 was incubated with FITC–BSA, and the fluorescence intensity (left panel) and fluorescence lifetime (right panel) were monitored at 1h, 5h, and 24h using two-photon fluorescence microscopy with time-resolved fluorescence lifetime imaging. As the probe was degraded within the cell, noticeable increases in fluorescence lifetime and intensity were detected. (Copyright Royal Microscopical Society, 1997. Reprinted with permission.)

Color Plates

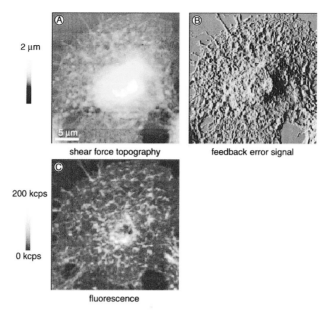

Fig. 12.3. Shear force topography (A), feedback error signal (B), and fluorescence (C) images of CHO cells expressing a fusion construct of the EGFR and GFP; λ_{ex} = 488 nm, 160 nW.

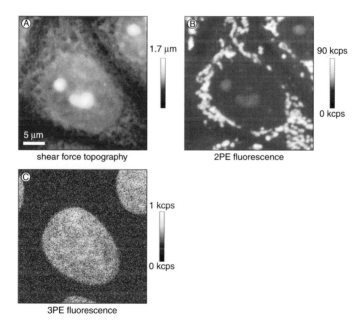

Fig. 12.7. 2PE and 3PE SNOM images of MCF7 cells. (A) Shear force topography, (B) 2PE fluorescence signal of the MitoTracker Orange-labeled mitochondria, and (C) 3PE fluorescence of the BBI-342-labeled nucleus. Scan parameters: 10 s/line, 128 lines, 256 points/line, excitation: 51 mW at 1064 nm.

Color Plates

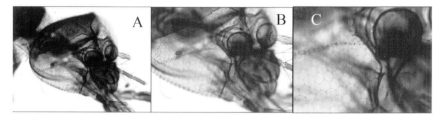

Fig. 13.4. Example of a zoomed image. Here, a mosquito head is imaged using the microscopes 10x objective with no zoom (A), zoom at 1.5x (B), and zoom at 3x (C).

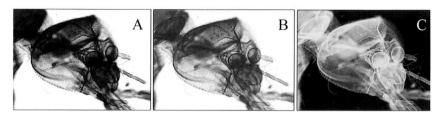

Fig. 13.6. Image enhancement features of COSMIC. The normal-mode image of Figure 4 (A), the image in contrast suppression mode (B), and the image in color-inverted mode (C). This mode is particularly useful to identify thin structures in cells where the color inversion highlights the objects.

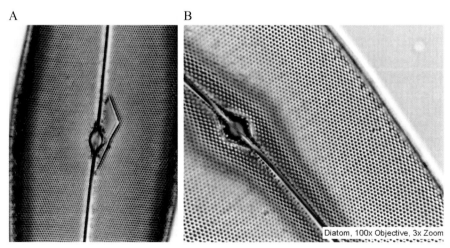

Fig. 13.8. Superresolution effect is demonstrated with an image of the diatom *Pleurasigma angulatum*. The holes in the diatoms are approximately 0.25 μm in diameter and hexagonal in shape. The full-color images were taken in white light. Obtaining such high-resolution images is very difficult with most other systems. Figure 13.8 (A) was taken with a 40x (0.75NA) objective. Figure 13.8 (B) was taken with a 100x (1.3NA oil) objective. However, the sample is in air under the cover slip, so the maximum usable NA for the 100x objective is less than 1.0. The condensor had an NA of 1.4 (oil).

signal processing becomes increasingly complex (multifrequencies), these electronics are a source of errors. By replacing the analog processing electronics with a computer-controlled digital processing and acquisition system (Feddersen et al., 1989), multifrequency phase/modulation lifetime data are processed by software. This digital acquisition technology is not a simple substitution of analog processing, but it provides a new approach and capabilities for measuring multiple excited-state lifetimes by frequency-domain methods.

Instrumentation

A Spectra Physics model 2025 5W argon laser (Spectra Physics, Mountain View, CA) serves as the excitation source and the modulator is a model 350 KDP dc-50 MHz bandwidth unit (Conoptics, Danbury, CT) (see Fig. 8.3) in the single-frequency, fluorescence lifetime flow cytometer, which has been described in detail (Steinkamp et al., 1993). A model 3335A RF signal synthesizer (Hewlett Packard, Palo Alto, CA) is used as the sine wave generator for the modulator drive electronics and as the reference frequency source for the homodyne signal. The laser beam is focused by crossed cylindrical lenses onto the cell stream in a Coulter Biosense flow chamber. The two-color fluorescence detector is a modified version of a multicolor unit (Steinkamp et al., 1987). An $f/0.95$ [0.53 numerical aperture (NA)] lens collects and collimates the modulated fluorescence emission to a second lens that focuses the emitted light onto a pinhole spatial filter located in front of a PMT. A Burle Industries model 4526 PMT/Comlinear model 401 high-speed operational amplifier configured in the transimpedance mode serves as the photodetector.

Forward-scattered light from cells is focused by a lens onto the photocathode of a PMT detector (see Fig. 8.3). The light-scatter signals are demodulated using a low-pass electronic filter set at 180 kHz (bandwidth) to obtain the cw-excited laser excitation light-scatter signal. The conventional cw-excited fluorescence signals are obtained by using a low-pass electronic filter set at 180 kHz, followed by an amplifier/integrator or logarithmic amplifier (see Fig. 8.4A). The phase-sensitive detection electronics (single-channel) consist of a band-pass electrical filter having a center frequency corresponding to the laser modulation frequency (ω), a passive phase detector (multiplier), and a low-pass electronic filter set at 180 kHz (see Fig. 8.4B). Switchable delay lines are used to shift the phase of the reference signal with respect to the modulated fluorescence signals. The PSD output signals are amplified or integrated. Logarithmic amplifiers may also be employed to amplify phase-resolved fluorescence signals. The two-phase ratio detector for making fluorescence lifetime measurements is shown in Figure 8.4C. A signal splitter is used to divide the fluorescence detector output signal to two PSD circuits. A 2–32-MHz quadrature phase hybrid module supplies two reference signals which are 90° out of phase with each other to the PSD circuits for generating outputs $v_{(\phi-90)}(t)$ and $v_{(\phi)}(t)$, the ratio that is directly proportional to the lifetime. The conventional light scatter and fluorescence, phase-resolved fluorescence, and lifetime signals are recorded as listmode data for display as frequency distribution histograms or as bivariate dot and contour diagrams using our LACEL computer-based data acquisition system (Hiebert et al., 1981). This system recently has been replaced by DiDAC, a new generation data acquisition system (Buican et al., 1991; Parson et al., 1993).

The longest lifetime that can presently be used in phase-resolved fluorescence measurements depends on the lowest usable excitation frequency, which is about 0.5 MHz. This corresponds to a 318-ns lifetime (45° phase shift). For lower modulation frequencies, the cw-excited, low-frequency signal component interferes with the 0.5-MHz high-frequency signal. The shortest lifetime depends on the maximum usable excitation frequency, which is about 35 MHz (bandwidth of the switchable delay lines) and corresponds to 0.5–1.0 ns. The longest and shortest measurable lifetimes depend on the usable frequency range of the phase comparators. Presently, the limiting component is the 2–32-MHz quadrature phase hybrid module. Higher frequency models are commercially available.

Cell Preparation, Staining, and Lifetime Measurement Procedure

Cultured human lung fibroblasts (HLFs) (primary cell line HFL-1; ATCC, Manassas, VA) were harvested before confluence from tissue culture flasks, washed in phosphate-buffered saline (PBS), and fixed in absolute methanol, absolute ethanol, 1.0% formaldehyde, 1.0% paraformaldehyde, and 0.1% glutaraldehyde for analysis of autofluorescence lifetime using 488-nm laser excitation, a 29-MHz modulation frequency, and a Schott OG515 long-pass (Melles Griot; Irvine, CA)/530 band-pass (Omega Optical, Brattleboro, VT) filter combination in the fluorescence detector. Viable (unfixed) HLFs (controls) were also analyzed under the same conditions. Murine thymocytes were obtained from deeply anesthetized C3H/HEJ mice after intraperitoneal injection of pentobarbital followed by excision of the thymus. Each thymus was washed in PBS; the thymuses were pooled and minced into small pieces and passed (syringed) 4–5 times through an 18-gauge needle to disperse cells. Pooled samples were then filtered through a 100-μm nylon mesh filter to remove large debris and clumps. The cells were centrifuged and the pellet was suspended in 1.0% bovine serum albumin (BSA), washed in PBS, and resuspended in 50 μl of BSA for antibody labeling.

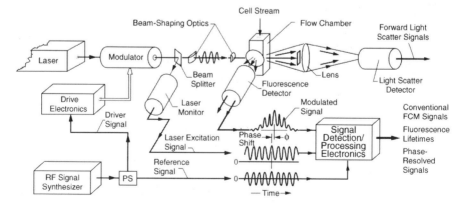

Fig. 8.3. System block diagram of the fluorescence lifetime flow cytometer illustrating the laser, modulator/drive electronics, beam splitter, laser monitor, flow chamber, forward light scatter and fluorescence detectors, synthesized sine wave signal generator, and signal detection/processing electronics.

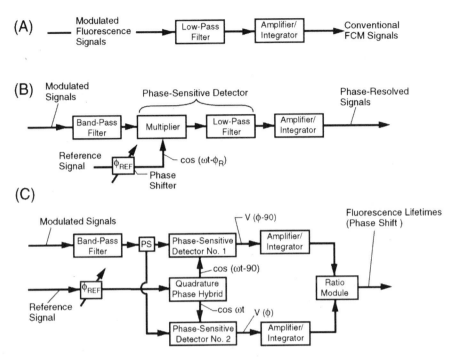

Fig. 8.4. Block diagram of the signal-processing electronics for obtaining conventional FCM signals by low-pass filtering (A); phase-resolved signals by phase-sensitive detection (B); and fluorescence lifetimes by the two-phase ratio method (C).

Murine thymus cells were labeled with anti-mouse Thy 1.2 antibody [anti(α)Thy 1.2] directly conjugated to FITC (FITC–αThy 1.2); PE (PE–αThy 1.2); Red 613, a PE/Texas Red tandem conjugate (Glazer and Stryer, 1983) (Red 613-αThy 1.2) (Gibco BRL, Grand Island, NY); and Quantum Red, a PE/Cy 5 tandem conjugate (Quantum Red-αThy 1.2) (Sigma Chemical, St. Louis, MO) and incubated for 30 min on ice. The labeled thymus cells were centrifuged, washed in PBS containing 1.0% BSA, and resuspended in PBS for flow analysis using 488-nm laser excitation and a 29-MHz modulation frequency. An OG515 filter was used to measure FITC- and PE-labeled antibodies, and Schott RG610 and RG645 long-pass filters (Melles Griot, Irvine, CA) were used to measure Red 613– and Quantum Red–labeled antibodies, respectively. Fluorescence quenching of FITC–αThy 1.2 antibody labeling of murine thymus cells at F/P [fluorescence-to-protein (antibody)] ratios of 4.5, 8.7, and 13.7 (Pharmingen, San Diego, CA) was demonstrated with murine thymocytes labeled at an FITC–αThy 1.2 antibody dilution of 1 : 100 using 488-nm laser excitation, a 29-MHz modulation frequency, and an OG515/530 filter. Thymocytes were also labeled with FITC–αThy 1.2 antibody (4.5 F/P ratio) at a 1 : 1000 dilution.

Chinese hamster cells (line CHO; ATCC, Manassas, VA) maintained in exponential growth in suspension culture were harvested, fixed in 70% ethanol (1 h), and centrifuged, and the ethanol was removed by aspiration prior to staining with DNA-binding fluorochromes. Ethanol-fixed CHO cells were stained with Hoechst 33342 (0.5 µg/ml; Calbiochem, La Jolla, CA), 4',6-diamidino-2-phenylindole (DAPI) (1.0 µg/ml; Molecular Probes, Eugene, OR), mithramycin (50.0 µg/ml; Pfizer Laboratories, Groton, CT) and 7-amino actinomycin D (7-AAD) (12.7 µg/ml; Molecular Probes). Hoechst- and DAPI-stained cells were analyzed using UV laser excitation, a 29-MHz modulation frequency, and a Schott GG400 long-pass filter. Mithramycin- and 7-AAD-stained cells were analyzed using 457-nm laser excitation, a 29-MHz modulation frequency, and a Schott GG495 long-pass filter and 514-nm laser excitation, a 29-MHz modulation frequency, and a Schott OG550 long-pass filter, respectively. Nuclei isolated from CHO cells (Vindelov et al., 1983) were stained with TOTO (5.2 µg/ml) and ethidium homodimer II (6.5 µg/ml; Molecular Probes) and analyzed using 496-nm laser excitation, a 29-MHz modulation frequency, and an OG515/530 filter combination and 488-nm laser excitation, a 6-MHz modulation frequency, and an OG515 filter, respectively. Fixed CHO cells treated with 50.0 µg/ml RNase (Worthington, Freehold, NJ) for 30 min at 37°C were stained with propidium iodide (15.0 µg/ml), ethidium monoazide (5.0 µg/ml), and ethidium bromide (15.0 µg/ml; Molecular Probes). Propidium iodide– and ethidium bromide–stained cells were analyzed using 488-nm laser excitation, a 6-MHz modulation frequency, and an OG515 filter, and ethidium monoazide–stained cells were analyzed using 488-nm laser excitation, a 10-MHz modulation frequency, and an OG515 filter. Chromosomes obtained from line clone Chinese hamster embryo (CCHE) cells [American Type Culture Collection (ATCC), Manassas, VA] growing in culture were prepared and stained with propidium iodide (50.0 µg/ml) (Aten et al., 1980) and analyzed using 488-nm laser excitation, a 6-MHz modulation frequency, and an OG515 filter.

In a final set of experiments, mouse thymus cells were labeled with anti-mouse Thy 1.2 antibody [anti(α)Thy 1.2] directly conjugated to Red 613 as described above. Sample tube 1 (first control) contained only Red 613–αThy 1.2 labeled thymocytes, sample tube 2 (second control) contained unlabeled thymocytes to which PI was added for a final concentration of 1 µg/ml, and sample tube 3 contained Red 613–αThy 1.2–labeled thymocytes to which PI was added for a final concentration of 1 µg/ml. All samples were kept on ice prior to analysis by flow cytometry using 488-nm laser excitation, a 10-MHz modulation frequency, and an RG610 filter. A fourth sample tube containing Red 613–αThy 1.2–labeled thymocytes and PI (1 µg/ml) was placed in a 37°C, 5% CO_2 incubator for 6 h prior to flow analysis.

The ability to quantify fluorescence decay times on particles and cells labeled with fluorescent probes by direct phase shift measurement in flow cytometry is illustrated in Figure 8.4C. The outputs of PSDs 1 and 2, that is, sin ϕ and cos ϕ output, were initialized by removing the long-pass barrier filter in the fluorescence detector, adjusting the reference phase shift (ϕ_R) to zero (null), and maximizing the sin ϕ and cos ϕ outputs, respectively, using nonfluorescent microspheres. The barrier filter was then replaced, and fluorescent-labeled particles of known lifetime (Deka et al., 1994), for example, Flow-Check fluorospheres (Coulter, Miami, FL), lifetime ~7.0 ns, were analyzed at the same PMT and PSD amplifier/integrator gain settings. The ratio module gain was adjusted to position the Flow-Check microspheres, typically in channel

70 of the histogram, prior to analyzing labeled cell samples at fixed gain settings. Neutral density filters were used in the fluorescence detector to compensate for differences in light scatter (nonfluorescent particles) and fluorescence (DNA Check microspheres and labeled cells) signal intensities when required to maintain the PMT voltage constant during lifetime measurement.

RESULTS

The results of experiments designed to compare the autofluorescence lifetimes of viable and fixed HLFs are shown in Figure 8.5A. The histograms show that the lifetimes of ethanol-, methanol-, paraformaldehyde-, and formaldehyde-fixed HLFs are slightly increased compared to unfixed HLFs, the exception being glutaraldehyde-fixed HLFs. The problem this presents is that the broadened lifetime histogram coefficients of variation (CVs) of viable and ethanol-, methanol-, paraformaldehyde-, and formaldehyde-fixed fixed cells partially overlap the lifetimes of the FITC-labeled antibody probe, thus making phase-resolved measurements based on lifetime differences difficult to accomplish. To alleviate this problem, we have de-

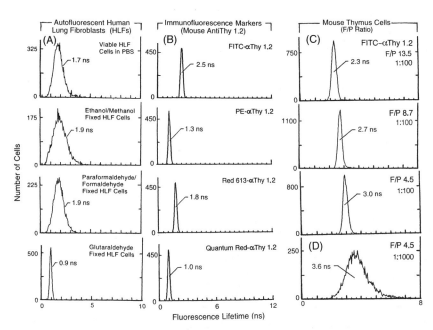

Fig. 8.5. Fluorescence lifetime frequency distribution histograms recorded on cultured viable human lung fibroblasts (HLFs), ethanol/methanol-fixed HLFs, paraformaldehyde/formaldehyde-fixed HLFs, and glutaraldehyde-fixed HLFs (A) and on murine thymus cells labeled with anti(α)Thy 1.2 antibody conjugated to FITC, PE, Red 613, and Quantum Red (B), FITC–αThy 1.2 having F/P ratios 13.5, 8.7, and 4.5 (all at 1 : 100 antibody dilution) (C), and FITC–αThy 1.2 having an F/P ratio of 4.5 (1 : 1000 antibody dilution) (D). Since the individually measured lifetime distributions for ethanol and methanol (and paraformaldehyde and formaldehyde) were identical, only one of the histograms for each pair is shown in (A).

veloped an approach in which low-concentration glutaraldehyde is used as the primary cell fixative for cell surface immunofluorescence measurements (Steinkamp et al., 1999b). When compared to viable HLFs and HLFs fixed in the other fixatives, the fluorescence lifetime is considerably shorter, and the histogram CV is smaller. We have also measured the lifetimes of FITC-labeled microspheres (simulated FITC-labeled antibody fluorescence) having fluorescence intensities similar to the glutaraldehyde-fixed HLF cells and determined that the lifetime histogram does not overlap the simulated FITC probe histogram (data not shown). The use of glutaraldehyde as a fixative to lower the lifetime of autofluorescent cells may facilitate the detection of FITC-labeled cell surface fluorescence by phase-resolved fluorescence emission measurements.

The measurement of fluorescence lifetimes on murine thymus cell surface markers labeled with FITC, PE, Red 613, and Quantum Red conjugated to anti(α)–Thy 1.2 antibody is illustrated in Figure 8.5B. The modal channel values of the lifetime histograms are less than the expected values listed in Table 1 (antibody solution measurements). Since these data suggest fluorescence quenching, studies using microspheres labeled with varying numbers of FITC molecules were performed

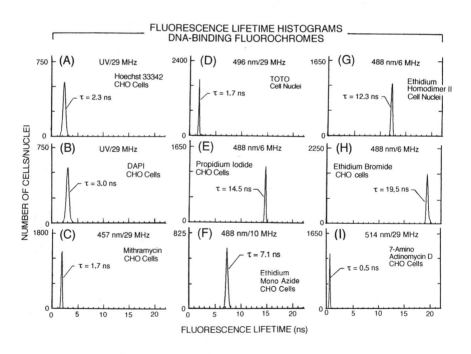

Fig. 8.6. Fluorescence lifetime frequency distribution histograms recorded on CHO cells and isolated CHO cell nuclei that were stained with DNA-binding fluorochromes: Hoechst 33342 (A), DAPI (B), mithramycin (C), TOTO (D), propidium iodide (E), ethidium mono azide (F), ethidium homodimer II (G), ethidium bromide (H), and 7-amino actinomycin D (I).

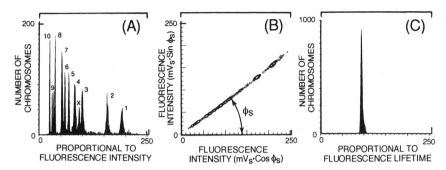

Fig. 8.7. Fluorescence intensity (amplified/integrated) frequency distribution histogram recorded on PI-stained chromosomes obtained from line CCHE cells (A) and corresponding bivariate contour diagram of $mV_s \sin \phi$ and $mV_s \cos \phi$ fluorescence signal intensities (B) and fluorescence lifetime histogram (C).

(Steinkamp et al., 1996). Lifetime values ranged from 2.0 to 3.8 ns, thus illustrating the self-quenching of FITC bound to microspheres. To demonstrate the FITC self-quenching phenomenon as a function of antibody labeling dilution and F/P ratio, fluorescence lifetime histograms were recorded on murine thymus cells labeled with FITC–αThy 1.2 having F/P ratios of 13.7 (maximum quenching), 8.7, and 4.5 (minimum quenching) (Deka et al., 1996) (see Fig. 8.5C). In Figure 8.5D the FITC–αThy 1.2 (F/P 4.5) antibody was diluted by 1 : 1000. The lifetime distribution peak modal value increased to 3.6 ns and was in close agreement with the value listed in Table 1, but with a broader CV due to lower fluorescence intensity (photon measurement statistics).

Figure 8.6 shows the fluorescence lifetime histograms measured on fluorochromes having DNA labeling specificity. The modal channel lifetime values of the histograms recorded for Hoechst 33342 and DAPI (A–T base pair specificity) and mithramycin (G–C base pair specificity) were slightly shorter (see Figs. 8.6A–C) than the reported values listed in Table 1. The lifetime histogram for TOTO (high-sensitivity dimer of thiazole orange) is shown in Figure 8.6D. TOTO is thought to intercalate into DNA without base-pair specificity. The modal channel value of 1.7 ns is the same as the lifetime value previously reported by time-resolved solution measurements (Castro et al., 1993). The lifetime histograms for the propidium iodide, ethidium monoazide, ethidium homodimer II, and ethidium bromide DNA–intercalating fluorochromes, which lack base-pair specificity, are shown in Figures 8.6E–H. Ethidium homodimer II is thought to exhibit A–T base-pair labeling preference. Figure 8.6I shows the lifetime histogram measured for 7-AAD-labeled cells having G–C base-pair preference. The lifetime values measured for propidium iodide, ethidium bromide, and 7-AAD are essentially the same as those given in Table 1. These data illustrate the capability to precisely measure lifetimes of DNA-binding probes and will serve as the basis for resolving fluorescent marker emissions that are to be quantified in combination with these fluorochromes. Sailer et al. have further characterized the fluorescence intensities and lifetimes of DNA-binding fluorochromes for labeling normal and apoptotic cells (Sailer et al., 1996, 1997a, 1998a, 1998b).

The ability to simultaneously measure fluorescence intensity and lifetime on PI-labeled chromosomes (from line CCHE cells) is illustrated in Figure 8.7. The flow karyotype fluorescence intensity histogram is shown in Figure 8.7A. Excellent measurement resolution was achieved with each chromosome peak defined. In Figure 8.7B, $mV_m\sin\phi$ and $mV_m\cos\phi$ [corresponding to $V(\phi - 90)$ and $V(\phi)$ of Fig. 8.4C] are plotted in bivariate form to show the lifetime measurement linearity. The slope of this line is equal to $\tan\phi$, which is directly proportional to the lifetime, and the corresponding lifetime distribution is unimodal (see Fig. 8.7C). These data demonstrate the measurement precision of the technology and will serve as the basis for future studies involving human chromosomes.

An example illustrating phase-resolved measurements on murine thymus cells labeled with Red 613–αThy 1.2 and stained with PI ("dead cells") (Stöhr and Vogt-Schaden, 1980) using two PSD channels operating in parallel (Steinkamp and Crissman, 1993) is shown below. Since the emission spectra of Red 613 and PI overlap (see Fig. 8.8A), the separation of Red 613 and PI signals cannot be achieved using conventional methods employing electronic compensation (Loken et al., 1977). Also, the PI labeling intensity is 42.5 times greater than the Red 613–αThy 1.2 fluorescence. The measured fluorescence lifetime histograms of Red 613 and PI individually labeled murine thymus cells are well separated (see Fig. 8.8B). Based on the lifetimes of Red 613 (2.0 ns) and PI (16.8 ns), murine thymus cells were labeled separately with Red 613–αThy 1.2 and PI (controls), and the phase shifts of the two phase-sensitive detector channels (see Fig. 8.9A) were adjusted to (1) null Red 613 signals in the PSD channel 2 output and (2) null PI signals in the PSD channel 1 output, as illustrated in Figure 8.9B. Thymus cells labeled with both Red 613–αThy 1.2 and PI

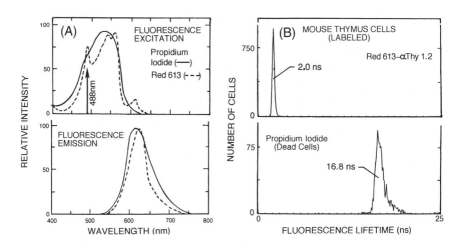

Fig. 8.8. Fluorescence excitation and emission spectra of PI and Red 613 (A) and fluorescence lifetime histograms recorded on unfixed murine thymus cells labeled separately with Red 613–αThy 1.2 and PI ("dead cells") (B).

Results

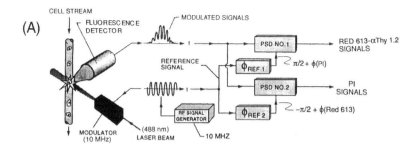

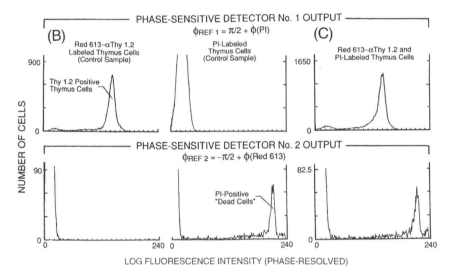

Fig. 8.9. Log phase-resolved fluorescence analysis of Red 613 and PI signals from murine thymus cells labeled with Red 613–αThy 1.2 and PI ("dead cells") by two PSD channels operating in parallel (A). In (B), the reference phase shift (ϕ_{REF1}) of PSD channel 1 was first adjusted [$\phi_{REF1} = \pi/2 + \phi(PI)$] to null signals from thymus control cells stained only with PI and to pass cells labeled with Red 613–αThy 1.2. The reference phase (ϕ_{REF2}) of PSD channel 2 was next adjusted [$\phi_{REF2} = -\pi/2 + \phi(Red\ 613)$] to null signals from thymus control cells labeled only with Red 613–αThy 1.2 and to pass cells stained with PI. In (C), cells labeled with Red 613–αThy 1.2 and stained with PI were analyzed, and the PSDs 1 and 2 signal output histograms were recorded.

were then analyzed at the same reference phase shift and gain settings, and the phase-resolved histograms (see Fig. 8.9C) and corresponding bivariate contour diagram (see Fig. 8.10A) were recorded. Approximately 96.5% of the total thymus cell population was labeled with Red 613–αThy 1.2 and the remaining 3.5% were PI positive. Essentially none of the Red 613–labeled thymus cells were PI positive when analyzed within a half hour after staining (at 4°C on ice) with PI (see Fig. 8.10A). In Figure 8.10B, thymus cells were labeled with Red 613–αThy 1.2 on ice, placed in PBS at 37°C for 6 with PI, and then analyzed. This resulted in a greater portion of Red 613-labeled cells initially taking up PI, followed by a loss in Red 613–αThy 1.2 labeling

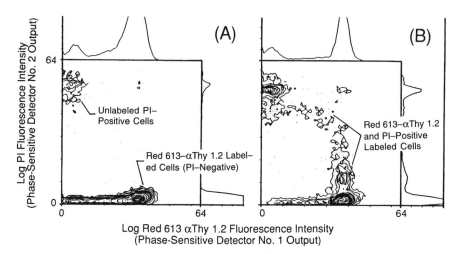

Fig. 8.10. Bivariate contour diagrams of murine thymus cells labeled with Red 613–αThy 1.2 and PI (from Fig. 8.9C) (A) and on perturbed cells labeled with Red 613–αThy 1.2 and placed in PBS at 37° with PI (6 h) (B).

intensity of the maximally PI positive cells. These results demonstrate the ability to resolve highly overlapping fluorescence emissions, differing in intensity by greater than 40 times, based on differences in lifetimes.

SUMMARY

Fluorescence lifetime flow cytometry is so new that many of its potential applications have not been fully explored or developed. This technology will add a new dimension to multiparameter flow cytometric analyses through the development of techniques for rapidly measuring fluorescence lifetime of probes bound to macromolecular complexes in cells, increase the range of fluorescent markers that can be used in multilabeling applications, reduce background interferences, and thus enhance measurement precision to yield more accurate results. In the past, procedures were limited in some cases by the availability of fluorescent markers with common excitation regions (so that a single laser excitation source could be used) and emission spectra that were sufficiently separated using optical color-separating filters. Because the lifetime-based sensing technology can separate fluorescence emissions electronically (and also optically), quantify fluorescence lifetimes directly, and make conventional flow cytometric measurements, it has a wide range of technically possible applications. The technology will significantly expand researchers' understanding of biological processes at the cellular, subcellular, and molecular level, and through clinical and biomedical research, we envision that it will contribute to improving diagnoses, treatment, and understanding the underlying mechanisms of human diseases. In addition, it can readily be added to existing commercial flow cytometry systems, where it can be used for virtually any clinical or research application involving the analysis of cells, cell function, or subcellular components through the use of fluorescent markers directed to specific targets.

ACKNOWLEDGMENTS

This work was performed at the Los Alamos National Laboratory, Los Alamos, New Mexico, under the sponsorship of the U.S. Department of Energy, the Los Alamos National Flow Cytometry Resource (National Institutes of Health Grant P41-RR013150), and National Institutes of Health Grant R01-RR07855. I thank Nancy M. Lehnert, Carolyn Bell-Prince, and Harry A. Crissman for their assistance in cell preparation and staining.

REFERENCES

Alcala JR, Gratton E, Jameson DM (1985): A multifrequency phase fluorometer using the harmonic content of a mode-locked laser. Anal Instrum 14:225–250.

Aten JA, Kipp JBA, Barendsen GW (1980): Flow cytofluorometric determination of damage to chromosomes from X-irradiated Chinese hamster cells. In Laerum OD, Lindmo T, Thorud E (eds). *Flow Cytometry IV. Proceedings of the Fourth International Symposium on Flow Cytometsry*. Oslo: Universitetsforlaget, pp 485–491.

Beavis AJ, Pennline KJ (1994): Simultaneous measurement of five-cell surface antigens by five-color immunofluorescence. Cytometry 15:371–376.

Beisker W, Klocke A (1997): Fluorescence lifetime measurement in flow cytometry. Proc SPIE 2982:436–446.

Berndt KW, Gryczynski I, Lakowicz JR (1990): Phase-modulation fluorometry using a frequency doubled pulsed laser diode light source. Rev Sci Instrum 61:1816–1820.

Blair DP, Sydenham PH (1975): Phase sensitive detection as a means to recover signals buried in noise. J Phys E Sci Instrum 8:621–627.

Buican TN, Habbersett RC, Martin JC, Naivar MA, Parson JD, Wilder ME, Jett JH (1991): A new FCM data acquisition and sorting system. Cytometry Suppl. 5:136.

Buican TN (1990): Real-time Fourier transform spectroscopy for fluorescence imaging and flow cytometry. Proc SPIE 1205:126–133.

Castro A, Fairfield FR, Shera EB (1993): Fluorescence detection and size measurement of single DNA molecules. Anal Chem 65:849–852.

Crissman HA, Steinkamp JA (1982): Rapid, one step staining procedures for analysis of cellular DNA and protein by single and dual laser flow cytometry. Cytometry 3:84–90.

Deka C, Steinkamp JA (1996): Time-resolved fluorescence-decay measurement and analysis on single cells by flow cytometry. Appl Opt 35:4481–4489.

Deka C, Lehnert BE, Lehnert NM, Jones GM, Sklar LA, Steinkamp JA (1996): Analysis of fluorescence lifetime and quenching of FITC-conjugated antibodies on cells by phase-sensitive flow cytometry. Cytometry 25:271–279.

Deka C, Cram LA, Habbersett R, Martin JC, Sklar LA, Steinkamp JA (1995): Simultaneous dual-frequency phase-sensitive flow cytometric measurements for rapid identification of heterogeneous fluorescence decays in fluorochrome-labeled cells and particles. Cytometry 21:318–328.

Deka C, Sklar LA, Steinkamp JA (1994): Fluorescence lifetime measurements in a flow cytometer by amplitude demodulation using digital data acquisition technique. Cytometry 17:94–101.

Demas JN (1983): *Excited State Lifetime Measurements*. New York: Academic Press.

Durack G, Yu W, Mantulin W, Gratton E (1998): Fluorescence lifetime flow cytometry: A phase-sensitive detection system utilizing a cross correlation technique and digital signal processing. Cytometry Suppl. 9:39.

Feddersen BA, Piston DW, Gratton E (1989): Digital parallel acquisition in frequency domain fluorometry. Rev Sci Instrum 60:2929–2936.

Gadella J, Van Hock A, Visser A (1997): Construction and characterization of a frequency-domain fluorescence lifetime imaging microscopy system. J Fluores 7:35–43.

Gadella TWJ, Jovin TM, Clegg RM (1993): Fluorescence lifetime imaging microscopy (FLIM): Spatial resolution of microstructures on the nanosecond time scale. Biophys Chem 48:221–239.

Glazer AH, Stryer L (1983): Fluorescent tandem phycobiliprotein conjugates. J Biophys 43:383–386.

Gratton E, Jameson DM, Rosato N, Weber G (1984): Multifrequency cross-correlation phase fluorometer using synchrotron radiation. Rev Sci Instrum 55:486–494.

Gratton E, Limkeman M (1983): A continuously variable frequency cross-correlation phase fluorometer with picosecond resolution. Biophys J 44:315–324.

Hiebert RD, Jett JH, Salzman GC (1981): Modular electronics for flow cytometry and sorting: The LACEL system. Cytometry 1:337–341.

Jameson DM, Gratton E, Hall RD (1984): The measurement and analysis of heterogeneous emissions by multifrequency phase and modulation fluorometry. Appl Spectros Rev 20:55–106.

Keating SM, Wensel TG (1991): Nanosecond fluorescence microscopy: Emission kinetics of Fura-2 in single cells. Biophys J 59:1186–202.

Keij JF, Bell-Prince C, Steinkamp JA (1999): Simultaneous analysis of relative DNA and glutathione content in viable cells using phase-sensitive flow cytometry. Cytometry 35:48–54.

Lakowicz JR (1983): *Principles of Fluorescence Spectroscopy*. New York: Plenum Publishing Corp.

Lakowicz JR, Szmacinski H, Nowaczy KK, Berndt KW, Johnson L (1992): Fluorescence lifetime imaging. Anal Biochem 202:316–330.

Lakowicz JR, Laczko G, Gryczynski I (1986): Two-GHz frequency-domain fluorometer. Rev Sci Instrum 57:2499–2506.

Lakowicz JR, Maliwal BP (1985): Construction and performance of a variable-frequency phase-modulation fluorometer. Biophys Chem 21:61–78.

Lakowicz JR, Laczko G, Cherek H, Gratton E, Limkeman M (1984): Analysis of fluorescence decay kinetics from variable-frequency phase shift and modulation data. Biophys J 46:463–477.

Lakowicz JR, Cherek H (1981): Resolution of heterogeneous fluorescence from proteins and aromatic amino acids by phase-sensitive detection of fluorescence. J Biol Chem 256:6348–6353.

Loken MR, Parks DR, Herzenberg LA (1977): Two-color immunofluorescence using a fluorescence activated cell-sorter. J Histochem Cytochem 25:899–907.

Meade ML (1982): Advances in lock-in amplifiers. J Phys E Sci Instrum 15:395–403.

Mitchell G, Swift K (1989): The 48000 MHF, A dual-domain fourier transform fluorescence lifetime spectrofluorometer. Proc SPIE 1204:270–274.

Parson JD, Olivier TL, Habbersett RC, Martin JC, Wilder ME, Jett JH (1993): Characterization of digital signal processing in the DiDAC data acquisition system. Cytometry Suppl. 6:40.

Pinsky BG, Ladasky JJ, Lakowicz JR, Berndt K, Hoffman RA (1993): Phase-resolved fluorescence lifetime measurements for flow cytometry. Cytometry 14:123–135.

Roederer M, De Rosa S, Gerstein R, Anderson M, Bigos M, Stoval R, Nozaki T, Parks D, Herzenberg L, Herzenberg, L (1997): 8 color, 10-parameter flow cytometry to elucidate complex leukocyte heterogeneity. Cytometry 29:328–339.

Sailer BL, Steinkamp JA, Crissman HA (1998a): Flow cytometric fluorescence lifetime analysis of DNA-binding probes. Eur J Histochem 42:19–28.

Sailer BL, Valdez JG, Steinkamp JA, Crissman HA (1998b): Apoptosis induced with different cycle perturbing agents produces differential changes in the fluorescence lifetime of DNA-bound ethidium bromide. Cytometry 31:208–216.

Sailer BL, Nastasi AJ, Valdez JG, Steinkamp JA, Crissman HA (1997a): Differential effects of deuterium oxide on the fluorescence lifetimes and intensities on dyes with different modes of binding to DNA. J Histochem Cytochem 45:165–175.

Sailer BL, Valdez JG, Steinkamp JA, Darzynkiewicz Z, Crissman HA (1997b): Monitoring uptake of ellipticine and its fluorescence lifetime in relation to the cell cycle by flow cytometry. Exp Cell Res 236:259–267.

Sailer BL, Nastasi AJ, Valdez JG, Steinkamp JA, Crissman HA (1996): Interactions of intercalating fluorochromes with DNA analyzed by conventional and fluorescence lifetime flow cytometry utilizing deuterium oxide. Cytometry 25:164–172.

So PTC, French T, Yu WM, Berland KM, Dong CY, Gratton E (1995): Time-resolved fluorescence microscopy using two-photon excitation. Bioimaging 3:49–63.

So PTC, French T, Gratton (1994): A frequency domain time-resolved microscope using a fast-scan CCD camera. Proc SPIE 2137:83–92.

Spencer RD, Weber G (1969): Measurements of subnanosecond fluorescence lifetimes with a cross-correlation phase fluorometer. Ann NY Acad Sci 158:361–376.

Steinkamp JA, Keij JF (1999): Fluorescence intensity and lifetime measurement of free and particle-bound fluorophore in a sample stream by phase-sensitive flow cytometry. Rev Sci Instrum 70:4682–4688.

Steinkamp JA, Lehnert BE, Lehnert NM (1999a): Discrimination of damaged/dead cells by propidium iodide uptake in immunofluorescently labeled populations analyzed by phase-sensitive flow cytometry. J Immunol Meth 226:59-70.

Steinkamp JA, Lehnert NM, Keij JF, Lehnert BE (1999b): Enhanced immunofluorescence measurement resolution of surface antigens on highly autofluorescent, glutaraldehyde-fixed cells analyzed by phase-sensitive flow cytometry. Cytometry 37:275–283.

Steinkamp JA, Parson JD, Keij JF (1998): Progress towards combined phase and amplitude demodulation fluorescence lifetime measurements by flow cytometry. Proc SPIE 3260:236–244.

Steinkamp JA, Lehnert BE, Keij JF (1997): Phase-sensitive detection as a means to recover fluorescence signals from interfering backgrounds in analytical cytology measurements. Proc SPIE 2982: 447–455.

Steinkamp JA, Deka C, Lehnert BE, Crissman HA (1996): Fluorescence lifetime as a new parameter in analytical cytology measurements. Proc. SPIE 2678:221–230.

Steinkamp JA, Crissman HA (1993): Resolution of fluorescence signals from cells labeled with fluorochromes having different lifetimes by phase-sensitive flow cytometry. Cytometry 14:210–216.

Steinkamp JA, Yoshida TM, Martin JC (1993): Flow cytometer for resolving signals from heterogeneous fluorescence emissions and quantifying lifetime in fluorochrome-labeled cells/particles by phase-sensitive detection. Rev Sci Instrum 64:3440–3450.

Steinkamp JA, Habbersett RC, Hiebert RD (1991): Improved multilaser/multiparameter flow cytometer for analysis and sorting of cells and particles. Rev Sci Instrum 62:2751–2764.

Steinkamp JA, Habbersett RC, Stewart CC (1987): A modular detector for flow cytometric multicolor fluorescence measurements. Cytometry 8:353–365.

Stöhr M, Vogt-Schaden M (1980): A new dual staining technique for simultaneous flow cytometric DNA analysis of living and dead cells. In: Laerum OD, Lindmo T, Thorud E (Eds). *Flow Cytomery*, Vol. 4. Bergen, Norway: Universitets-forlaget.

Thompson RB, Frisoli JK, Lakowicz JR (1992): Phase fluorometry using a continuously modulated laser diode. Anal Chem 64:2075–2078.

Veselova TV, Cherkasov AS, Shirokov VI (1970): Fluorometric method for individual recording of spectra in systems containing two types of luminescent centers. Opt Spectrosc 29:617–618.

Vigo J, Salmon JM, Viallet P (1987): Quantitative microfluorometry of isolated living cells with pulsed excitation: Development of an effective and relatively inexpensive instrument. Rev Sci Instrum 58:1432–1438.

Vindelov LL, Christensen IJ, Nissen NI (1983): A detergent-trypsin method for the preparation of nuclei for flow cytometric DNA analysis. Cytometry 3:323–327.

Wilkerson CW, Goodwin PM, Ambrose WP, Martin JC, Keller RA (1993): Detection and lifetime measurement of single molecules in flowing sample streams by laser-induced fluorescence. Appl Phys Lett 62:2030–2032.

Zarrin F, Bornhop DJ, Dovichi NJ (1987): Laser doppler velocimetry for particle size determination by light scatter within the sheath flow cuvette. Anal Chem 59:854–860.

9

Application of Fluorescence Lifetime and Two-Photon Fluorescence Cytometry

Donald J. Weaver, Jr.,* Gary Durack, Edward W. Voss, Jr., and Anu Cherukuri[†]
University of Illinois at Urbana-Champaign, Urbana, Illinois

INTRODUCTION

CD4[+] T-lymphocytes recognize fragments of antigenic proteins in association with major histocompatibility complex (MHC) class II molecules on the surface of cells known as antigen-presenting cells (i.e., macrophages, B cells, and dendritic cells). These antigenic proteins enter the antigen-presenting cell via several processes, including phagocytosis, receptor-mediated endocytosis, and pinocytosis (Lanzavecchia, 1990; Fig. 9.1). Upon internalization, the foreign protein is transported through the endocytic system of the cell (Watts, 1997). As transport occurs, the foreign molecule encounters a variety of biochemical phenomena that serve to denature and degrade the antigen (Germain and Marguiles, 1993; Fig. 9.1). Although alterations in pH and reduction of disulfide bonds facilitate this process, proteolytic cleavage of antigenic proteins into peptides is a fundamental requirement for this pathway (Fineschi and Miller, 1997). Several of the proteolytic enzymes involved in this process have been characterized (Diment, 1993; Fineschi and Miller, 1997). Termed

---

* Current address: Department of Microbiology and Immunology, University of North Carolina-Chapel Hill, Chapel Hill, North Carolina.
[†]Current address: Department of Biochemistry, Northwestern University, Evanston, Illinois.

Emerging Tools for Single-Cell Analysis, Edited by Gary Durack and J. Paul Robinson.
ISBN 0-471-31575-3   Copyright © 2000 Wiley-Liss, Inc.

cathepsins, these enzymes belong to both the cysteine and aspartic families of proteases, and although the precise sequence and site of action of these enzymes have not been clearly elucidated, data suggest that the processing of antigenic proteins by cathepsins is a sequential series of events that results in the formation of 16–21 amino acid fragments capable of interacting with the MHC class II molecule and initiating an immune response to the original protein (Diment, 1993; Fig. 9.1).

As the exogenous protein is degraded, MHC class II molecules are synthesized within the endoplasmic reticulum (ER) of the cell and associate with the nonpolymor-

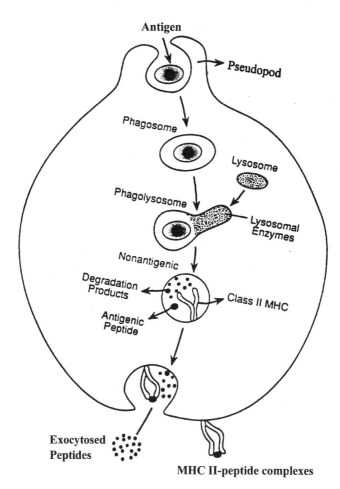

Fig. 9.1. Classical depiction of the antigen-processing pathway in murine macrophage. Antigenic proteins are internalized into the cell and immediately surrounded by an intracellular vacuole. The antigen-containing vacuole proceeds through the endocytic system of the cell. As this occurs, the pH of the vacuole decreases and the disulfide bonds of the protein are reduced. Finally, the protein is degraded into small peptides by the cathepsin family of enzymes. A select few of these peptides are bound by the MHC class II molecule which is also expressed within the endocytic system. Upon peptide binding, the MHC class II-peptide complex is transported to the surface of the cell for recognition by the appropriate T-lymphocyte receptor.

phic invariant chain as nonameric complexes consisting of three MHC class II heterodimers bound to an invariant chain trimer (Cresswell, 1994; Fig. 9.1). Upon exiting the ER, these complexes are transported to the Golgi and trans-Golgi network for carbohydrate modification prior to entering the cell's endocytic system (Germain and Marguiles, 1993; Fig. 9.1). Following carbohydrate modification, these complexes are targeted to the endosomal/lysosomal compartments of the cell via a signal sequence in the cytoplasmic tail of the invariant chain and the β chain of the MHC class II molecule (Watts, 1997). Intersection of MHC II molecules with the endocytic system provides an opportunity for MHC II molecules to sample a vast array of peptides that have been internalized from various extracellular sources (Watts, 1997). However, the invariant chain must be removed from the MHC class II binding groove prior to peptide binding, and several experiments indicate that this process is dependent upon the same proteases that are responsible for cleavage of antigenic proteins as well as the H-2M chaperonin protein (Cresswell, 1994). Once a stable MHC II–peptide complex is formed, the complex is transported to the plasma membrane for T-cell recognition (Ward and Qadri, 1997; Fig. 9.1). Following stimulation, the activated T cell proceeds to secrete interleukin-2 and upregulate several surface molecules such as CD40 ligand. The ultimate result of these events is the differentiation and expansion of antigen-specific $CD4^+$ T-cell clones and subsequent initiation of the immune response.

DEVELOPMENT OF THE PROBLEM

Several questions regarding the intracellular pathways described above remain unanswered. First, little is known about the precise mechanisms that generate antigenic fragments presented by MHC class II molecules (Pieters, 1997). For example, it is unclear if these enzymes act simultaneously or as a sequential cascade (Pieters, 1997). In particular, it is possible that the endopeptidase activity displayed by all cathepsins initiates antigen processing and generates larger fragments containing antigenic epitopes. Subsequently, exopeptidases may be responsible for trimming epitopes to their final size following MHC II binding to ensure preservation of epitopes recognized by T cells. Second, the exact intracellular location for these processing events remains unknown. Numerous research groups have demonstrated that a specialized organelle termed the MHC class II containing compartment (MIIC) is a major component of the processing pathway (Tulp et al., 1994; Qui et al., 1994). However, experiments by Amigorena et al. (1994) demonstrated that a second type of endocytic vesicle termed the CIIV (class II vesicle), which is distinct from the MIIC based on both morphology and marker proteins, is the major compartment for MHC II peptide loading (Amigorena et al., 1994). To reconcile these experiments, several investigators have begun a systematic characterization of the endocytic system of antigen-presenting cells (Kleijmeer et al., 1997). However, since these studies represent a limited number of model antigens and cell types, more rigorous studies are required. Finally, it is imperative that the mechanisms responsible for the transport of these complexes to the surface of the cell be delineated. Initial efforts by Wubbolts et al. (1996) to visualize this process have met with limited success. In short, several issues involved in this pathway remain unresolved. Therefore, a novel experimental approach was required to provide additional insight into this pathway. Since previous experimental strategies relied solely on invasive strategies including electron microscopy and subcellular fractionation, new noninvasive techniques could contribute significantly toward elucidation of this pathway. For this purpose, a novel fluorescent antigenic probe was synthesized and employed.

FLUORESCEIN-DERIVATIZED BOVINE SERUM ALBUMIN AS AN EXOGENOUS ANTIGEN

The probe consists of the small fluorescent hapten fluorescein-5-isothiocyanate (FITC) conjugated to the protein bovine serum albumin (BSA). In the highly derivatized form, fluorescein molecules are relatively nonfluorescent due to their close spatial proximity, orientation, and the short Stoke's shift between absorption and emission of fluorescein (Fig. 9.2). However, if the spatial constraint is alleviated due to unfolding or proteolytic degradation, a significant increase in fluorescence intensity results (Fig. 9.2). Therefore, FITC-derivatized BSA can be used as a multifunctional probe to monitor several intracellular events, including protein unfolding, disulfide bond reduction, and enzymatic cleavage (Voss et al., 1996). Of major concern in this study was the selection of both the protein carrier and fluorophore. Similar substrates were developed by Rothe et al. (1990) in which synthetic substrates for the enzymes cathepsin L and elastase were conjugated with rhodamine 110 (Rothe et al., 1990; Rothe and Valet, 1993a,b). Bowser and Murphy (1990) conjugated a substrate for cathepsin B with 4-methoxy-β-napthylamine, and upon cleavage, the nonfluorescent substrate was converted to a fluorescent product (Bowser and Murphy, 1990). Although similar in principle, these synthetic substrates had several critical differences from the fluorescein–BSA probe mentioned above. BSA is susceptible to cleavage by a variety of enzymes. Therefore, FITC-derivatized BSA can be used as a proteolytic probe for a vast array of enzymes as opposed to a few specific examples. The antigenic and biochemical properties of BSA are well defined, and BSA typically produces a T-cell-dependent response eliciting high-affinity IgG antibodies (Benjamin et al., 1984). Several studies have also demonstrated that T cells cloned from mice of the H-2^d haplotype such as BALB/c were high responders to antigen-presenting cells incubated with BSA (Benjamin et al., 1984). As a result, BSA was an ideal carrier due to its ability to elicit a strong immune response.

The immunogenicity and antigenicity of fluorescein in the context of a carrier protein is well documented (Voss, 1984). Fluorescein typically elicits high-affinity IgG antibodies (Kranz and Voss, 1981; Voss, 1990). In addition, fluorescein possesses several unique and advantageous spectral properties. Specifically, since the fluorescence lifetime of fluorescein is sensitive to alterations in pH, FITC–BSA can be used as a probe to monitor the pH gradient encountered during the endocytic trafficking of proteins with the cell (Fig. 9.3; Ohkuma and Poole, 1978). The fluorescent properties of fluorescein also allow for the measurement of several parameters, including fluorescence polarization, lifetime, and intensity. Therefore, a more complete under-

A. FITC-BSA

B. Non-Reduced

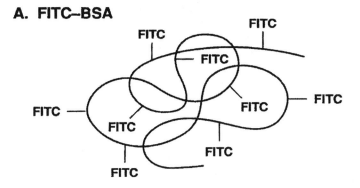

C. Thiol-Reduced

D. Proteolysis

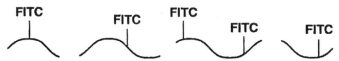

Fig. 9.2. Principle of FITC-derivatized BSA and generation of a kinetic probe. Upon spatial separation of fluorescein molecules via unfolding or proteolytic cleavage of the protein carrier, a significant increase in fluorescence intensity results that can be detected using several fluorescence techniques, including flow cytometry and two-photon fluorescence microscopy.

standing of the intracellular microenvironment can be achieved. Finally, the fluorescent properties of the antigen are amenable to noninvasive methodologies to quantitate and monitor alterations in fluorescence parameters within live cells. Therefore, this novel fluorescent probe provided some of the first real-time kinetic measurements of intracellular uptake and processing within antigen-presenting cells, specifically macrophages.

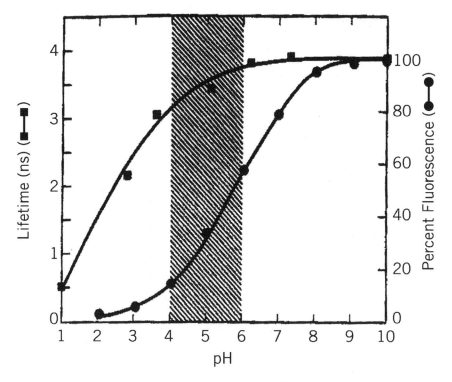

Fig. 9.3. Representation of the pH dependence of fluorescein. As pH decreases, a noticeable decrease in the fluorescence intensity of fluorescein is detected. The shaded region represents the pH range observed within the endocytic system. (Copyright Royal Microscopical Society, 1997. Reprinted with permission.)

TWO-PHOTON FLUORESCENCE MICROSCOPY

Prior to flow cytometry, J774 macrophages were incubated with the antigenic probe, and both fluorescence lifetime and fluorescence intensity of single cells were measured using two-photon fluorescence microscopy with time-resolved lifetime imaging. Two-photon fluorescence microscopy is a technique involving the excitation of a fluorophore using two photons each containing one-half the energy required for transition from the ground state to the excited state (French et al., 1997). Therefore, only regions of high photon density will result in excitation (French et al., 1997). The practical application of this phenomenon is that over 80% of the total fluorescence intensity derives from a 1-μm-thick region about the focal point, allowing for the construction of three-dimensional images similar to those observed in confocal microscopy with several additional advantages (French et al., 1997). Namely, two-photon fluorescence allows for better separation of excitation and emission light while localizing photobleaching and photodamage to submicrometer regions at the focal point (French et al., 1997). Two-photon fluorescence microscopy also provides for generation of three-dimensional images without using confocal pinholes (French et al., 1997). This is especially important in the context of the current application due to the high spatial density of fluorescent endocytic vesicles (French et al., 1997; Weaver et al., 1997). Because of the three-dimensional imaging capabilities of two-photon fluorescence microscopy, out-of-focus endocytic vesicles contributed minimal fluorescence to lifetime images and did not affect experimental results. In short, each of the parameters listed above was critical for the successful application of two-photon fluorescence microscopy in studying antigen processing within viable cells (French et al., 1997; Weaver et al., 1996).

To perform two-photon experiments, macrophage cells were incubated with FITC–BSA for various periods of time at 37°C, and the cells were immediately ob-

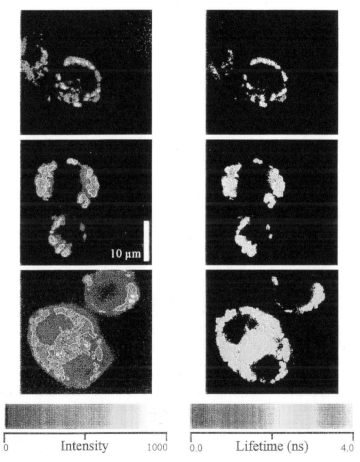

Fig. 9.4. Two-photon imaging of macrophage cell line J774 incubated with FITC–BSA for various periods of time. Macrophage cell line J774 was incubated with FITC–BSA, and the fluorescence intensity (left panel) and fluorescence lifetime (right panel) were monitored at 1h, 5h, and 24h using two-photon fluorescence microscopy with time-resolved fluorescence lifetime imaging. As the probe was degraded within the cell, noticeable increases in fluorescence lifetime and intensity were detected. (Copyright Royal Microscopical Society, 1997. Reprinted with permission.) *See color plates.*

served microscopically (French et al., 1997). Increases in both fluorescence lifetime and intensity were observed throughout the experiment, indicative of the proteolysis of FITC–BSA (Fig. 9.4). In a similar experiment, increased fluorescence intensity was not observed with the nondegradable polymer FITC–poly-D–lysine, verifying that the increases in both fluorescence lifetime and intensity observed with FITC–BSA were due to specific enzymatic degradation. Interestingly, the average fluorescence lifetime of intracellular FITC–BSA was only 2.2 ns after 24 h, which was significantly lower than expected, even when accounting for the acidic environment of the endocytic system (French et al., 1997; Fig. 9.4). This discrepancy can be explained in part by the fact that the reported values are averages, and single cells contain vacuoles with totally degraded protein as well as vacuoles with partially degraded antigen. Moreover, a macromolecule such as the MHC II could be interacting with fluoresceinated peptide fragments and causing fluorescence quenching (French et al., 1997). Regardless of the mechanism, these results proved that FITC–BSA was localized to an acidic environment and enzymatically processed over a period of 24 h within the endocytic system of murine macrophages (French et al., 1997). More importantly, these data represented the first kinetic measurements of antigen processing within living cells and served as a foundation for flow cytometry experiments.

APPLICATION OF FLOW CYTOMETRY FOR MONITORING ANTIGEN PROCESSING AND PRESENTATION

Fluorescence Intensity Measurements

Once two-photon microscopy confirmed the validity of the antigenic probe, flow cytometry fluorescence intensity measurements were utilized to characterize the processing pathway, and these experiments provided crucial details in regard to both the mechanism of internalization and the kinetics of degradation. However, it is important to mention that several fluoresceinated derivatives of BSA were studied both in vitro and in vivo to determine the most suitable derivative for flow cytometry measurements (Voss et al., 1996). In contrast to other probes, $FITC_{10}BSA$ exhibited an important balance of spectral properties. The probe produced a sufficient level of fluorescence in the flow cytometer even at low concentrations (2–10 µg/ml), and the fluorescence was sufficiently quenched such that $FITC_{10}BSA$ was sensitive to events such as unfolding and proteolysis (Voss et al., 1996). Additionally, enzyme-linked immunosorbent assay (ELISA) with anti-BSA antibodies revealed that the antigenic structure of the BSA carrier was left relatively unaffected by FITC conjugation (French et al., 1997). Based on these studies, 10 FITC groups per molecule of BSA were selected as the optimum degree of substitution to monitor antigen processing in flow cytometry measurements (Weaver et al., 1996).

Flow cytometry fluorescence intensity measurements proved invaluable in dissecting the mechanism of uptake by murine macrophages. Specifically, the ability to monitor the internalization of antigen by live cells provided important kinetic information that ultimately resulted in the identification of a novel macrophage receptor possessing specificity for fluorescein and other aromatic structures (Cherukuri et al., 1997a). Using methods developed by Sklar et al. (1984), flow cytometry measurements revealed that as fluorescein substitution of BSA increased, a concomitant increase in the rate of uptake was observed (Cherukuri et al., 1997a; Sklar and Finney, 1982; Fig 9.5). Rates of $4.2 \pm 0.2 \times 10^6$, $9.6 \pm 0.1 \times 10^6$, and $1.9 \pm 0.1 \times 10^7 M^{-1}$ min^{-1} were observed for FITC$_5$BSA, FITC$_{10}$BSA, and FITC$_{22}$BSA, respectively (Cherukuri et al., 1997a). The same study also demonstrated that receptor-mediated endocytosis was the primary mechanism for internalization of FITC–BSA by macrophages (Cherukuri et al., 1997a). One obvious conclusion derived from these studies was that a receptor possessing specificity for fluorescein and other aromatic structures was involved in antigen internalization (Cherukuri et al., 1997a). To con-

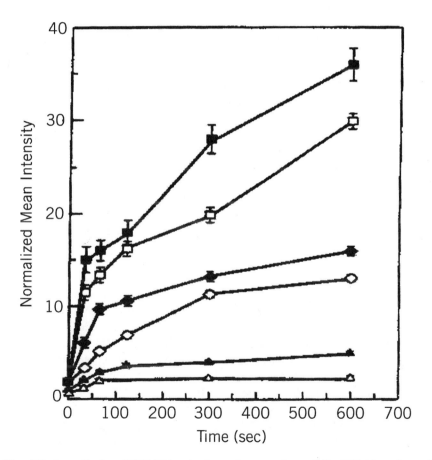

Fig. 9.5. Internalization of FITC$_x$BSA probes by murine macrophage cell line J774. Macrophage cells were incubated with 10 μg of FITC$_3$BSA (△), FITC$_5$BSA (▲), FITC$_8$BSA (◇), FITC$_{10}$BSA (◆), FITC$_{15}$BSA (□), or FITC$_{22}$BSA (■) for the specified period of time. Fluorescence intensity was monitored on the flow cytometer. Data represented averages ± SEM of triplicate experiments. (Copyright Elsevier Science, 1997. Reprinted with permission)

firm this conclusion, a series of model synthetic probes consisting of fluorescein labeled poly-D-lysine were developed for use with flow cytometry (Cherukuri et al., 1997b). The probes possessed several characteristics essential for measuring intracellular uptake. Most importantly, as a D-isomer, FITC–poly-D-lysine could be neither enzymatically degraded nor recognized by receptors (Cherukuri et al., 1997b). Therefore, the effects of carrier molecules on cellular uptake were negligible, and the role of the hapten could be analyzed in a more controlled system (Cherukuri et al., 1997b). Using the methods of Sklar et al. (1984), these studies confirmed that hapten recognition was critical for the optimal uptake of fluorescein-derivatized probes by J774 macrophages as well as peritoneal macrophages (Cherukuri et al., 1997b; Cherukuri and Voss, 1998; Sklar et al., 1984).

In addition to identifying the receptor, flow cytometry offered a simple technique for examining the specificity of the receptor. Cells were either preincubated or coincubated with various monovalent ligands at several concentrations (Cherukuri and Voss, 1998). These samples were analyzed by flow cytometry, and the percent inhibition of fluorescence intensity was calculated (Cherukuri and Voss, 1998). These data revealed that unconjugated phenyl ring structures (i.e., L-phenylalanine) possessed higher inhibitory efficiency relative to derivatized phenyl ring structures such as fluoresceinamine (Cherukuri and Voss, 1998). Once the specificity of the receptor was determined, purification of the novel protein was required, and although subsequent purification involved biochemical techniques, flow cytometry fluorescence intensity measurements facilitated the identification and initial characterization of this novel receptor (Cherukuri et al., 1997a, 1998; Cherukuri and Voss, 1998). The obvious benefits of flow cytometry in the context of this system were rapid data collection, existence of theoretical models for monitoring receptor–ligand interactions, consistent quantitative measurements, and statistically significant analyses of the entire cell population (Sklar and Finney, 1982; Cherukuri et al., 1997a).

Flow cytometry fluorescence intensity measurements also provided a powerful tool for analyzing the processing of the antigenic probe within the endocytic system of the cell. In these experiments, two protocols were available. First, the cells were added to the flow cytometer, and once the level of autofluorescence was established, antigen was added to the cell sample directly on the flow cytometer followed by a brief mixing step. The fluorescence intensity of cells was then monitored over extended periods of time. The second experiment involved incubating cells in tissue culture with antigen and removing an aliquot of cells at specified time points for analysis by flow cytometry. Each of these experiments provided important information related to processing of antigen for MHC II loading. As evident in Figure 9.6, the average fluorescence intensity of cells was relatively homogeneous, suggesting that the processing of antigen by these cells was a synchronous process within the cell population (Weaver and Voss, 1998a; Fig. 9.6). In the context of the second protocol, flow cytometry offered a novel method to investigate the requirements for antigen processing by applying various inhibitors of the processing pathway to the cells. Inhibitors of adenosine triphosphate (ATP) synthesis and endocytic trafficking were utilized throughout these studies and proved to be invaluable (Weaver et al., 1996;

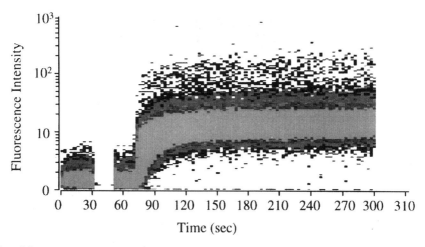

Fig. 9.6. Homogeneous uptake and processing of $FITC_{10}BSA$ by murine peritoneal macrophages. Cells were monitored on the flow cytometer for several seconds prior to the addition of $FITC_{10}BSA$ to establish the level of autofluorescence. $FITC_{10}BSA$ at 10 µg/ml was then added directly to the cells on the flow cytometer, and fluorescence intensity was monitored as a function of time. (Copyright Editions Scientifiques et Medicales Elsevier, 1998. Reprinted with permission.)

Weaver and Voss, 1998b). As a specific example, protease inhibitors were also incubated with the cells prior to the addition of $FITC_{10}BSA$. Interestingly, the kinetics of processing were altered in the presence of these inhibitors (Weaver et al., 1996; Weaver and Voss, 1998a). Inhibition by thiol and aspartyl protease inhibitors was observed at slightly different time points early in the pathway (Fig. 9.7; Weaver et al., 1996; Weaver and Voss, 1998a). This was consistent with previous reports which demonstrated the importance of cathepsin B, a thiol protease, and cathepsin D, an aspartyl protease, in the generation of T-cell epitopes (Fineschi and Miller, 1997). In the case of serine and metalloprotease inhibitors, inhibition was observed at 120 min. (Fig. 9.7; Weaver et al., 1996). The conclusion derived from these experiments was that the enzymatic processing of antigen within these cells was a sequential series of events. Therefore, the localization and activity of these enzymes provide an important regulatory mechanism for this pathway (Fineschi and Miller, 1997). This is important since the goal of processing is to generate peptides that induce T-cell stimulation and protective immunity (Weaver et al., 1996). Moreover, if this process were not regulated, many antigenic determinants could be destroyed (Weaver et al., 1996). Although discussed in the context of other systems, this idea represented an important conceptual advance in the area of antigen processing and presentation (Weaver et al., 1996). For this reason, the application of the novel fluorescent probe to flow cytometry intensity measurements proved to be a significant methodology for offering new insights into these intracellular events.

At this point, it is important to note the similarity in kinetics between flow cytometry and two-photon fluorescence microscopy. This was significant because the 3-dimensional resolution afforded by two-photon fluorescence verified that only fluorescence localized within the vacuoles of the cells was observed, and the apparent similarity in data validated the use of flow cytometry as a method to monitor the kinetics of intracellular, endocytic events (Fig. 9.8; French et al., 1997). Flow cytometry had additional advantages, specifically the ability to gate on a particular population of cells, thereby eliminating the contribution of dead cells to total fluorescence. Moreover, compared to two-photon fluorescence microscopy, the flow cytometry analysis provided a statistically significant analysis of the entire cell population over an extended period of time.

Besides detecting intracellular events, flow cytometry intensity measurements were also useful for detecting surface-expressed MHC II–peptide complexes. Subcellular fractionation experiments revealed that fluorescein was associated with peptides bound in the context of the MHC II molecule (Weaver and Voss, 1998b). Furthermore, the fluorescyl moieties were accessible for binding by high-affinity antifluorescein antibodies (Weaver et al., 1998b). Therefore, both antifluorescein and anti-MHC II antibodies could be used to stain for MHC II–fluoresceinated peptide complexes on the cell surface. Namely, these antibody reagents should provide a useful nonradioactive alternative for calculating the number of specific MHC II–peptide complexes on the cell surface.

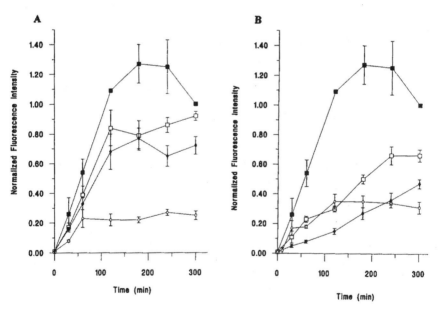

Fig. 9.7. Effect of specific protease inhibitors on the processing of $FITC_{10}BSA$. (A) Prior to incubation with 10 μg/ml of $FITC_{10}BSA$, macrophage cells were preincubated with 15 μM TPCK (□), 250 μM E-64 (◇), and 45 μM 1,10 phenanthroline (◆). (B) Cells were also preincubated with 350 μM Pepstatin A (□), 50 μM TLCK(◆), or 500 μM leupeptin (◇) prior to addition of 10 μg/ml of $FITC_{10}BSA$. Data represented averages ± SEM of triplicate experiments. Fluorescence intensity was normalized to the fluorescence of cells incubated for 300 min. Cells incubated with 10 μg/ml of $FITC_{10}BSA$ (■). TPCK: N-tosyl-L-phenylalanine chloromethyl ketone. E-64: trans-epoxysuccinyl-L-leucylamido-(4-guanidino) butane. TLCK: Nα-p-tosyl-L-lysine chloromethyl ketone.

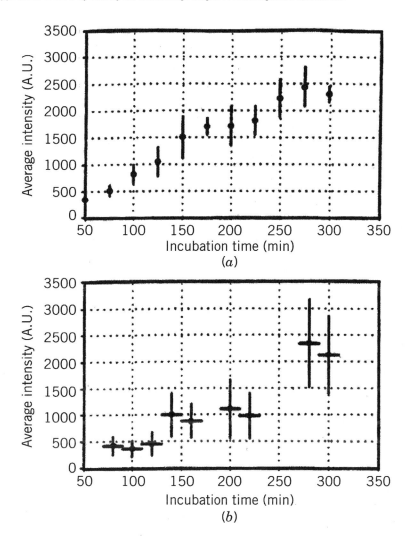

Fig. 9.8. Kinetics of intracellular degradation of FITC$_{10}$BSA by murine macrophage cell line J774. Cells were preincubated with 10 μg/ml of FITC$_{10}$BSA and incubated for the specified period of time at 37°C. After several washes, fluorescence intensity of cells was monitored by two-photon fluorescence microscopy (lower panel) or flow cytometry (upper panel). Data represented averages ± SEM of triplicate experiments.

Fluorescence Polarization Measurements

In order to monitor fluorescence polarization, a fluorophore is excited with plane-polarized light, and the resulting fluorescence is detected in the planes parallel and perpendicular to the excitation light. Assuming that the rotational lifetime of the fluorophore is less than the fluorescence lifetime, the emission will have a different polarization than the excitation, making it possible to gain information about the fluorophore's local environment (Lakowicz, 1983; Weber, 1952). In the present system, fluorescence polarization analysis by flow cytometry provided a useful method for monitoring the local environment and intracellular processing of $FITC_{10}BSA$ within the cell. However, in these experiments, calibration of the flow cytometer was critical. Therefore, high- and low-fluorescence polarization controls were monitored prior to the experiment. For the low-fluorescence polarization control, $FITC_{10}BSA$ was digested with Pronase for 35 min. These peptides were then incubated with macrophages for 30 min, and fluorescence polarization was monitored on the flow cytometer. As expected, due to extensive rotational motion of these peptides, a fluorescence polarization value of 0.04 ± 0.01 was observed (Fig. 9.9). As a high-fluorescence polarization control, the fluorescence polarization of Immunobrite III beads was also monitored on the flow cytometer (Fig. 9.9). Upon addition of the beads, a fluorescence polarization value of 0.38 ± 0.01 was determined (Figure 9.9). Since these results verified that the flow cytome-

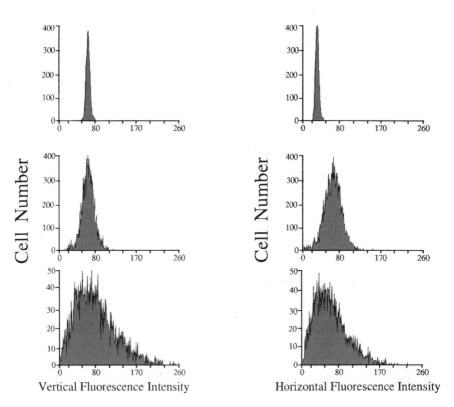

Fig. 9.9. Representative fluorescence intensity histograms of murine macrophage cell line J774 preincubated with various fluorescent probes. The upper panel represents the fluorescence intensity of Immunobrite level III beads. The middle panel represents the fluorescence intensity of Pronase-digested $FITC_{10}BSA$ peptides. The lower panel represents the fluorescence intensity of cells incubated with 10 μg/ml $FITC_{10}BSA$. In all examples, fluorescence intensity in the both the vertical and horizontal planes was detected.

Application of Flow Cytometry for Monitoring Antigen Processing and Presentation

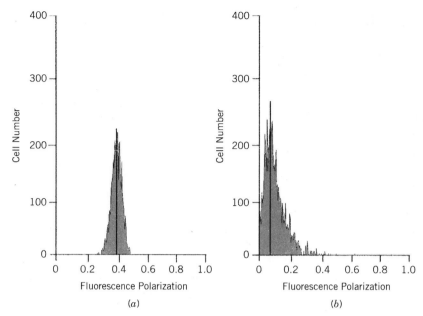

Fig. 9.10. Fluorescence polarization histograms of Immunobrite III beads or J774 macrophages preincubated with fluoresceinated peptides. (a) The fluorescence polarization of Immunobrite beads was monitored. (b) Cells were incubated with fluoresceinated peptides and fluorescence polarization was monitored.

ter was calibrated accurately, the cells were incubated with $FITC_{10}BSA$, and fluorescence polarization was monitored. Figure 9.9 represents various fluorescence intensity histograms depicting both the vertical and horizontal fluorescence components of the cells and illustrates several important points concerning the measurements. First, in terms of the processing of the probe, the cells were relatively homogeneous, producing Gaussian distributions that facilitated data analysis (Fig. 9.9). However, as evident from the data, within the population there was a wide range of intensities, which indicated that the cells were at slightly different stages of processing (Fig. 9.9). For example, the low fluorescence intensities could be attributed to cells that had only recently begun to degrade the probe while the higher fluorescence intensities could be the result of cells in which extensive proteolytic degradation had already occurred (Fig. 9.9). As described in the Introduction, several different events occur as proteins proceed through the endocytic compartment. Since the probe employed in these studies was sensitive to each of these steps, it was not unexpected to see variability in the intensity data (Fig. 9.9). In short, these results simply reflected the wide variety of events that occurred within the cells rather than improper calibration of the instrument (Fig. 9.9, 9.10). Furthermore, to ensure that the results were accurate, no less than 90% of the gated population was represented in each calculation, and both the horizontal and vertical gains were adjusted throughout the experiment to ensure an accurate representation of the dynamic population.

Results of fluorescence polarization measurements of macrophage cells incubated with $FITC_{10}BSA$ for various periods of time are depicted in Figure 9.11. A steady increase in fluorescence polarization was observed for the first 100 min of the experiment (Fig. 9.11). The fluorescence polarization then remained constant for the next 100 min, and finally, a decrease in fluorescence polarization was observed after 200 min (Fig. 9.11). Decreased fluorescence polarization at 200 min was the result of proteolytic cleavage of $FITC_{10}BSA$ and generation of relatively small fluoresceinated peptides; yet as indicated in fluorescence polarization histograms, the degree of degradation varied slightly from cell to cell (Fig. 9.11; Weaver et al., 1996, 1997). Increased fluorescence polarization could be caused by several factors, including: (1) an increase in the viscosity of the endocytic compartment, (2) a decrease in the pH of the local environment of the fluorophore, and (3) binding by a macromolecule. However, control experiments demonstrated that these data reflected the kinetics of endosomal transport by macrophages. To test this hypothesis more directly, the fluorescence polarization of cells treated with the weak base ammonium chloride was monitored as well (Poole and Ohkuma, 1981). When the fluorescence polarization of the cells was monitored, the increase in fluorescence polarization was not observed; instead the fluorescence polarization decreased, reaching a plateau at ~0.10. Extensive studies by Ohkuma and Poole (1978) have shown that upon addition of ammo-

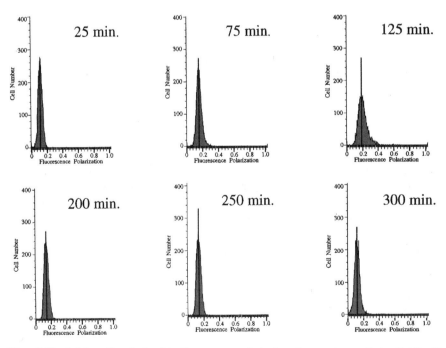

Fig. 9.11. Cell-by-cell analysis of the fluorescence polarization of macrophage cells incubated with $FITC_{10}BSA$. Cells incubated with 10 μg/ml $FITC_{10}BSA$ for various periods of time were analyzed on a cell-by-cell basis and plotted as a function of fluorescence polarization.

nium chloride the endocytic environment reached a pH of 6.5. Therefore, a polarization value of ~0.10 would be expected. In short, the flow cytometry polarization analysis provided further evidence that $FITC_{10}BSA$ was localized to the endocytic environment as well as provided a method to observe endocytic transport.

Fluorescence Lifetime Measurements

The recent development of a flow cytometer capable of measuring fluorescence lifetime at the University of Illinois Flow Cytometry Facility has provided an important tool for dissecting the intracellular microenvironment of macrophage. Specifically, fluorescence lifetime measurements offer a powerful tool for uniting both structural and functional details of biological processes. In the case of the present system, fluorescence lifetime measurements proved invaluable for measuring both intracellular trafficking and degradation of the antigenic probe. However, in the application of fluorescence lifetime measurements to antigen-processing studies, two important factors will affect data. First, as indicated in Figure 9.3, fluorescein's fluorescence lifetime is exquisitely sensitive to pH. Therefore, as the antigenic probe migrates through the endocytic system, a decrease in fluorescence lifetime will be detected. Second, when the protein carrier is in the globular form, fluorescein's fluorescence lifetime is quenched due to the close spatial proximity and orientation of neighboring fluorescein molecules (Fig. 9.4). Therefore, as the spatial constraint is alleviated via proteolytic cleavage of the protein carrier, an increase in fluorescence lifetime will also be detected.

To perform fluorescence lifetime experiments, several standards were used to calibrate the instrument, including acridine orange, ethidium bromide, and fluorescein-5-isothiocyanate with corresponding fluorescence lifetime values of 1.7, 1.6, and 4.0 ns, respectively. As indicated in Table 1, several trends were observed in the lifetime experiments. First, slight variations were observed when comparing measurements of fluorescein-derivatized probes using different reference standards. For example, when cells were incubated with $FITC_{20}BSA$ for 2 h, fluorescence lifetime values of 2.4, 2.2, and 2.1 ns were observed using acridine orange, ethidium bromide, and fluorescein reference standards, respectively (Table 1). In the case of $FITC_{10}BSA$, values of 2.6 2.4, and 2.1 ns were observed for acridine orange, ethidium bromide, and FITC reference standards, respectively (Table 1). Therefore, selection of the appropriate standard is imperative. Second, lifetime experiments verified previous fluorescence intensity and polarization experiments that suggested the probe localized to acidic compartments within the cell since low fluorescence lifetime values were observed even after 24 h (Table 1). More importantly, as time progressed, increased fluorescence lifetime values were observed in the case of fluorescein-derivatized BSA, indicative of degradation within the endocytic system (Table 1). However, when fluorescein-derivatized poly-D-lysine was incubated with the cells, increased fluorescence lifetime values were not detected, which was expected since poly-D-lysine is nondegradable (Table 1). Therefore, flow cytometry fluorescence lifetime measurements further validated the use of FITC–BSA as a tool for monitoring the endocytic processing of proteins.

TABLE 1. Fluorescence Lifetime Flow Cytometry Measurements of Fluorescein-Derivatized Probes Following Endocytosis and Processing by Murine Macrophage Cell Line J774

Antigenic Probe	Incubation Time (min)	Lifetime (ns)
Acridine Orange Reference Standard (1.7 ns)		
FITC$_{10}$BSA	60	2.6
FITC$_{20}$BSA	5	1/2
FITC$_{20}$BSA	10	1.2
FITC$_{20}$BSA	60	2.6
FITC$_{20}$BSA	120	2.4
FITC$_{20}$BSA	240	2.1
FITC$_5$PDL	60	1.9
FITC$_5$PDL	120	1.9
FITC standard		4.0
Ethidium bromide standard		1.8
FITC Reference Standard (4.0 ns)		
FITC$_{10}$BSA	60	2.1
FITC$_{20}$BSA	60	2.1
FITC$_{20}$BSA	120	2.2
FITC$_{20}$BSA	240	2.2
FITC$_5$PDL	60	1.6
FITC$_5$PDL	120	1.6
Acridine orange standard		1.4
Ethidium bromide standard		1.5
Ethidium Bromide Reference Standard (1.6 ns)		
FITC$_{10}$BSA	60	2.4
FITC$_{20}$BSA	5	1.3
FITC$_{20}$BSA	10	1.4
FITC$_{20}$BSA	60	2.4
FITC$_{20}$BSA	120	2.2
FITC$_{20}$BSA	240	2.4
FITC$_5$PDL	60	1.9
FITC$_5$PDL	120	1.9
FITC standard		4.1
Acridine orange standard		1.6

Note: PDL, poly-D-lysine. Subscripts designate the moles of fluorophore per mole of protein carrier. Reference standards represent the compound used to calibrate the instrument prior to performing experiments.

When comparing fluorescence lifetime values obtained through flow cytometry and two-photon fluorescence imaging, similar values were observed (French et al. 1997). For example, following a 24-h incubation, a fluorescence lifetime of 2.4 ns was detected using flow cytometry while a value of 2.3 ns was detected using two-photon fluorescence microscopy (French et al., 1997). In the case of FITC–poly-D-lysine, values of 1.2 and 1.3 ns were observed for flow cytometry and two-photon fluorescence microscopy, respectively (French et al., 1997). The similarity in data validates the accuracy of flow cytometry fluorescence lifetime measurements. One obvious advantage of using two-photon fluorescence microscopy is that the three-dimensional imaging capabilities afforded by two-photon excitation can be used to monitor specific regions within the cells (French et al., 1997). However, in certain experimental systems, flow cytometry may offer an alternative since flow cytometry provides information on populations of cells in a short period of time, although total cell-associated fluorescence is detected.

CONCLUSIONS

The complexities of antigen processing and presentation require the development of novel technologies capable of providing new insights into these cellular processes. Therefore, the goal of this chapter is to emphasize the significance of a novel antigenic probe and its application to fluorescence microscopy and flow cytometry (Voss et al., 1996; Weaver et al., 1996). The fluorescent properties of FITC-derivatized BSA provide for analysis of intracellular events via multiple parameters, including fluorescence lifetime, intensity, and polarization. Integration of these fluorescent techniques ensures that the inherent biases of individual measurements do not influence experiments. For example, fluorescence intensity and polarization are biased toward the most intense signals within a cell. Therefore, signals derived from small or extensively degraded peptides will predominate results while signals derived from partially degraded peptides will be masked. However, in the case of fluorescence lifetime, all signals are registered independent of intensity, resulting in an average, albeit a weighted one. Thus, to determine the kinetics of intracellular pathways, all three parameters are required to establish an accurate representation of events. Moreover, this probe has also been used in combination with invasive techniques such as electron microscopy and subcellular fractionation, and the ability of each of these methodologies, including flow cytometry, two-photon fluorescence microscopy, and subcellular fractionation, to complement one another represents a powerful system for the elucidation of various intracellular phenomena (Weaver et al., 1998a,b). To illustrate this, a model for the processing and presentation of MHC II–peptide complexes is presented here. Moreover, this model is based solely on data obtained by combining the novel fluorescent antigen with the techniques described above.

The antigenic protein entered the cell through multiple processes including pinocytosis and receptor-mediated endocytosis (Fig. 9.12; Cherukuri et al., 1997a; Cherukuri and Voss, 1998). Upon entering the cell, the antigen probe was internalized into an acidic environment, and within 5 min, the protein localized to early endocytic organelles (Weaver and Voss, 1998b). More importantly, degradation of the protein was also evident by 5 min. As time progressed, the protein was detected in late endosomal/lysosomal organelles by 10 min (Weaver and Voss, 1998b). Flow cytometry experiments suggested that thiol and aspartyl proteases were the primary enzymes involved in the cleavage of the probe (Weaver and Voss, 1998b). MHC II–peptide loading of antigenic fluoresceinated peptides was observed by 15 min in transferrin receptor–positive, LAMP-1–positive, and cathepsin D–positive organelles, whereas loading of unlabeled BSA peptides occurred in transferrin receptor–negative organelles (Weaver and Voss, 1998b). In addition, MHC II–fluoresceinated peptide complexes utilized transferrin receptor–positive organelles for transport to the plasma membrane at 30 min (Weaver and Voss, 1998b). Therefore, macrophage cells utilized

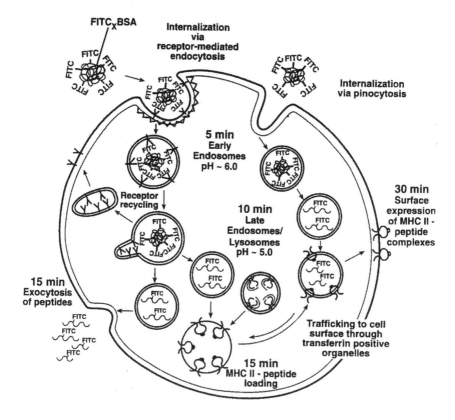

Fig. 9.12. Kinetic model for MHC II–peptide loading and surface expression in murine macrophages. The antigenic protein enters the cell through receptor-mediated endocytosis or pinocytosis. Upon entering the cell, the antigenic protein encounters an acidic environment at 5 min. Also at 5 min, proteolytic degradation of the protein begins, and by 10 min, the accumulation of antigenic peptides within the late endosomal/lysosomal environment is observed. Exocytosis of fluoresceinated peptides begins at 15 min, but more importantly, MHC II–peptide loading of fluoresceinated peptides occurs in transferrin receptor–positive or –negative late endosomal organelles. Finally, MHC II–fluoresceinated peptide complexes migrate to the plasma membrane via a retrograde pathway for surface expression at 30 min.

a retrograde pathway through the endocytic system to transport MHC II–peptide complexes to the cell surface (Weaver and Voss, 1998b; Fig. 9.12). This technique is not restricted to proteolytic processing by macrophages. Using various fluorescent probes and carrier proteins, the microenvironment of a wide variety of macromolecules could be monitored noninvasively in viable cells, facilitating kinetic analysis and providing insight into numerous cellular processes. For example, several of the downstream effector molecules responsible for programmed cell death were recently identified (Boldin et al., 1996; Muzio et al., 1996). Termed caspases, these enzymes are believed to act as a signal cascade to initiate apoptosis (Wang and Lenardo, 1997). With the recent development of membrane-permeable caspase substrates, the construction of fluorescent substrates similar to FITC–BSA is possible and the development of a fluorescent system to monitor the kinetics of caspase activity in live cells is also feasible. As indicated in the macrophage system, directly measuring the kinetics of intracellular processes in live cells can offer several advantages over conventional techniques and provide new insights into the mechanisms responsible for those kinetics.

REFERENCES

Amigorena S, Drake J, Webster P, Mellman I (1994): Transient accumulation of new class II MHC molecules in a novel endocytic compartment in B lymphocytes. Nature 369:113–120.

Benjamin DC, Berzofsky JA, East IJ, Gurd FRN, Hannum C, Leach SJ, Margoliash E, Michael JG, Miller A, Prager EM, Secarz EE, Smith-Gill SJ, Todd PE, Wilson AC (1984): The antigenic structure of proteins: reappraisal. Annu Rev Immunol 2:67–102.

Boldin MP, Goncharov TM, Goltsev YV, Wallach D (1996): Involvement of MACH, a novel MORT1/FADD-interacting protease, in FAS/APO-1 and TNF receptor induced cell death. Cell 85:803–815.

Bowser R, Murphy RF (1990): Kinetics of hydrolysis of endoctyosed substrates by mammalian cultured cells: early introduction of lysosomal enzymes into the endocytic pathway. J Cell Physiol 143:110–117.

Cherukuri A, Voss EW, Jr. (1998): Ligand binding specificity of a macrophage surface receptor utilized by the fluorescein hapten for uptake into the endocytic pathway. Mol Immunol 35:115–125.

Cherukuri A, Nelson J, Voss EW, Jr (1998): Biochemical purification and partial characterization of a macrophage surface receptor possessing specificity for small aromatic moieties including fluorescein. J Mol Recognition 12:94–102.

Cherukuri A, Durack G, Voss EW, Jr (1997a): Evidence for hapten recognition in receptor-mediated intracellular uptake of a hapten-protein conjugate by murine macrophage. Mol Immunol 34:21–32.

Cherukuri A, Frye J, French T, Durack G, Voss E.W., Jr. (1997b): FITC-poly-D-lysine conjugates as fluorescent probes to quantify hapten-specific macrophage receptor binding and uptake kinetics. Cytometry 31:110–124.

Cresswell P (1994): Assembly, transport, and function of MHC class II molecules. Annu Rev Immunol 12:259–293.

Diment S (1993): Proteolytic processing for MHC class II antigen presentation. Ciencia Cultura 45:305–312.

Fineschi B, Miller J (1997): Endosomal processing and antigen processing. Trends Biol Sci 22:377–382.

French T, So PTC, Weaver DJ, Jr, Coelho-Sampaio T, Gratton E, Voss EW, Jr, Carrero J (1997): Two-photon fluorescence lifetime imaging microscopy of macrophage-mediated antigen processing. J Microsc 185:339–353.

Germain RN, Marguiles DH (1993): The biochemistry and cell biology of antigen processing and presentation. Annu Rev Immunol 11:403–450.

Kleijmeer MJ, Morkowski S, Griffith JM, Rudensky AY, Geuze HJ (1997): Major histocompatibility complex class II compartments in human and mouse B lymphoblasts represent conventional endocytic compartments. J Cell Biol 139:639–649.

Kranz DM, Voss EW, Jr (1981): Partial elucidation of an anti-napter repertoire in BALB/c mica: comparative characterization of several monoclonal anti-fluoresay antibodies. Mol Immunol 18:889–898.

Lanzavecchia A (1990): Receptor-mediated antigen uptake and its effect on antigen presentation to class II-restricted T lymphocytes. Annu Rev Immunol 8:773–793.

Lakowicz JR (1983): *Principles of Fluorescence Spectroscopy*. New York: Plenum Press.

Muzio M, Chinnaiyan AM, Kischkel FC, O'Rourke K, Shevchenko A, Ni J, Scaffidi C, Bretz JD, Zhang M, Gentz R (1996): FLICE, a novel FADD-homologous ICE/CED-3-like protease, is recruited to the CD95 (FAS/APO-1) death-inducing signal complex. Cell 85:817–827.

Ohkuma S, Poole B (1978): Fluorescence probe measurement of intralysosomal pH in living cells and the perturbation of pH by various agents. Proc Natl Acad Sci USA 75:3327–3331.

Pieters J (1997): MHC class II restricted antigen presentation. Curr Opin Immunol 9:89–96.

Poole B, Ohsuma S (1981): Effect of weak bases on the intralysosonal pH in mouse peritoneal macrophages. J Cell Biol 90:665–669.

Qui Y, Xu X, Wandinger-Ness A, Dalke DP, Pierce SK (1994): Separation of subcellular compartments containing distinct forms of MHC class II. J Cell Biol 125:595–605.

Rothe G, Valet G (1993a): Measurement of neutrophil elastase activity with (N-benzyloxycarbonyl-Ala-Ala)$_2$-Rhodamine 110). In Robinson JP (ed). *Handbook of Flow Cytometry Methods*. New York: Wiley, pp 200–201.

Rothe G, Valet G (1993b): Measurement of mononuclear phagocyte cathepsin B/L activity with (N-benzyloxycarbonyl-Arg-Arg)$_2$-Rhodamine 110). In Robinson JP (ed). *Handbook of Flow Cytometry Methods*. New York: Wiley, pp 202–203.

Rothe G, Oser A, Assfalg-Machleidt I, Machleidt W, Mangel WF, Valet G (1990): Cathepsin B activity measured with (Z-Phe-Arg)$_2$-Rhodamine 110 as a new flow cytometric marker of monocyte/macrophage activation. Cytometry Suppl. 4:77.

Sklar LA, Finney DA (1982): Analysis of ligand-receptor interactions with the fluorescence activated cell sorter. Cytometry 3:161–165.

Sklar LA, Finney DA, Oades ZG, Jesaitis AJ, Painter RG, Cochrane CG (1984): The dynamics of ligand-receptor interactions. Real-time analyses of association, dissociation, and internalization of an N-formyl peptide and its receptors on the human neutrophil. J Biol Chem 259:5661–5669.

Tulp A, Verwoerd D, Dobberstein B, Ploegh HL, Pieters J (1994):Isolation and characterization of the intracellular MHC class II compartment. Nature 369:120–126.

Voss EW, Jr. (1984): *Fluorescein Hapten: An Immunological Probe*. Boca Raton, FL: CRC Press.

Voss EW, Jr (1990): Anti-fluorescein antibodies as structure function models to examine fundamental immunochemical and spectroscopic principles. Comments Mol Cell Biophys 6:197–221.

Voss EW, Jr, Workman CJ, Mummert ME (1996): Detection of protease activity using a fluorescent-enhancement globular substrate. BioTechniques 20:286–291.

Ward ES, Qadri O (1997): Biophysical and structural studies of TCRs and ligands: implication for T cell signaling. Curr Opin Immunol 9:97–106.

Watts C (1997): Capture and processing of exogenous antigens for presentation on MHC molecules. Annu Rev Immunol 15:821–850.

Wang J, Lenardo MJ (1997): Molecules involved in cell death and tolerance. Curr Opin Immunol 9:818–825.

Weaver DJ, Jr, Voss EW, Jr (1998a): Analysis of rates of receptor-mediated endocytosis and exocytosis of a fluorescent hapten-protein conjugate in murine macrophage: implications for antigen processing. Biol Cell 90:169–181.

Weaver DJ, Jr, Voss EW, Jr (1998b): A novel macrophage receptor enhances MHC II-peptide loading and surface expression of a hapten-protein conjugate. Biol Cell. 90:427–438.

Weaver DJ, Jr, Voss EW, Jr (1999): Kinetics and intracellular pathways required for major histocompatibility complex II-peptide loading and surface expression of a fluorescent hapten-protein conjugate in murine macrophage. Immunology 96:557–568.

Weaver DJ, Jr, Durack G, Voss EW, Jr (1997): Analysis of the intracellular processing of proteins: application of fluorescence polarization and a novel fluorescent probe. Cytometry 28:25–35.

Weaver DJ, Jr, Cherukuri A, Carrero J, Coehlo-Sampaio T, Durack G, Voss EW, Jr (1996): Macrophage-mediated processing of an exogenous antigenic fluorescent probe: Time-dependent elucidation of the processing pathway. Biol Cell 87:95–104.

Weber G (1952): Polarization of the fluorescence of macromolecules. I. Theory and experimental methods. Biochem J 51:145–155.

Wubbolts R, Fernandez-Borja M, Oomen L, Verwoerd D, Janssen H, Calafat J, Tulp A, Dusseljee S, Neefjes J (1996): Direct vesicular transport of MHC class II molecules from lysosomal structures to the cell surface. J Cell Biol 135:611–622.

Xu X, Song W, Cho H, Qiu Y, Pierce SK (1995): Intracellular transport of invariant chain MHC II complexes to the peptide loading compartment. J Immunol 155:2984.–2992.

10

Probing Deep-Tissue Structures by Two-Photon Fluorescence Microscopy

Chen-Yuan Dong, Ki Hean Kim, Christof Buehler, Lily Hsu, Hyun Kim, and Peter T. C. So
Massachusetts Institute of Technology, Cambridge, Massachusetts

Barry R. Masters
University of Bern, Bern, Switzerland

Enrico Gratton
University of Illinois at Urbana-Champaign, Urbana, Illinois

Irene E. Kochevar
Massachusetts General Hospital, Boston, Massachusetts

INTRODUCTION

Two-photon fluorescence microscopy is increasingly becoming an important part of modern optical techniques. Excellent axial depth discrimination, limitation of photodamage to focal volume, good depth penetration, and more efficient fluorescence detection are some of the key advantages of two-photon microscopy over conventional techniques. This chapter begins with an overview of two-photon microscopy relevant to applications in deep-tissue imaging. Basic principles, deep-tissue models, and

---

Emerging Tools for Single-Cell Analysis, Edited by Gary Durack and J. Paul Robinson.
ISBN 0-471-31575-3 Copyright © 2000 Wiley-Liss, Inc.

instrumentation are discussed. From there, recent advances in two-photon deep-tissue imaging are addressed: (1) the application of a blind deconvolution algorithm in further improving image quality and providing point spread function (psf) information in deep tissue, (2) the development of video-rate two-photon instrumentation to extend the imaging technology to potential clinical applications, and (3) the acquisition and analysis of two-photon spectroscopic data as complementary information to structural imaging.

Confocal versus Two-Photon Fluorescence Microscopy

Confocal microscopy has been traditionally the most promising technology for the study of three-dimensional (3D) deep-tissue structure with subcellular resolution (Raijadhyaksha et al., 1995; Corcuff et al., 1993; Masters et al., 1997c; Masters, 1996; Rummett et al., 1994). The basic concept behind confocal microscopy is straightforward (Wilson and Sheppard, 1984; Pawley, 1995). Excitation light deflected by a beam splitter is focused through the objective onto the sample. The excitation source generates scattered light (reflected mode) or fluorescence (fluorescence mode) throughout the hour-glass-shaped excitation volume. To obtain an optical section, the light collected by the objective is refocused onto a pinhole placed at a telecentric plane. At this plane, only the photons originating from the objective focal volume will be focused through the aperture whereas the off-focal photons are rejected. High-resolution 3D image stacks from human skin and eyes have been obtained in vivo (Raijadhyaksha et al., 1995; Corcuff et al., 1993; Masters et al., 1997b, 1997c; Masters, 1990, 1996; Piston et al., 1995). This technique provides spatial resolution with radial and depth resolution on the order of 0.2 and 0.5 μm, respectively. The penetration depth of confocal techniques varies depending on the experimental conditions. In the reflected light mode, the penetration depth of a confocal microscope is about 200 μm. However, in the fluorescence mode, the penetration distance is much shorter. With near-UV excitation, the penetration depth in human skin is reduced to only 30 μm. Furthermore, the one-photon excitation used in typical confocal microscopy results in signal generation throughout the excitation volume and can result in unnecessary photodamage in off-focal regions.

In the past decade, development in two-photon microscopy, initiated by Denk et al. (1990), bypasses many of the drawbacks of conventional confocal microscopy and emerges as a new and powerful technique in applications involving deep-tissue imaging. In this method, chromophores can be excited by the simultaneous absorption of two photons each having half the energy needed for the excitation transition (Denk et al., 1990; Göppert-Mayer, 1931). Unlike one-photon excitation, two-photon absorption occurs most significantly in regions with high excitation photon flux density. The high photon flux density can be easily achieved by focusing lasers with short pulse widths using microscope objectives with high numerical aperture (NA). In this manner, excitation photons arrive at the spatially confined focal volume within a narrow temporal window, thus creating the high instantaneous photon flux density necessary for efficient two-photon excitation of fluorescent molecules. While two-photon excitation can be achieved using continuous wave (CW) laser excitation, tissue thermal damage is a serious concern because of the high average power needed (Booth and Hell, 1998). The efficiency of two-photon excitation under different experimental conditions can be examined by calculating n_a, the number of photon pairs absorbed per laser pulse per chromophore. For a focused, mode-locked laser source with an average power p_0, repetition rate f_p, pulse width τ_p, and wavelength λ, n_a can be estimated from (Denk et al., 1990)

$$n_a \approx \frac{p_0^2 \delta}{\tau_p f_p^2} \left(\frac{\pi (\mathrm{NA})^2}{hc\lambda} \right)^2$$

where NA is the numerical aperture of the focusing lens, c is the speed of light, h is Planck's constant, and δ is the two-photon cross section, typically on the order of 10^{-50}–10^{-49} cm^4 photon^{-1} molecule^{-1}.

There are a number of advantages to implementing two-photon-induced absorption in microscopy:

1. Since excitation occurs appreciably only at the focal region, excellent axial depth discrimination is achieved. For one-photon excitation in a spatially uniform fluorescent sample, significant fluorescence is generated from different axial sections above and below the focal plane. On the other hand, since two-photon excitation is nonlinear in nature (quadratic dependence on excitation power), over 80% of the total fluorescence intensity originates from a 1-μm-thick region near the focal point for a NA 1.25 objective. The confinement of excitation volume allows 3D imaging with excellent depth discrimination to be achieved. An added benefit of the restricted excitation volume is the accompanying reduction in photodamage. Since only fluorophores near the focal volume are excited, sample photodamage is limited to the focal region, and this prolongs sample viability.

2. A second major advantage is the achievable greater penetration depth. The typical absorbancy of infrared photons is more than an order of magnitude less than that of near-UV or blue-green light. Furthermore, since Rayleigh scattering is inversely proportional to the fourth power of wavelength, the red photons used in two-photon microscopy scatter much less than the blue photons used in one-photon excitation (Jackson, 1975). Therefore, the red/near-IR excitation light source used in two-photon excitation can be focused to a greater depth, and optically thick samples can be examined without disruptive procedures.

3. Large-area photodetectors that can be used in two-photon microscopy allow more efficient photon collection than confocal detection. In confocal microscopy, the emission pinhole is used to reject out-of-focus light. However, inside deep tissue, scattering of signal photons is inevitable and can lead to significant loss of these photons at the confocal pinhole. The detection geometry for the fluorescence photons is less critical in the two-photon case, and most of the forward-scattered photons can be retained (Fig. 10.1).

4. Two-photon excitation wavelengths are typically about twice that used in one-photon techniques. This wide separation between excitation and emission spectra ensures easy optical rejection of the excitation light and Raman scattering without significant attenuation of the fluorescence. The result is a more complete collection of the signal photons.

Tissue Autofluorescence

The importance of autofluorescence imaging has been recognized because of its use for noninvasive monitoring of tissue cellular metabolism by redox fluorometry (Masters, 1990; Chance and Thorell, 1959; Chance, 1976; Chance et al., 1979; Bennett et al., 1996; Masters and Chance, 1993). Intrinsic fluorescent probes include pyridine nucleotides and flavoproteins. The pyridine nucleotide NAD(P)H can be excited in the 365-nm region and fluorescence emission occurs in the 400–500 nm region. Spectroscopic properties of the flavoproteins are more red shifted with excitation

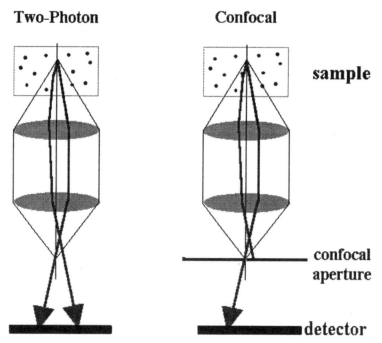

Fig. 10.1. A comparison between two-photon and confocal detection geometry. In a highly scattering specimen, a large fraction of emitted photons can be scattered before they are collected by the objective. In the two-photon system, most of these scattered photons can be collected by a large area detector. In the confocal system, these scattered photons can be blocked by the confocal pinhole aperture and escape detection. (Reprinted with permission from IEEE. IEEE Engineering in Medicine and Biology Magazine 18:23–30, 1999.)

around 450 nm and fluorescence between 500 and 600 nm. Typically, the changes in fluorescence intensity follow changes in cell and tissue oxidative metabolism. Fluorescence imaging of cellular redox metabolism has been successfully performed in the cornea and in the skin.

Recent studies by Webb's group at Cornell have demonstrated that two-photon microscopy is capable of imaging other tissue components such as collagen/elastin using their endogenous fluorescence (Maiti et al., 1997; Zipfel, 1997). Optical properties of these extracellular matrix components are known. Tropocollagen and collagen fibers typically have absorption in the range of 280–300 nm and emission ranges from 350 to 400 nm (LaBella and Gerald, 1965; Dabbous, 1966; Hoerman and Balekjan, 1966). Elastin can be excited by wavelengths of 340–370 nm and fluoresces from 400 to 450 nm with excitation (Thomas et al., 1963; LaBella, 1961; LaBella and Lindsay, 1963).

SKIN AND CORNEA AS DEEP-TISSUE IMAGING MODELS

Two important physiological models for deep-tissue studies are skin and cornea. Both are epithelial tissues and can be easily accessed by a two-photon microscope.

Anatomy of Skin

Typical mammalian skin (Fig. 10.2a) can be divided into two main layers: the epidermis and the dermis. The outer layer, the epidermis, is a squamous, multilayered structure composed primarily of keratinocytes. Epidermal tissue continuously renews itself by mitotic cell division from its deepest stratum, the basal layer. As the cells gradually migrate to the surface, they progressively mature by the process of keratinization, which results in the synthesis of the fibrous protein keratin. The basal monolayer is composed of cuboidal cells adhering to the basement membrane from which anchoring fibrils extend into the superficial dermis (papillary dermis). In a light microscope, the dermal-epidermal junction is observed as an undulating pattern of ridges. Cells produced by basal-cell division form the outer layers as they migrate toward the surface. Stratum corneum, the surface layer, is composed of hexagonal-shaped, fully keratinized dead cells without nuclei and cytoplasmic organelles. The dermis is composed mainly of collagen/elastin fibers mixed with a sparse population of fibroblasts. In human, the epidermal thickness varies from 50 μm on the eyelids to 1.5 mm on the palms and soles. In mouse skin, the epidermis is typically 30 μm thick. In the case of a mouse's ear, cartilage forms the structural support for the mouse ear below the skin.

Anatomy of Cornea

As an epithelial tissue, cornea has great structural similarity to skin (Fig. 10.2b). The corneal epithelium is about 50 μm thick. The epithelium consists of a single layer of basal epithelial cells, an adjacent layer of wing cells, and additional poorly organized wing cell strata up to the surface. The cells in the uppermost layer are the superficial epithelial cells. The region below the epithelium is called the stroma, which is structurally similar to the dermal layer in the skin, with a typical thickness of about 350–450 μm, and consists of collagen fibers interspersed with stromal keratocyte cells. The cell bodies of the stromal keratocytes are about 1 μm thick and form long extended processes in the plane of the nuclei.

TWO-PHOTON FLUORESCENCE MICROSCOPE

Setting up a Two-Photon Fluorescence Microscope

A typical two-photon microscope previously presented is shown in Figure 10.3 (So et al., 1995; Masters et al., 1997a). Central to this microscope is an ultrafast, mode-locked titanium–sapphire (Ti–sapphire) laser (Mira 900, Coherent, Santa Clara, CA). The typical pulse width of the Ti–sapphire laser is about 150 fs, which is ideal for inducing two-photon excitation. The excitation wavelength of the laser can be

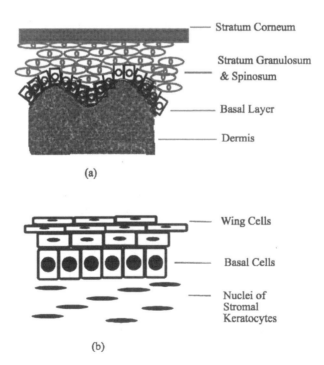

Fig. 10.2. Anatomy of (a) skin and (b) cornea. (Reprinted with permission from IEEE. IEEE Engineering in Medicine and Biology Magazine 18:23–30, 1999.)

tuned between 700 and 1000 nm. Optical components such as a Glan–Thomson polarizer can be used to control the excitation laser power. The beam-expanded laser light is directed into the microscope via a galvanometer-driven x–y scanner (Cambridge Technology, Watertown, MA). Images are generated by raster scanning the x–y mirrors. The excitation light enters Zeiss microscopes (Zeiss, Thornwood, NY) via a modified epiluminescence light path. The scan lens is positioned such that the x–y scanner is at its eye point while the field aperture plane is at its focal point. Since the objectives are infinity corrected, an excitation tube lens is positioned to recollimate the excitation light. The scan lens and the excitation tube lens function together as a beam expander that overfills the back aperture of the objective lens. The dichroic mirror reflects the excitation light to the objective. The dichroic mirrors are custom-made short-pass filters (Chroma Technology, Brattleboro, VT) that maximize reflection in the infrared and transmission in the blue-green region of the spectrum. Various high-NA objectives can be used for the experiments. High-NA objectives ensure tight focusing and result in highly efficient two-photon excita-

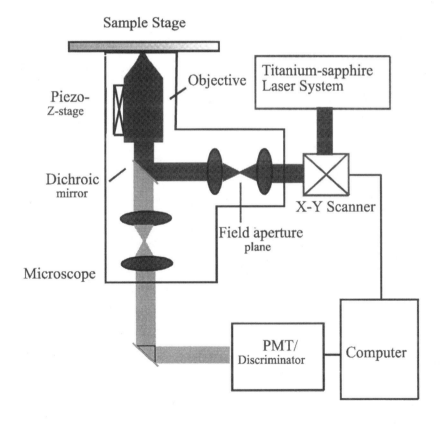

Fig. 10.3. A two-photon fluorescence microscope.

tion. Two commonly used objectives are Zeiss c-Apochromat 40X (NA 1.2) and Zeiss Fluar 40X (NA 1.3). A piezo-driven device, which is interfaced to a computer, controls the objective axial position. The field of view is about 50–100 μm on a side. Typical image acquisition time for a frame is about 1 s.

The fluorescence emission is collected by the same objective and transmitted through the dichroic mirror along the emission path. An additional barrier filter is needed to further attenuate the scattered excitation light because of the high excitation intensity used. Since two-photon excitation has the advantage that the excitation and emission wavelengths are well separated (by 300–400 nm), suitable short-pass filters such as 2 mm of BG39 Schott glass filter (CVI Laser, Livermore, CA) eliminate most of the residual scatter with a minimum attenuation of the fluorescence. A descan lens can be inserted after the tube lens. A single-photon counting signal detection scheme is commonly used to detect and process the fluorescence photons. Fluorescence signal at each pixel is first detected by an R5600-P photomultiplier tube (PMT; Hamamatsu, Bridgewater, NJ), which is a compact single-photon counting module with high quantum efficiency. Then the signal is conditioned by a low-noise preamplifier and a photon discriminator (Advanced Research Instrument, Boulder, CO) before being transferred to the memory of the data acquisition computer.

Imaging of Mouse Ear

To demonstrate the power of two-photon imaging in deep tissue, we use an ex vivo mouse ear as a model imaging system. Images at five layers of the tissue were acquired (Fig. 10.4), and distinct physiological features at each layer can be readily resolved. At the tissue surface, that is, the stratum corneum, large, hexagonal-shaped cells were observed. These highly fluorescent cells in the stratum corneum typically have lost their nuclei. Cells in the process of sloughing were also observed. Two layers of living cells were observed between the stratum corneum and the basal layer. Quantitatively, the fluorescence intensity originating from these layers was one-half to one-third that of the stratum corneum. The cells in these layers were fairly large (about 15 μm across) and well organized. The fluorescence of these cells originates mainly from the cytoplasm. From previous studies (Albota et al., 1998), the major chromophore in the cytoplasm responsible for the observed fluorescence is NAD(P)H. The cellular nuclei with a negligible NAD(P)H concentration were clearly observed as large oblong dark regions. The basal layer lies above the epidermal–dermal junction and cellular autofluorescence from the basal layer is slightly more intense as compared to the upper cell layers. Cell packing in this layer was less well organized. These cells were smaller in cross section but were thicker as compared with cells in the upper epidermal layers. In the dermal layer, we observed fiberlike structures likely to be collagen or elastin fibers. Below the dermis, we observed honeycomblike structures with a loose hexagonal packing arrangement. Fluorescence originates from the walls of these honeycomb structure. The morphology of this layer

Fig. 10.4. Sections of mouse ear structures obtained by two-photon deep-tissue microscopy. From left to right, the images are of stratum corneum, epidermal cell layer, basal cell layer, dermal structure, and cartilage. The depth of these images is indicated in the upper left corner of the each frame. The bar scale is 20 μm. (Reprinted with permission from OSA. Optics Express 3:339–350, 1998.)

is consistent with the cartilage structure below the dermis, which is the primary structural support of the mouse ear.

APPLYING A BLIND DECONVOLUTION ALGORITHM TO IMPROVE TWO-PHOTON IMAGES

Blind Deconvolution Algorithm

Although two-photon microscopy is an excellent tool for obtaining structural information in deep tissues, greater image distortion in the axial axis is expected due to the larger dimension of the psf along that direction. This discrepancy in radial and axial psf means that out-of-plane noises can contribute significantly to distortion in image quality, resulting in the presence of "haze" in the acquired images. To rectify this problem, 3D deconvolution techniques can be used to further enhance spatial resolution and therefore the image quality of the specimen. In addition to improving image quality, deconvolution can also provide information about the psf in deep tissue. For example, a 3D blind deconvolution algorithm based on the principle of maximum likelihood may not require the psf to be known. In fact, the AutoDeblur software (AutoQuant, Watervliet, NY) we use actually returns the psf as a parameter. This software is based on the blind deconvolution technique developed by Holmes (1992) and is extremely important for several reasons:

1. Inside inhomogeneous samples such as tissues, it is very difficult to derive the psf theoretically. Since the psf in a given tissue specimen is generally unknown, a blind deconvolution algorithm that does not require prior knowledge of psf is a more general technique.

2. The ability of the blind deconvolution algorithm to produce psf at different depths in the sample allows us to characterize optical properties inside the complex, inhomogeneous specimen and can lead to a better understanding of photonic processes in such samples.

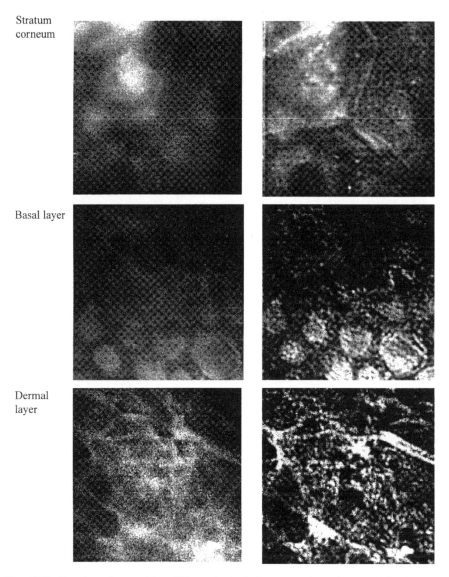

Fig. 10.5. Two-photon human skin at different depths. Left: raw two-photon images. Right: deconvoluted results. The blind deconvolution algorithm also reveals the psf. FWHM's of the psf in the order of stratum corneum, basal layer, and dermal layer are: radial 0.49, 0.45, and 0.43 μm; axial: 2.5, 2.3, and 2.2 μm.

Applying Deconvolution to Deep-Tissue Images of Human Skin

To realize the power of deconvolution in improving image quality and derive information on the psf inside deep tissue, we applied the blind deconvolution software AutoDeblur to different layers inside human skin imaged by two-photon fluorescence microscopy. A 100x oil-immersion objective (Zeiss Fluar, NA 1.3) was used to image three distinct regions inside a sample of human skin: (1) stratum corneum near the surface, (2) basal layer at the epidermal–dermal junction, and (3) filamentous structures inside the dermal layer. These images along with the deconvoluted results are shown in Figure 10.5. First, the images clearly show a qualitative improvement in image quality. Finer details are revealed in the deconvoluted images. Furthermore, the psf's determined from the deconvolution show that the full widths at half maximum (FWHMs) did not change significantly even though these three layers are located at different depths inside the skin. To be specific, the radial FWHMs varied from 0.49 μm in the stratum corneum to 0.45 μm in the basal layer and is 0.43 μm in the dermal layer. Axial FWHMs showed the same trend. The FWHMs are 2.5, 2.3, and 2.2 μm in the respective layers. This is consistent with results published by Centonze and White (1998) in which latex particles were imbedded in tissue phantoms.

VIDEO-RATE TWO-PHOTON IMAGING

Implementing Video-Rate Two-Photon Microscopy

Although two-photon methodology can be applied to in-depth imaging of thick tissue, challenges remain in implementing this technology in the clinical setting. A major difficulty is associated with the relatively slow imaging speed of typical two-photon microscopes. The typical frame rates of two-photon and confocal microscopes are between 0.5 to 10 s. In order to acquire a three-dimensional image of a 200-μm-thick tissue with 500 optical sections, data acquisition time on the order of hours is needed. In the clinical setting, such long acquisition time is impractical for patient diagnosis. In addition, the slow data acquisition rate can present problems in image acquisition throughout the 3D image stack. For long acquisition times, the images acquired are more likely to be subject to motional artifacts caused by the patients. Two different approaches have been demonstrated in bringing two-photon imaging speed to video rates (about 20–30 frames per second) (Brakenhoff et al., 1996; Guild and Webb, 1995; Bewersdorf et al., 1998).

The first technique is based on the line scanning approach. A line scanning approach reduces image acquisition time by covering the image plane with a line instead of a point (Brakenhoff et al., 1996; Guild and Webb, 1995). Line focusing is typically achieved by using a cylindrical element in the excitation beam path. The resulting fluorescent line image is acquired with a spatially resolved detector such as a charge-coupled device (CCD) camera. The main drawback associated with line scanning is the degradation of the image psf, especially in the axial direction.

A second approach, which has been termed *multiphoton multifocal microscopy*, is analogous to Nipkow disk-based confocal systems (Bewersdorf et al., 1998). This approach is based on a custom-fabricated scan lens consisting of a specially designed lenslet array that focuses the incident laser into multiple focal spots at the field aperture plane. The lenslet array is arranged similarly to the traditional Nipkow design. Upon the rotation of the scan lens, the projected focal spots of the lenslet array will uniformly cover the field aperture plane. The CCD cameras are used to register the spatial distribution of the resulting fluorescent spots and integrate them into a coherent image. The ability to image multiple sample regions simultaneously reduces total data acquisition time. The technique suffers much less from resolution degradation and has the added advantage of being extremely robust.

A High-Speed, Video-Rate Point Scanning System

A third method optimized for high-speed deep-tissue imaging is based on raster scanning of a single diffraction-limited spot utilizing a high-speed polygonal mirror. This instrument is adapted from a very successful two-photon microscope designed for deep-tissue imaging (So et al., 1998). Since a single spot is scanned, this system retains the diffraction-limited resolution of a standard two-photon microscope. Furthermore, since fluorescence is generated at only one location at any time, a large-area single-point detector can be used. The spatial information is encoded by the timing of the raster scan pattern, as in a typical confocal microscopy. The image resolution can also be further improved by replacing the CCD camera with a large single-pixel detector, to remove the dependence on the emission psf. This is particularly important in turbid specimens where the scattered fluorescence signal is not confined in a single pixel of the CCD camera and results in degradation of the image resolution.

In our implementation, a fast rotating polygonal mirror (Lincoln Laser, Phoenix, AZ) is used for high-speed line scanning along the x-axis and a slower galvanometer-

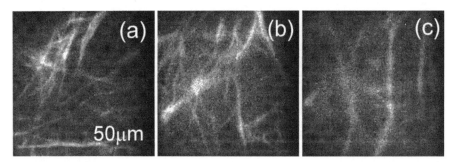

Fig. 10.6. Collagen/elastin fibers imaged in the dermal layer of an ex vivo human skin sample. Each picture was acquired in 90 ms. Images at depths of (a) 80, (b) 100, and (c) 120 μm are shown. (Reprinted with permission from SPIE. Proc SPIE 3604:60–66, 1999.)

driven scanner with 500-Hz bandwidth (Cambridge Technology, Watertown, MA) correspondingly deflects the line-scanning beam along the y-axis. The spinning disc of the polygonal mirror comprises 50 aluminum-coated facets (2 × 2 mm) arranged contiguously around the disc's perimeter. The facets repetitively deflect the laser beam over a specific angular range, correspondingly scanning a line 50 times per revolution. Rotation speeds of either 10,000, 15,000, 20,000, or 30,000 rpm are available.

The fluorescence is recorded by an intensified, frame transfer CCD camera (Pentamax; Princeton Instrument, Trenton, NJ). The 12-bit data of the 512 × 512 pixel CCD chip can be read out at 5 MHz. The maximum achievable image transfer rate is about 10 frames per second for 256 × 256 pixel images (2 × 2 pixel binning). Consequently, the polygonal mirror spinning speed is adjusted to 10,000 rpm, and the CCD exposure time is correspondingly set to 90 ms. In order to perform 3D volume scans, the objective is mounted on a computer-controlled piezoelectric objective translator (P-721.00, Physik Instrumente, Waldbronn, Germany).

A laser diode (1 mW at 632 nm; Thorlab, Newton, NJ) and a photodiode detector (Thorlab, Newton, NJ) are used to encode the polygonal mirror position and to generate a reference signal to synchronize the x, y scanners, the objective translator, and the CCD camera.

To demonstrate the acquisition of clinically relevant images using a video-rate two-photon microscope, we have studied the dermal structures in ex vivo human skin. The collagen/elastin fiber structures in the dermal layer were imaged (Fig. 10.6). One hundred images were taken at depths between 80 and 120 μm below the skin surface. The frame rate was 90 ms and the entire stack of images was acquired in 9 s. The filamentous structures of the collagen/elastin fibers can be clearly resolved from the figure. Representative images of the fiber structures were shown in Figure 10.5. Due to the lower fluorescence intensity of the NAD(P)H components in the epidermal keratinocytes, they are more difficult to observe. The basal cells can be visualized; however, their high melanin content makes thermal damage due to one-photon absorption a major concern. Thermal damage to the basal layer was occasionally observed. We are investigating a method to mitigate this problem (Masters et al., submitted).

TWO-PHOTON FLUORESCENCE SPECTROSCOPY IN DEEP TISSUE

Importance of Spectroscopic Information in Deep-Tissue Imaging

In addition to structural information, two-photon fluorescence microscopy can be used to assess tissue functional state by spectroscopic study of the induced fluorescence. Information such as emission wavelength, fluorescence lifetime, and emission polarization can be used to monitor the tissue microenvironment. Fluorescence spectroscopic techniques can be applied in a number of ways. First, different structures in the tissue can be distinguished either by their endogenous spectroscopic properties or by extrinsic fluorophores labeling the tissue. Second, fluorescence spectroscopy can be used to study tissue biochemical states and the concentration of metabolites such as calcium and oxygen. For example, cellular metabolism can be monitored optically and non-invasively by the technique of redox fluorometry (Masters, 1990; Chance and Thorell, 1959; Masters et al., 1993). A class of common endogenous probes is the pyridine nucleotides. The relative concentration of NAD (nonfluorescent) versus NAD(P)H (fluorescent) in the tissue varies depending on cellular metabolic conditions. Therefore, tissue metabolism can be monitored by tracking fluorescence intensity changes in cells and tissues.

Biomedical Applications of Fluorescent Spectroscopic Information in Two-Photon Microscopy

A study was performed to quantify the biochemical species responsible for the fluorescence observed in the cornea of a freshly excised rabbit eye. Sample preparation procedures were similar to those used in Piston et al. (1995). The emission spectra of the various cell layers in the ex vivo rabbit cornea (wing cells, basal cells, and stro-

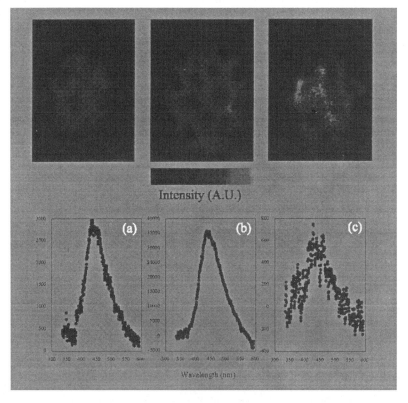

Fig. 10.7. Representative images and emission spectra of three cell layers in the rabbit cornea. (a) wing cells, (b) basal cells, and (c) stroma keratocytes. (Reprinted with permission from IEEE. IEEE Engineering in Medicine and Biology Magazine 18:23–30, (1999).)

mal keratocytes) with two-photon excitation at 730 nm are shown in Figure 10.7. These spectra show a peak in the region of 450 nm and an FWHM of about 100 nm. The shape and wavelengths of the emission spectra are consistent with those of the NAD(P)H fluorescence.

The maximum intensities of the fluorescence emission occur at 450 nm and show a large variation between the various cell layers. The highest fluorescent intensity was recorded for the basal epithelial cells (35,000 AU). The peak intensity recorded from the wing cells (3000 AU) was more than an order of magnitude less than that from the basal epithelial cell. The lowest fluorescent intensity (600 AU) was recorded from the stromal keratocytes. These results support the fact that the cells in the basal region are the most metabolically active.

CONCLUSION

Two-photon fluorescence microscopy is a powerful technique for studying deep-tissue physiology. In this chapter, we discussed three recent complementary developments in the field with significant impacts in advancing two-photon, deep-tissue technology. We have shown that a blind deconvolution algorithm can not only further improve quality of two-photon images but also provide psf information in inhomogeneous specimen. The additional psf information can help to understand two-photon-induced photonic interaction in deep tissue. We have also seen the importance of video-rate microscopy in extending two-photon technology to potential clinical applications. In the section on video-rate, two-photon technology, we saw that video-rate, two-photon microscopy can provide fast structural information in deep tissue. Future development clearly will extend the faster imaging technique to clinical diagnosis of diseases. In addition to structural information, two-photon spectroscopic data can be acquired in conjunction with images. The biochemical information acquired can be complemented to structural images, providing a new dimension to the two-photon technique.

These developments represent significant advances in two-photon, deep-tissue microscopy. In the future, it can be envisioned that they will act in unison to address important physiological problems in deep-tissue studies.

REFERENCES

Albota M, Beljonne D, Bredas JL, Ehrlich JE, Fu JY, Heikal AA, Hess SE, Kogej T, Levin MD, Marder SR, McCord-Maughon D, Perry JW, Rockel H, Rumi M, Subramaniam G, Webb WW, Wu XL, Xu C (1998): Design of organic molecules with large two-photon absorption cross sections. *Science* 281:1653–6.

Bennett BD, Jetton TL, Ying G, Magnuson MA, Piston DW (1996): Quantitative subcellular imaging of glucose metabolism within intact pancreatic islets. *J Biol Chem* 271:3647–3651.

Bewersdorf J, Pick R, Hell SW (1998): Mulitfocal multiphoton microscopy. *Opt Lett* 23:655.

Booth ML, Hell SW (1998): Continuous wave excitation two-photon fluorescence microscopy exemplified with the 647-nm ArKr laser line. *J Microsc* 190:298–304.

Brakenhoff GJ, Squier J, Norris T, Bliton AC, Wade WH, Athey B (1996): Real-time two-photon confocal microscopy using a femtosecond, amplified Ti:sapphire system. *J Microsc* 181(Pt 3):253.

Centonze VE, White JG (1998): Multiphoton excitation provides optical sections from deeper within scattering specimens than confocal imaging. *Biophys J* 75(4):2015–2024.

Chance B (1976): Pyridine nucleotide as an indicator of the oxygen requirements for energy-linked functions of mitochondria. Circ Res Suppl 1 38:I-31–I-38.

Chance B, Schoener B, Oshino R, Itshak F, Nakase Y (1979): Oxidation reduction ratio studies of mitochondria in freeze-trapped samples. J Biol Chem 254:4764–4711.

Chance B, Thorell B (1959): Localization and kinetics of reduced pyridine nucleotide in living cells by microfluorometry. J Biol Chem 234:3044–3050.

Corcuff P, Bertrand C, Leveque L (1993): Morphometry of human epidermis in vivo by real-time confocal microscopy. Arch Dermatol Res 285:475–481.

Dabbous MK (1966): Inter- and intramolecula cross-linking in tyrosinase-treated tropocollagen. J Bio Chem 241:5307–5312.

Denk W, Strickler JH, Webb WW (1990): Two-photon laser scanning fluorescence microscopy. Science 248:73–76.

Göppert-Mayer M (1931): Über Elementarake mit zwei Quantensprungen. Ann Phys (Leipzig) 5:273–294.

Guild JB, Webb WW (1995): Line scanning microscopy with two-photon fluorescence excitation. *Biophys J* 68:290a.

Hoerman KC, Balekjian AY (1966): Some quantum aspects of collagen. Fed Proc 25:1016–1021.

Holmes T (1992): Blind devoncolution of quantum-limited incoherent imagery: maximum-likelihood approach. J Opt Soc Am A 9(7):1052–1061.

Jackson D (1975): *Classical Electrodynamics*. New York: Wiley.

LeBella FS (1961): Studies on the soluble products released from purified elastic fibers by pancreatic elastase. Arch Biochm Biophys 93:72–79.

LaBella FS, Gerald P (1965): Structure of collagen from human tendon as influence by age and sex. J Gerontol 20:54–59.

LaBella FS, Lindsay WG (1963): The structure of human aortic elastin as influence by age. J Gerontol 18:111–118.

Maiti S, Shear JB, Williams RM, Zipfel WR, Webb WW (1997): Measuring serotonin distribution in live cells with three-photon excitation. Science 275:530–532.

Masters BR (1996): Three-dimensional confocal microscopy of human skin *in vivo*: Autofluorescence of normal skin. Bioimages 4:13–19.

Masters BR (1990): *In vivo* corneal redox fluorometry. In Masters BR (ed). *Noninvasive Diagnostic Techniques in Ophthalmology*. New York: Springer.

Masters BR, Chance B (1993): Redox confocal imaging: intrinsic fluorescent probes of cellular metabolism. In Mason WT (ed). *Fluorescent and Luminescent Probes for Biological Activity*, London: Academic Press.

Masters B, So PT, Dong CY, Buehler C, Gratton E (submitted): The use of a laser pulse picker to mitigate two-photon excitation damage to living specimen. *J Rev. Sci. Instr.*

Masters BR, So PTC, Gratton E (1997a): Multiphoton excitation fluorescence microscopy and spectroscopy of in vivo human skin. Biophys *J* 72:2405–2412.

Masters BR, So PTC, Gratton E (1997b): Multiphoton excitation fluorescence microscopy and spectroscopy of in vivo human skin. Biophys *J* 72:2405–2412.

Masters BR, Gonnord G, Corcuff P (1997c): Three-dimensional microscopic biopsy of in vivo human skin: a new technique based on a flexible confocal microscope. J Microsc 185:329–338.

Masters BR, Kriete A, Kukulies J (1993): Ultraviolet confocal fluorescence microcopy of the in vitro cornea: redox metabolic imaging. Appl Opt 32:592–596.

Pawley JB (1995): *Handbook of Biological Confocal Microscopy*. New York: Plenum Press.

Piston DW, Masters BR, Webb WW (1995): Three-dimensionally resolved NAD(P)H cellular metabolic redox imaging of the *in situ* cornea with two-photon excitation laser scanning microscopy. J Microsc 178:20–27.

Raijadhyaksha M, Grossman M, Esterowitz D, Webb RH, Anderson RR (1995): In vivo confocal scanning laser microscopy of human skin: melanin provides strong contrast. J Invest Dermat 6:946–954.

Rummelt V, Gardner LMG, Folberg R, Beck S, Knosp B, Moninger TO, Moore KC (1994): Three-dimensional relationship between tumor cells and microcirculation with double cyanine immunolabelling, laser scanning confocal microscopy, and computer assisted reconstruction: An alternative to cast corrosion preparation. J Histochem Cytochem 42:681–686.

So PT, Kim H, Kochevar IE (1998): Two-photon deep tissue ex vivo imaging of mouse dermal and subcutaneous structures. Opt Exp 3:339.

So PTC, French T, Yu WM, Berland KM, Dong CY, Gratton E (1995): Time-resolved fluorescence microscopy using two-photon excitation. Bioimaging 3:49–63.

Thomas J, Elsden DF, Partridge SM (1963): Degradation products from elastin. Nature 200:651–652.

Wilson T, Sheppard C (1984): *Theory and Practice of Scanning Optical Microscopy.* New York: Academic Press).

Zipfel W (1997): Multi-photon Excitation of Intrinsic Fluorescence in Cells and Intact Tissue. Presented in Application of multi-photon excitation imaging, Pre-Microscope Society of America Symposium, Cleveland, OH, Aug 9–10.

11

Limits of Confocal Imaging

James B. Pawley
University of Wisconsin, Madison, Wisconsin

INTRODUCTION

What Limits?

Perhaps a measure of the utility of the confocal imaging principle is the fact that it seems to have been independently invented at least seven times since the late 1950s (Minsky, 1988; Egger and Petran, 1967; Slomba et al., 1972; Sheppard and Choudhury, 1977; Brakenhoff et al., 1979; Carlsson et al., 1985; White et al., 1987). The optical principle is not complex: focus a point light source into or onto the specimen and arrange for light emitted or scattered from this point on the specimen to be focused onto a point detector [usually an aperture in front of a photomultiplier tube (PMT)]. Photons emerging from features in the specimen that are not in the plane of focus will not be focused into a point at the plane of the detector aperture and, consequently, most of them are excluded. The exclusion of out-of-focus light from the data stream gives the confocal its most characteristic feature: the ability to make optical sections. The price for this sectioning ability is that most laser confocal microscopes at any one time image only a single point on the specimen, and a two- or three-dimensional (2D, 3D) image can be produced only by scanning the focused spot over the specimen or vice versa. This sampling approach allows an image to be built up from a number of individual measurements that reflect optical properties within specific regions of the sample. Any such measurements involve not only the optical limitations inherent in

Emerging Tools for Single-Cell Analysis, Edited by Gary Durack and J. Paul Robinson.
ISBN 0-471-31575-3 Copyright © 2000 Wiley-Liss, Inc.

focusing and imaging systems but also counting photons, a process that implies important limitations on data rate and statistical accuracy.

The task of the confocal light microscope is to measure optical or fluorescent properties within a number of small, contiguous subvolumes of the specimen (Fig. 11.1; Pawley and Centonze, 1994). Because confocal microscopy is probably the most sensitive method for imaging living cells and because imaging such cells places the greatest demands on the instrumentation and technique, we will generally consider the specimen to be a living cell. The fundamental limits on this process are related to the quantitative accuracy with which these measurements can be made, a factor that depends on the number of photons that pass into, n_1, and out of, n_2, the subvolume; its size (δx, δy, δz); and its position (x, y, z). Additional limitations are imposed on the rate at which these measurements can be made by the effects of photodamage to the specimen, source brightness, and fluorescence saturation. Finally, limitations are imposed by the fact that the continuous specimen must be measured in terms of discrete volume elements called voxels (a voxel is the 3D equivalent of a pixel, which is the smallest element of a 2D image). Although, for simplicity in discussion, these factors are usually treated separately, in practical microscopy they almost always interact. The discussion that follows will often highlight the interactions.

This chapter will outline the factors that ultimately limit the accuracy with which these measurements can be made. Although most of the discussion should be applicable to any form of confocal microscope, we will assume that scanning is accomplished by moving a single beam of laser light.

The data recorded from a confocal laser scanning microscope (CLSM) are usually a set of intensity values matched to every pixel of a 2D optical section or to every voxel throughout a 3D volume in the specimen. Ideally, these values represent the concentration of fluorophore as a function of position. In fact, many other factors can produce unintended contrast. In fluorescence confocal microscopy, the rate at which data can be produced is limited by fluorescence saturation. As a result, the statistical quality of the data is often related inversely to the scanning speed, the instability of the fluorophore, and the size of the raster.

Counting Statistics: The Importance of *n*

The accuracy of any particular measurement involving fundamental, quantum interactions (such as counting photons) is limited by Poisson statistics, which ensure that the 1σ error in a measurement of n photons is $\sqrt{n}$. While similar considerations limit the performance of all types of microscopical measurements, they are more explicit in their effect on confocal microscopy, where peak signal levels often produce signal levels of only about 8 photons per 2-μs pixel.

The uncertainty associated with counting quantum-mechanical events is the source of *intrinsic* statistical noise. Confocal data sets can also contain *extrinsic* noise introduced by detector dark current, electronic noise, or interference or produced by stray or out-of-focus light. Unlike intrinsic noise, extrinsic noise can be reduced by careful technique and technological improvements. Indeed, the first test of a confocal microscope should be to collect signal from a "blank field." To do this, form an image

Introduction

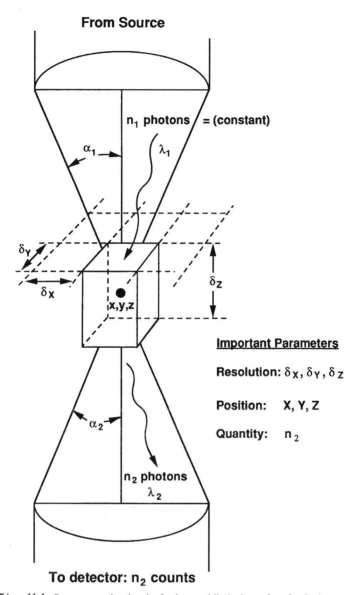

Fig. 11.1. Parameters related to the fundamental limitations of confocal microscopy.

of a nonfluorescent object such as pure water and set the PMT gain as you would for "normal" operation for a weak specimen and the dc offset so that the average recorded signal is about 10 digital units (assuming 8-bit, 256-level digitization). A statistical analysis of a single-scan data set should allow you to differentiate pixels with one dark count from those with none. A count rate of more than 10,000 "bright" pixels per 1-s scan should be a cause for concern. Perhaps the PMT photocathode is being accidentally warmed by nearby equipment. Alternatively, you may have a light leak. To localize the problem, try comparing one detector channel with another, turning off the room lights, or obscuring the laser.

To check for "fixed-pattern" noise, accumulate data from a large number of dark frames by Kalman averaging. The appearance of any sort of pattern on the screen as more data are acquired may also indicate a problem. However, one must realize that fixed-pattern noise always exists at some level. It is only a problem if it is similar in size to the signal level, and even then its effect can sometimes be reduced by subtracting a "noise-only" pattern from a "signal-plus-noise" pattern.

In a well-designed CLSM, the major noise source is intrinsic noise. This fact highlights the importance of making sure that as many as possible of the available photons are recorded as part of the signal.

Specimen Response: Dye Saturation. In "normal microscopy" it is safe to assume that photons interact with the specimen in a manner that is independent of the intensity of the illumination and that output signal is always directly proportional to excitation. However, this linear response is not characteristic of any laser-based fluorescence confocal microscopes operated with >1 mW in a 0.5-μm-diameter spot in the specimen. Not only may absorption in the specimen cause sufficient warming to produce damage or motion, but the electric field strength of the focused light may also become sufficient to produce a variety of nonlinear responses such as fluorescence saturation.

Saturation occurs when fluorescent molecules are excited by a flux of exciting illumination so intense that, at any instant, a significant fraction of the fluorescent molecules is in the excited state. As excited molecules no longer absorb light at the usual wavelength (λ), this has the effect of lowering the effective dye concentration. This saturation threshold can easily be reached at flux levels around 10^6 W/cm^2. Saturation is related to absorption cross section and fluorescence lifetime. As can be seen from Equations (1) and (2), the problem is more severe when using dye molecules with long fluorescent decay times τ_f:

$$\alpha = \frac{I\sigma}{h\nu} \qquad (11.1)$$

where α is the excitation rate in reciprocal seconds at low intensity, σ is excitation cross section in centimeters squared, I is the excitation intensity in watts per centimeters squared, and $h\nu$ is the energy of the absorbed excitation photon. Then

$$N_1 = \frac{\alpha}{\alpha + 1/\tau_f} \qquad (11.2)$$

where N_1 is the fraction of dye molecules in the first singlet state, near saturation. With most dyes, singlet-state saturation occurs at power levels of about 1 mW if focused into a diffraction-limited spot by a numerical aperture (NA) 1.3 lens. As lower NA produces a larger spot area, an NA 0.65 would not saturate until about four times this power. If the spot is not diffraction limited because of specimen effects, then more power can be used because the light is spread over a larger area. It is recommended to stay at least a factor of 10 below saturation, preferably a factor of 100.

Effects of Singlet-State Saturation

Because fluorescence saturation affects any dye molecules in the crossover, those in high-concentration (bright) pixels are affected the same as those in low-concentration (dim) pixels. As a result, the contrast of the final image may seem unaffected: it does not look "saturated" in the same way that a video image does when many pixels are recorded as 256 because the signal level is too high. Therefore, one cannot feel confident that one is avoiding fluorescence saturation by the casual evaluation of collected image data.

Rather than distorting the intensity contrast, fluorescence saturation causes the signal from *any particular* pixel to become a function of variables other than dye concentration. These variables include the local fluorescent decay time, which can depend strongly on the molecular environment, and local defocus effects that may affect the peak intensity in the focused spot and hence the degree of saturation. It follows from this last point that, as saturation occurs mainly in the focused spot, relatively more signal will be produced (and detected) from *adjacent* planes when one operates near saturation. This effect will marginally reduce the z-resolution.

However, the most important reason that saturation should be avoided is that it exposes the specimen to additional damaging illumination without producing proportionally more data. In fact, this bleaching effect may be more than linearly proportional to the "wasted" illumination: it may change the damage mechanism. Although, photodegradation mechanisms are as yet poorly understood, it now seems likely that the absorption of a second photon by a molecule still in the singlet-excited state may turn out to be a common bleaching mechanism (Sanchez et al., 1997). If this occurs, then it follows that above some intensity level, the bleach rate will be proportional to the square of the light flux, and operation near saturation would increase the effective bleach rate/excitation photon as well as reducing the signal-to-illumination ratio.

Fluorescence saturation is a fundamental limitation on the rate at which confocal fluorescent data can be acquired. It can be side-stepped only if the beam is focused into a large spot or if the illuminating light is formed into more than one focused spot on the specimen. These conditions are present only in the disk-scanning instruments or laser line scanners (Hell et al., 1996; Buist et al., 1998; White and Amos, 1995).

Problem of Interactions

Limits on the performance of the CLSM do not act alone but can combine in complex and subtle ways. To highlight these interactions, let us define a characteristic microscopical problem shown schematically in Figure 11.2. The specimen is a living cell in which some of the protein subunits making up bundles of cytoskeletal fibers have been replaced with fluorescent analogs. The diameter of each fiber is about 10 times smaller than the resolution limit of light optics, and excitation of the fluorescent dye causes it to bleach, a process that is toxic to the cell. Because the object is to observe the formation and movement of the linear cytoskeletal elements within the living cell, we will need to collect many images over time.

The important features of this example are as follows:

1. Observations will be improved by high spatial resolution in all three dimensions.
2. The ability to quantitate intensity of stain/unit length may permit determination of the number of subresolution, linear polymers bundled together to make up each visible fiber.
3. A number of images must be recorded at different times in order to show change/motion.
4. Each measurement will cause bleaching and cellular toxicity.

Clearly, these conditions contain an inherent contradiction: to obtain more quantitative temporal or spatial accuracy, we must pass more light through the sample, but this will produce more fading and cytotoxicity. As a result, the "improved" images may be of a dying cell rather than a living one, and, what we might call the *biological reliability* of the measurement may be inversely proportional to the *physical accuracy* set by counting statistics.

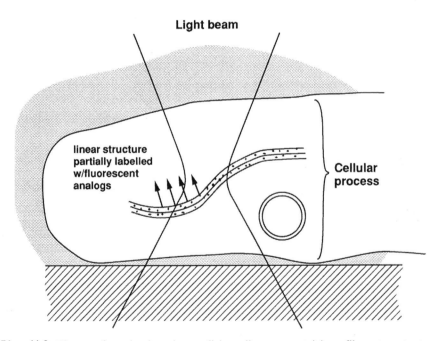

Fig. 11.2. Diagram of a notional specimen: a living cell process containing a filamentous structure smaller than the resolution limit and sparsely stained with fluorescent analog molecules.

Introduction

There are other interactions among these parameters. High spatial resolution and Nyquist sampling imply that the microvolumes (pixels) excited by the beam and sampled by the detector must be partially overlapping. The measurement at each pixel is really the detection of that small fraction (usually <1%) of the fluorescent excitations in the specimen that actually produce signal in the detector (Sandison et al., 1994).

Higher spatial resolution implies smaller pixels (or voxels in 3D) and hence the need to count more photons from any given volume of the specimen. In addition, smaller features show less contrast. Therefore, if small features are to be resolved, even more events must be detected in order to reduce the statistical noise enough to visualize lower contrast features.

The increases in detected signal required for improved resolution are not insignificant. Just maintaining the statistical accuracy of the measurements when the size of the voxels is reduced by a factor of 2 requires four times more signal from the same volume of specimen in order to image a single plane and eight times more if a 3D volume is to be sampled.

Although this calculation accounts for the increase in the number of voxels needed, it still grossly underestimates the increase in dose needed to visualize smaller features. Assuming that the contrast of small features is inversely proportional to their size, the signal required from each voxel to "visualize" a feature is therefore inversely proportional to the square of the contrast. Therefore, if one accounts for both the additional voxels and the reduced contrast, the dose of excitation required to both sample and "resolve" a small feature increases to the fourth power of its spacings.

Fortunately, the x–y resolution, $d_{x\text{-}y}$, is defined by the Abbé equation,

$$d_{x\text{-}y} = \frac{0.6\lambda}{NA} \qquad (11.3)$$

and the only way to actually increase it by two times is to increase the NA by two times. Because this also increases the fraction of the photons emitted from the specimen that contribute to the signal by a factor of $(\Delta NA)^2$, or four times, improved "resolution" can be attained with only a moderate dose penalty.

The most commonly used expression for resolution in the z-direction, d_z, is

$$d_z = \frac{2\lambda_0 \eta}{(NA_{obj})^2} \qquad (11.4)$$

This argument introduces the importance of keeping the pixel size appropriate to the operating resolution. Microscopists who normally record wide-field (WF) images on film may be less familiar with the idea that pixel size is an explicit experimental variable. It is important to remember that, on most commercial CLSMs, the "zoom" magnification factor used to change the size of the area scanned on the specimen usually does so by changing the pixel size (on the specimen). Therefore, although the ability to vary the display magnification by a factor of about 10 : 1 may seem to be a great convenience, *with a given specific λ and NA, only one zoom setting provides optimal Nyquist sampling* of the optical image data. All other zoom settings must necessarily either over- or undersample the object and produce either more beam damage or less resolution than they should.

Given the interrelated constraints highlighted above, the two parameters of a confocal microscope that must be optimized to approach its ultimate performance are:

1. *Photon Efficiency.* The system must count as many as possible of the photons scattered or emitted by the sample at the plane of focus.
2. *Spatial and Temporal Resolution.* Although, in practice, spatial resolution in fluorescence confocal microscopy is more likely to be limited by insufficient signal than by optics per se, it is still important to increase the x, y, and z optical resolution as much as possible by using the appropriate pixel size and the highest NA available and by scanning the beam no faster than is required to answer the biological question.

PRACTICAL PHOTON EFFICIENCY

Photon efficiency (γ) is the fraction of the signal photons generated by the action of the laser beam on the specimen that is accurately represented in the final image data set. Although photons can be lost both between the light source and the specimen or between the specimen and the detector, those lost before reaching the sample can usually be "replaced" with relative ease by an increase in laser power. However, photons lost after leaving the sample represent a fundamental loss: they carry information obtained at the expense of radiation damage to the specimen.

Light generated inside the specimen can be lost through several mechanisms:

1. absorption or scattering by either the objective lens or the medium that couples it to the focus plane or by the fixed and/or moving mirrors and transfer optics needed to scan the beam;
2. incorrect alignment of the optical system resulting in the improper placement or orientation of the pinhole;
3. low quantum efficiency (QE) of the photon detector; or
4. imprecise digitization of the output of the photon detector.

These subjects will be covered here as a background to the descriptions of practical tests of performance that follow.

Losses in the Optical System

Objectives. To a first approximation, fluorescent light proceeds from the site of its generation equally in all directions. As a result, only a small fraction even strikes the objective lens: ~30% of the total for NA 1.4 (assuming that the specimen is mounted in medium with an index of refraction of 1.515). The fraction of this light that emerges above the objective depends on its transmittance. Microscope manufacturers

Practical Photon Efficiency

TABLE 11.1. Transmission of Selected Representative Objectives at Various Wavelengths

Brand	Description	\	\	Wavelength, nm	\	\	\	Design
		320	350	400	500	600	700	
Nikon	CF 40×/1.3 Fluor oil	16	66	80	90	91%		160 mm
	CF 40×/1.2 Planapo water		56		77			160 mn
	CF 40×/1.15 Fluor water	20	56	72	80	83	81	160 mm
	CF 40×/1.3 Fluor oil	28	64	78	85	86		160 mm
	CF 60×/1.4 Planapo oil				80			160 mm
	CF 100×/1.3 Fluor Oil	30	58	67	80	87	89	160 mm
Zeiss	16×/.5 Plan Neofluar		50		86	80	76	Infinity
	40×/1.2 Planapo water	20	55		84	91	91	Infinity
	40×/1.3 Fluar oil	29	79	88	95	99		Infinity
	100×/1.0 Achroplan water		60	90	94	90	90	Infinity
Olympus	20×/.4 Planapo water		56	67	86	89		160 mm
	40×/.9 Planapo UVLSM water		56	68	88	90		160 mm
	100×/1.1 Planapo UVLSM water		60	73	90	92		160 mm

are now more willing to provide transmission data for certain modern lenses as a function of wavelength (Table 1). Instructions on making comparative measurements of the transmission of objectives can be found in Centonze and Pawley (1995).

Mirrors. A simple setup can be used to test the performance of the internal mirrors on most microscopes. All that is needed is a photodiode light sensor (one *without* an air space between it and the clear "window") linked to a sensitive current meter (or better, a basic photometer) and a front-surfaced mirror. After adjusting the instrument to measure backscattered light (BSL) with low laser power, align the instrument if necessary; then switch to a high zoom setting and measure the light emerging *from* the objective with the sensor. Call this light reading I_a. Be sure to prevent stray room light from reaching the sensor and to couple the lens directly to the sensor with immersion oil if appropriate.

Now place a front-surfaced mirror on the specimen stage and set up the microscope to image the reflecting surface using the BSL signal. Then the beam scanning is stopped (or made to scan a very small raster) when focused on the mirror surface (brightest). Be careful to keep the illumination very low so as not to damage the PMT in the microscope and to realign the instrument if reducing the laser power means you have to add any neutral density (ND) filters in front of the laser. Focus initially on dust on the mirror specimen; then adjust the focus and alignment to give maximum reflected signal.

If your microscope produces specular reflection artifacts (bright blobs having nothing to do with the specimen) when used in the BSL mode, make your measurements in a part of the image field unaffected by these reflections.

After turning off the PMT(!), make a second reading (P_a) with the same photodiode placed just in front of the pinhole, making sure that *all* the light strikes the sensitive part of the sensor. To check that the sensor is protected from room light, obscure or turn off the laser, which should make the reading go to zero.

The P_a/I_a ratio will usually be depressingly small (~10–20%), but the difference covers losses at the various mirror and filter surfaces, including the beamsplitter (up to 70% loss for one pass through a so-called 50/50 reflected light beamsplitter), as well as those at the eyepiece, the various transfer lenses, and the objective.

Though indirect and somewhat cumbersome, such measurements can be useful for two reasons: (1) as a rough method of comparing the performance of different instruments (or different types of mirrors fitted to the same instrument) and (2) to monitor performance of a specific instrument over time. Performance degrades as dust and/or hydrocarbons and other atmospheric vapors condense onto the surfaces of mirrors and other optical components. In instruments having up to 11 reflective surfaces in the detector chain, a change in reflectance of even 1% can have marked effects on total photon efficiency.

Pinhole. In a confocal microscope, the pinhole prevents fluorescent light from reaching the detector unless it originates from the plane of focus. The pinhole is mounted in an image plane, and if it is misaligned or if its size is reduced beyond that corresponding to a diffraction-limited spot in that plane, then it will severely reduce the number of photons reaching the detector while producing only a marginal improvement in *x–y* resolution. Making the pinhole larger than the diffraction-limited spot allows detection of more photons; but as almost all of those originating *from* the plane of focus were already being collected, *most of the additional signal comes from adjacent planes,* reducing the z-resolution. Choice of the appropriate size and shape for the pinhole is a sensitive function of λ and of the objective lens NA and magnification. However, even an "optimum" aperture will exclude at least the 18% of photons that are in the outer rings of the Airy disk.

Multiphoton Excitation. Now that we have considered the loss produced by the pinhole, it is perhaps appropriate to insert a brief discussion of multiphoton fluorescent excitation—the only method of laser scanning light microscopy that can produce optical section data *without* the use of a pinhole. Earlier we discussed singlet-state saturation, a problem that occurs only when the extreme optical brightness of the milliwatt laser is focused into a diffraction-limited spot. Two-photon or multiphoton microscopy depends for its operation on a peak light intensity almost one million times higher again (Denk et al., 1990; see Chapter 10).

Intensity levels this high can produce nonlinear interactions, that is, interactions in which the output event is not linearly dependent on the input intensity. One such interaction is two-photon excitation. In this process, the density of photons in the "vicinity" of a fluorescent molecule becomes so great that occasionally two are absorbed simultaneously, almost as if they were a single photon having half the wavelength and twice the energy. Because this process requires the presence of two photons, the likelihood of its occurring varies with the square of the intensity of the photon flux.

Assuming negligible absorption, the intensity of light at any level in the focused cone emerging from a high-NA objective varies with the square of the distance from the focus plane. Consequently, the probability of exciting a fluorescent photon by a two-photon process drops off with the fourth power of the distance from the focus plane, and light is effectively excited *only* from very near the focus plane.

This is a very useful characteristic. It means that the dye is not subject to bleaching above or below the focus plane and that optical sectioning occurs automatically, without the need for using a pinhole in front of the photodetector. It is well suited to looking deep into living tissue slices because the lack of a pinhole permits one to collect and utilize as a signal even fluorescent light that has been scattered between the focus plane and the objective lens. In addition, UV-excited dyes can now be excited using normal optics and near-IR illumination, the design of the dichroic is simplified by the large "Stokes shift" between the wavelength of the illumination and that of the fluorescent light, and the longer-wavelength, near-IR illumination is scattered less by cellular structures.

There are also some disadvantages. To obtain the required intensity without cooking the specimen, one normally uses a laser capable of making pulses about 10^{-13} s in duration at a repetition rate of about 100 MHz (i.e., the laser is on only about 0.001% of the time, but the peak power is about 10^5 times greater than the average power). Such pulses require complex and expensive lasers that usually must be coupled to the scan head directly rather than via an optical fiber. In addition, the linear resolution is about two times larger (and the volume resolution eight times larger) because of the longer wavelength used; because average laser power levels are often 10–100 times greater than those used in single-photon excitation, cells are likely to be severely damaged if they happen to contain any pigments that absorb at the illumination wavelength. Finally, although the total amount of bleaching is reduced because it occurs only at the plane of focus, it seems that two-photon excitation may be about twice as damaging per excited photon as single-photon excitation. (Sanchez et al., 1997).

In like manner, it is possible in some cases to use similar equipment to perform three-photon excitation (Hell et al., 1996).

Because the fluorescent light is confined to a single plane by the excitation process, multiphoton instruments can in principle be operated not only without pinholes but also without descanning the fluorescent light through the scanning mirrors at all (Wokosin et al., 1998). However, as with normal confocal, the final image is still dependent on the performance of the photodetector and the digitizing system employed.

Detection and Measurement Losses

The Detector. The detector characteristics of importance to 3D microscopy have been reviewed by several authors (Pawley, 1994; Sheppard et al., 1992; Sandison et al., 1994; Stelzer, 1997). The characteristics of most importance to photon efficiency are:

1. *Quantum Efficiency (QE).* The proportion of the photons arriving at the detector that actually contribute to its output signal. (This may be a strong function of the wavelength of the detected photons; see Figure 11.3.)

2. *Noise Level*. This includes both additive noise, in the form of electronic amplifier noise or statistical variations in the dark current of the PMT or charge-coupled device (CCD), and multiplicative noise, in the form of pulse-to-pulse variations in the PMT output produced by individual photoelectrons.

The PMT. Although the PMT is the most common detector used in the CLSM, it is not necessarily the ideal detector. While the QE of a PMT may be as high as 30% in

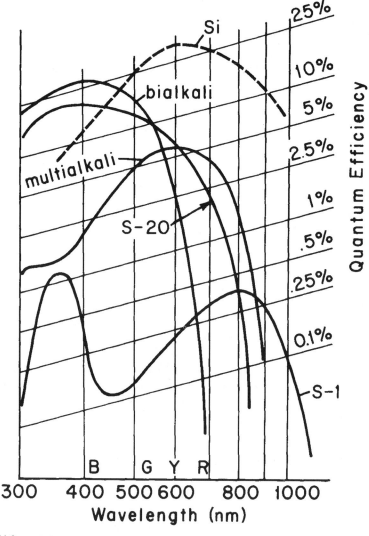

Fig. 11.3. Variation of quantum efficiency with wavelength for some representative photocathode materials (solid lines) and silicon photodiodes (dashed line). Unless they are thinned and rear illuminated, the performance of the Si sensors in a CCD is substantially (50%) less in the green and blue because of absorption in the charge transfer electrodes.

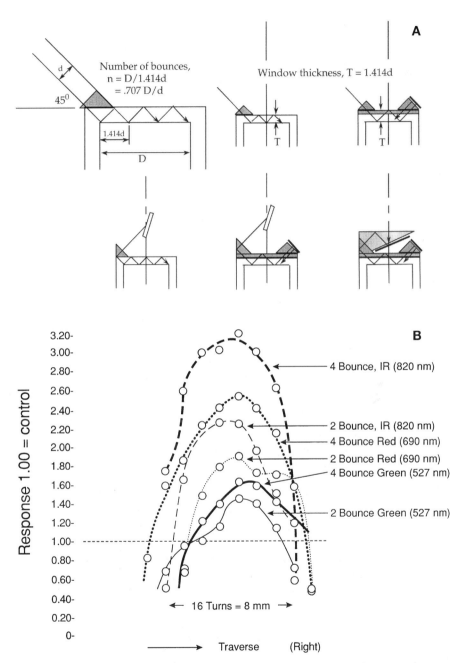

Fig. 11.4. (**A**) Optical designs for increasing the QE of end-window PMTs. Each system endeavors to reduce transmission and reflection losses by using total internal reflection to trap incoming light inside the window material so that it has a greater chance of reacting with the photocathode material and producing a photoelectron. (**B**) Performance improvement at 527, 690, and 820 nm with a prism arrangement that caused either two or four encounters with the photocathode. Such systems work well only with collimated light and have a limited acceptance aperture, in this case 8 mm for a 25-mm PMT.

the blue/green, this still means that 70% of the photons produce no signal. At 565 nm, only about 15% of photons striking the outer surface of a selected, S-20, end-window PMT are absorbed in the thin photocathode (PC) layer evaporated onto the inner surface of the window. The remaining 85% are either transmitted or reflected. As only absorbed photons can produce photoelectrons (PEs) to be subsequently amplified in the multiplier section, it is clear that any mechanism that reduces PC transmission or reflection losses may substantially improve PMT QE. This can be done by introducing collimated light into the PMT at an angle such that when it strikes the PC, it is totally internally reflected (Gunter et al., 1970; Fig. 11.4*A*). The light not absorbed when it first reaches the PC reflects off the outer surface of the window and then strikes the PC again. This process continues until the light either is absorbed or reaches the far side of the PC. Using such an optical enhancer can *increase* the QE by 60% at 520 nm, 180% at 690 nm, and 220% at 820 nm (Fig. 11.4*B*; Pawley et al., 1993b).

To take advantage of this optical effect, manufacturers have introduced tubes in which the PC material has been evaporated onto a "prismatic" surface made up of contiguous 45° pyramids. This allows a parallel, axial light beam to interact with the PC twice, a process that improves the average QE at a cost of pronounced (50%) variations in local, effective QE: depending on where on the pyramid a particular photon arrives, the resulting PE may or may not be able to avoid colliding with the walls of the "valleys."

Transmission losses can also be eliminated by employing a solid photocathode, and recent side-window PMTs such as the Hamamatsu R-3896 have markedly higher QE, especially in the red. Although a significant achievement, such improved PC performance in the red is usually accompanied by significantly ($10\times$) higher dark-count rates. While the dark current of a PMT is usually low enough to be inconsequential, it is a strong function of temperature. In the heated interior of some commercial instruments, the dark count may not always be small compared to the signal level of a weakly fluorescent sample, and this is especially true when excitation levels have been reduced to permit photon counting without pulse pileup (see below).

Of more concern is the fact that the PMT output pulses produced by individual PEs can vary in size by an order of magnitude because of statistical variations in the small number of particles present in the early stages of the electron multiplier (Fig. 11.5*A*). The distribution in pulse height of these single-PE pulses is shown for several representative PMTs in Figure 11.5*B*.

As long as all photons make an equivalent contribution to the output current, the current remains exactly proportional to the number of photons. However, in practice, there is no way of knowing if the current sensed during a single digitizing interval corresponds to (for instance) one photon that happened to have made a large pulse or three photons that have made smaller ones. This uncertainty contributes to an equivalent noise term, called multiplicative noise, that (depending on the single-photoelectron response of the tube) can add 15–30% to the intrinsic, statistical, or photon noise. If not remedied by the use of pulse-counting techniques, the effect of this additional noise can be removed only by increasing the counting time by $(1.15)^2 = 132\%$ to $(1.30)^2 = 169\%$. In other words, operating a PMT in the analog mode *reduces the effective QE of the PMT by between 25 and 45%*!

Practical Photon Efficiency

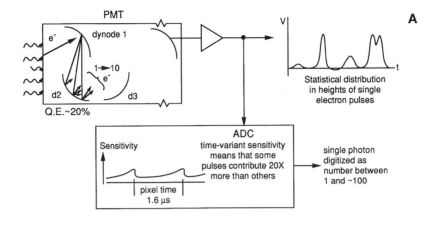

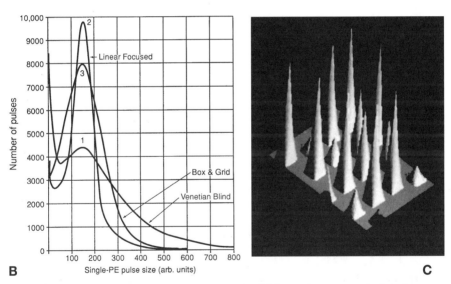

Fig. 11.5. (**A**) In microscope systems that do not employ full integration or operate in the pulse-counting mode, two physical processes can add uncertainty to the measurement of the photon signal. Statistical variations in the number of secondary electrons produced during the early stages of electron multiplication in the PMT have the effect that identical photons can produce pulses that vary in height by a factor of 10. In addition, inappropriate design of the digitizing circuitry can have the result that even identical pulses from the PMT will be recorded as numbers that may vary by a factor of 10. (**B**) Spectrum of single-PE pulse heights for three representative PMTs. The spread in the major peak is due to statistical variations in the gain of the early stages of the dynode multiplier. The peaks of smaller pulses represent PE produced thermally from the later dynodes. In pulse counting, these smaller pulses are eliminated by setting the discriminator threshold near the valley in the distribution. (**C**) An intensity surface representing the pulse heights actually recorded by such a system when digitizing identical pseudo-Gaussian pulses from a signal generator. The overall result of both these processes is to add considerable uncertainty as to the actual intensity of the signal from the specimen.

Cooled CCD. The only practical alternative photodetector to the PMT is the silicon detector, of which the cooled CCD is the optimal example. This detector has both different capabilities and different problems. On back-illuminated CCDs, the QE can be as high as 85% and extend well into the infrared (Fig. 11.6). Furthermore, as each photon is recorded as an identical amount of current, there is no multiplicative noise. Unfortunately, to keep the noise level acceptably low (± 3 photons per measurement), this detector must usually be cooled to -40 to $-80°C$ and read out at the relatively low rate of 25,000–250,000 pixels/s (vs. 600,000 pixels/s for a normal CLSM). This noise level is clearly too high if the *peak* signal level is only 10 photons/pixel, as it can be on many CLSMs. It is less serious when the signal from the *darkest* pixel is at least 10 photoelectrons because then statistical variations in the signal are similar in size to the measurement noise. These features make the cooled-CCD detector more suitable for slowly scanned images (10–100 s/frame) producing relatively high signal levels. So far single-pixel CCDs optimized for use in the confocal microscope have only just reached the prototype stage (Pawley et al., 1996).

In the disk-scanning and line-scanning confocal microscopes, the image data emerge as a real image rather than as a time sequence of intensity values from a single detector. Although this real image can be detected photographically or by eye, these sensors have fairly low QE. However, such confocal microscopes can surpass the photon efficiency of the CLSM if they incorporate a cooled-CCD sensor having detection performance similar to that described above. This combination is now implemented in some commercial instruments presently marketed by EG&G Wallach.

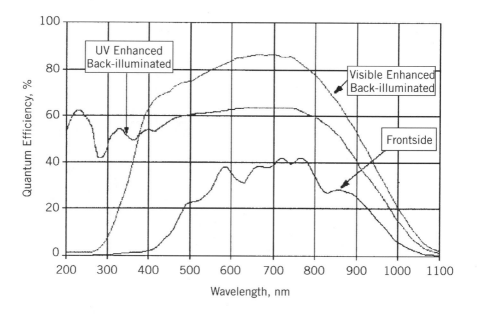

Fig. 11.6. Quantum Efficiency of front- and back-illuminated CCD sensors. (Courtesy of PixelVision Inc., Beaverton, OR.)

Of even more interest to live-cell microscopists are the microlens multiphoton excitation systems developed independently by Straub and Hell (1998) and Buist et al. (1998). By utilizing multiple excitation beams, these instruments avoid the data-rate constraints imposed on single-beam confocals by fluorescence saturation and do so in a fashion that confines excitation (and damage?) to the focus plane. In addition, they increase the data rate by coupling the high-speed disk-scanning system to a high-QE, wide-field detector such as a CCD.

Silicon photon detectors have potential practical advantages besides high QE. As the sensitive element in such a detector is typically very small (7–30 μm on a side), use of small planar arrays (3 × 3 or 5 × 5) could permit it to operate in the CLSM as a combination pinhole and detector (Pawley et al., 1996). Optical misalignment could be sensed electronically simply by comparing the signal from sensor elements on one side against those opposite during scan retrace. Likewise, the effective size of the detector could be adjusted on a scale of 10–50 μm, a size compatible with that of a pinhole operating at the intermediate image plane. Such a CCD detector could even acquire data at several effective pinhole sizes simultaneously.

Digitization. In the simplest CLSM system, the output of the PMT head amplifier is passed directly to the analog-to-digital converter (ADC), which samples the voltage for a few nanoseconds during the time of each pixel (T_p) and turns the sensed voltage into a digital number (Fig. 11.5a). As T_p is usually a few microseconds, it is important to ensure that the voltage present during the short sampling time is a good measure of the average signal level during T_p. This is usually accomplished by giving the amplifier immediately preceding the ADC a time constant of $T_p/4$. This simple approach effectively expands the sampling time from $\sim T_p/1000$ to $T_p/4$ without excessively blending the signal from each pixel with that of its neighbors. However, it still means that the system is only "counting" about 25% of the time, a circumstance that reduces the "effective" system QE by an additional 75%.

The situation can be improved if one uses a digitizer employing full integration. Such a system can be implemented by feeding the current from the PMT into a capacitor during the pixel time, then reading the capacitor voltage out to an ADC, and finally resetting the capacitor voltage back to zero (Fig. 11.7). The Bio-Rad MRC-600 and later instruments incorporate three circuits of this type in each digitizing channel. Three circuits are used so that one can accumulate while the second is being read out and the third is being reset.

The second method of implementing full integration is to feed the output of a high-bandwidth ($T_p/20$) head amplifier into a high-speed ADC running at, say, 10 times the pixel rate and then to utilize fast digital circuitry to average the 10 successive readings needed to produce a stored value characteristic of the whole pixel.

Compared to $T_p/4$ bandwidth limiting, either method of full integration effectively provides four times more useful signal for a fixed amount of light from the specimen. This matter is important enough for it to be worth determining the method used by any confocal instrument that you are considering purchasing.

Photon Counting. Obtaining a digital representation of optical data is ultimately a question of counting photons. This means not only using a high-QE detector but also recording a signal in which the contribution of each photon is equal. In the case of solid-state sensors, the uniformity condition is automatically met by the sensing process (1 photon → 1 PE). This condition can be met by the PMT only if it is operated in a pulse-counting mode.

In pulse counting, the object is not to measure the average level of the output current during the T_p, but rather to eliminate the effects of multiplicative noise by discriminating and then counting the individual output pulses resulting from the emission of individual PEs from the PC. To reduce the effect of the small noise pulses generated from the dynodes of the PMT, photon pulses are passed through a discriminator; each time the PMT output goes above some preset threshold, one pulse is counted (Fig. 11.7).

Unfortunately, problems arise when large numbers of photons must be counted in a short time because the PMT output does not immediately return to zero after each pulse. If a second pulse arrives before the first one is over, the second, or piled-up, pulse will be missed.

Suppose a laser-scanning instrument scans a 512 × 768 raster in 1 s and 35% of this time is used for retrace. That leaves $T_p = \sim 1.6$ μs/pixel. If we assume that each photon pulse occupies 20 ns (T_p), the maximum number of pulses that one could possibly count in each pixel would be 80, but because the photons arrive at random times, even at $1/10$ of this rate (eight photons), 10% of the photons will still arrive when the circuit is already busy.

Figure 11.8 shows the performance of the fast photon-counting circuitry in the Bio-Rad MRC-600 and MRC-1000. The response is linear up to about 10 counts/pixel with the former and about 20 counts/pixel with the newer model, which

Pulse counting circuitry

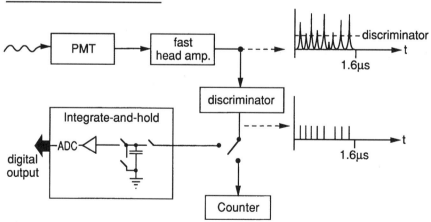

Fig. 11.7. Two alternative approaches to counting single-photon pulses. In both, the signal from a fast head amplifier is passed to a discriminator. Uniform pulses from the discriminator can be either counted with digital circuitry or integrated in a capacitor and then read out through an analog-to-digital converter (ADC).

Practical Photon Efficiency

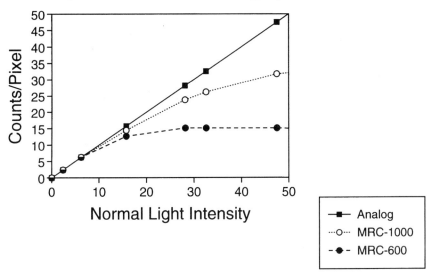

Fig. 11.8. Performance of the fast-photon counting circuitry in the MRC-600 and the MRC-1000 CLSMs. The response is linear up to about 10 counts/pixel with the MRC-600 and up to about 20 counts/pixel with the MRC-1000, which employs a faster PMT and head amplifier.

employs a faster PMT and head amplifier. This later instrument is also unusual for employing the optical techniques for QE enhancement mentioned above and for allowing the user to switch easily between the analog and photon-counting modes.

On first hearing, 20 counts/pixel may sound like a very low level of signal. In fact, however, considerable experience shows that much fluorescence confocal microscopy is performed at even lower signal levels. Owners of instruments capable of fast-photon counting can calibrate their PMT gain controls following instructions in Pawley (1995). Those able to follow this procedure may be surprised to find that when they are using "normal" PMT gain on a "normal" specimen, the "256" that they are storing in the memory in the analog mode corresponds to a signal of <10 photons.

Although the signal saturation that occurs when one exceeds the linear range of the fast-photon counting circuitry is of concern, it is also important to remember that one can correct for piled-up losses to some degree with a simple look-up table. Furthermore, such losses need not be reduced to zero but merely made small compared to the intrinsic statistical noise. Pile-up losses will become less important as manufacturers switch to faster pulse-counting PMTs and circuits; the Bio-Rad MRC-1024 saturates at 58 counts/pixel.

Digital counters are not strictly mandatory. Multiplicative noise can be avoided by clipping all the single-PE pulses to a uniform size and feeding them to a fully integrating ADC (Fig. 11.7). In fact, in some commercial instruments, much of the beneficial effect of photon counting has been incorporated into the analog digitization system by the simple expedient of arranging the electronic and PMT gain so that the single-PE pulses saturate a fast, high-gain amplifier installed between the PMT and the low-bandwidth amplifier feeding the ADC. Because this saturable amplifier is fast, to a reasonable extent, each pulse is clipped separately to a uniform height, thereby meeting the criterion that each photon contribute equally to the recorded signal.

Where Have All the Photons Gone? All present instruments embody design compromises that prevent them from obtaining the ultimate in photon efficiency throughout the four stages discussed above. Many systems employ more refractive optics than is absolutely necessary. Absorption and reflection losses in these produce optical transmission losses in the range of 50–75% (Table 11.1). Although the metal mirrors that produced *losses of 85%* in early confocal instruments have now been replaced with broadband, dielectric-multilayer mirrors, mirror losses still can reach 25% if many surfaces are involved.

It is also important to pay attention to the selection and adjustment of the PMT itself. While many recognize that any specific tube will operate best over only a narrow range of accelerating voltages and that PMTs with bialkali photocathodes on end-window tubes have lower dark current and higher QE in the green, while those with S-20, multialkali photocathodes are better in the red and near infrared, it is usually forgotten that the performance of individual tubes often varies from the mean for their type by a factor of more than 3. Therefore, selection of tubes for high performance and to couple the photocathode material to the wavelength to be detected can pay dividends.

As mentioned above, additional degradation is imposed on the data by multiplicative noise and poor digitizing circuitry, and of course, signal can be lost because of poor alignment. Finally, the improper choice of pinhole diameter may exclude as much as 90% of the useful signal from the detector in a vain attempt to extract an "imaginary" improvement in x–y resolution (imaginary because the resulting low signal levels prevent statistically useful information from being obtained before the specimen is destroyed).

Taken together, all these factors can add up to a factor of 100 times or more in photon efficiency between state-of-the-art and sloppy operation on poor equipment. Every action that results in more efficient use of the photons generated within the specimen should be thought of as being directly responsible for making it possible to collect a proportionally larger number of images (or images with better statistics).

As mentioned above, the use of multiphoton excitation with a non-descanned photodetector allows many of these losses (e.g., dichroic, alignment, and scanning mirrors) to be reduced or eliminated, a condition that greatly improves the photon efficiency of such systems. This advantage must be set against the other cost and performance penalties of the technique that are mentioned above.

Measuring Photon Efficiency. The PMT output signal is the only source of readily available data with which to measure the photon efficiency. Unfortunately, a large number of parameters can have a major effect on this single measurement. An incomplete list includes the following:

Practical Photon Efficiency

- laser power, which is a function of, for example, temperature, cavity gain, stability of stabilizing circuit, and relative power between lines of multiline lasers;
- the transmission at a particular λ of the ND and other filters used;
- the NA and transmission of the objective lens and other optics, usually a strong function of position in the field-of-view;
- reflectivity of the mirrors for the particular λ and polarization of the laser;
- the fraction of the laser beam that is actually accepted by the objective entrance pupil;
- pinhole diameter and alignment;
- the PMT itself: the specific tube in use, its QE as a function of wavelength, and its gain as a function of voltage setting;
- the PMT controls: voltage (gain), gamma, and black-level setting;
- staining density, dye type, and local environment; and
- focus level and refractive index of embedding media.

The number and diversity of these parameters make it difficult to measure the fraction of the photons leaving the focused spot that contribute to the stored image data.

What is needed is a stable point source of light of known intensity that can be mounted conveniently below the objective. One way to make such a source is by allowing a measurable amount of laser light to strike a stable phosphor. First measure the light *emerging* from the objective (as described above) and then adjust it to some standard level. Specimens that maintain a constant level of fluorescent efficiency (i.e., ones that do not bleach or change with time) include such inorganic phosphors as single crystals of CaF_2–Eu or YAG–Ce and uranyl glass. Unfortunately, although these materials are very stable under intense laser illumination, they also have a very high refractive index. Consequently, high-NA objectives are unable to form an aberration-free focus within them, and therefore, with a given pinhole setting, the signal that they generate at the PMT decreases rapidly as the focal plane moves into the material. However, such samples can be useful to those who normally use objectives of lower NA where refractive index effects are less serious. An alternative fluorescence standard can be fabricated by dissolving dye in immersion oil or water (depending on the correction of the objective), but the fluorescent efficiency of such specimens is seldom stable over long periods.

A more direct approach to measuring photon efficiency involves using the microscope simply to image a small light source such as a light-emitting diode (LED) or one formed by the microscope's normal transmission illumination system set up for Köhler illumination (Fig. 11.9). In the latter case, the arc or incandescent source must be provided with a measurable and regulated power supply. Once this has been set up, the only major variables remaining are the amount of metal deposited on the inside of the glass envelope surrounding the source, the bandpass effects of any filters that remain in the light path, the NA of the condenser, the pinhole diameter, and the PMT voltage. In many instruments, it is relatively easy to turn off or obscure the laser, remove all of the dichroic and bandpass filters, and let the light from the image plane pass directly to the PMT. Under these conditions, one should get a standard reading with a given

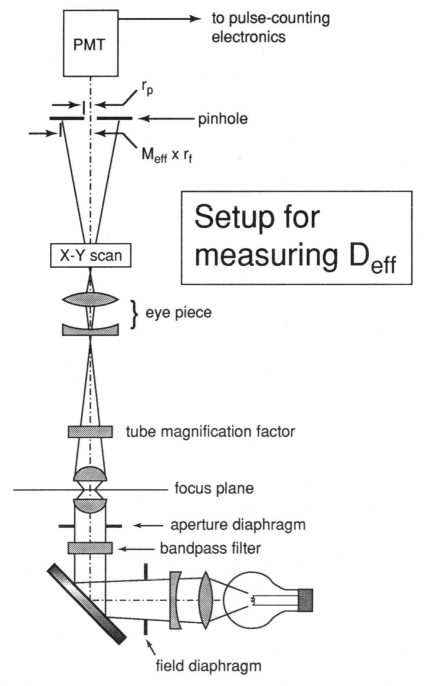

Fig. 11.9. Optical set-up for measuring the detection efficiency or the effective size of the pinhole using the internal transmitted illumination system of the microscope as a standard light source.

objective, pinhole size, and lamp power. Of course, the microscope is now a flying spot detector that collects photons from only one pixel at a time. This has the result that the effective intensity is about 400,000 times less than if the PMT were measuring the entire field, but even so, it will usually be necessary to place ND filters between the source and the condenser lens to permit the PMT to be operated at a voltage high enough to be stable. Line-frequency variations in the filament-heating current may be visible in the data, as will be any instability in the plasma of arc sources. For this reason, it may be more convenient to measure the PMT output with an analog dc voltmeter than with an image storage system of the confocal microscope.

By introducing ND filters below the stage, it is possible to reduce the intensity of the light to the level at which photon counting is appropriate. On those instruments that have this ability, you can easily make measurements to determine a rough ratio between the actual number of photons being detected (using photon counting) and analog intensity values stored in the memory at the same settings of the PMT gain. This is done by recording and then comparing the same blank, bright-field "image" in both analog and photon-counting modes. Such information should be used to reach a rational decision about when one can use photon counting without fear of pulse pileup. As noted above, with a PMT setting of 8.00 on the Bio-Rad MRC-600, an analog-mode, stored intensity of 256 is usually equivalent to the detection of <15 photons/pixel per 1-s frame.

Nothing increases the probability that normal operating procedures are optimal so much as practicing these techniques under test conditions such as these because one knows if one is getting the "right" answer only if one already knows what the answer should be (Pawley et al., 1993a).

OTHER ASPECTS OF RESOLUTION

Limitations Imposed by Spatial and Temporal Quantization

Although the image viewed in a disk-scanning or slit-scanning confocal microscope is, in principle, as continuous as that from a WF microscope, this distinction is lost when the image is finally sensed using a cooled CCD or a digitized photodetector. The fact that all digital confocal images must be recorded and treated in terms of measurements made within discrete pixels can limit the effective spatial resolution of the instrument in ways that may not be familiar to some who approach digital microscopy for the first time (Stelzer, 1997). In a sense, these limits are more practical than fundamental because if the microscope is operated in conformance to the rules of Nyquist sampling theory as discussed below, these limits should present no obstacle to recording good images. However, because the incautious use of the "zoom" magnification control present on all commercial CLSMs makes it relatively easy to operate these instruments outside the Nyquist conditions, a brief discussion of sampling theory is included here. It is mentioned here under the heading "Resolution" because it involves the ability to *record* "the separation of two closely spaced objects."

Spatial Frequencies and the Contrast Transfer Function

Sampling theory, like resolution itself, is easier to think of in the *spatial frequency domain,* where we consider not the size of objects themselves (millimeters per feature) but the inverse of their size (features per millimeter). In the spatial frequency domain, an image is seen as being composed of the *spacings* between features rather than the features themselves. The reason for using this seemingly obscure mental construct is that the ability of an optical system to transmit information depends almost entirely on the spatial frequency of this information. More specifically, all optical systems transmit the contrast of the high spatial frequencies (representing smaller spacings) less effectively than they transmit lower spatial frequencies (which represent larger features). This fact is made evident when one plots the contrast transfer function (CTF) of an optical system by measuring the contrast present in an image of a test object made up of regular arrays of black-and-white bars having specific spacings or frequencies (Oldenbourg et al., 1993). Such a CTF is shown by the solid line in Figure 11.10, the dotted line below it representing the image contrast produced by a test target in which the "black" lines are only 70% gray (i.e., 30% contrast).

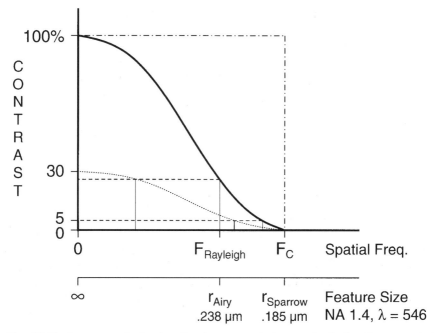

Fig. 11.10. Contrast transfer function of a microscope optical system showing how the contrast in the image of a high-contrast bar pattern varies with the spatial frequency of the pattern (spacings per micrometer). The dotted line represents the response measured if the dark bars in the test target reflect (or transmit) 70% of the incident light rather than 0%. The dashed line represents the imaginary situation of a CTF in which all spatial frequencies up to the "resolution limit" really have equal contrast. The diagram makes clear why high-frequency features of a test object are more likely to remain visible above any arbitrary noise level (5% of 26.5% shown) if they have 100% contrast rather than 30%.

CTF and Resolution

Although in common parlance we often refer to spatial "resolution" as though it were an independent parameter, Figure 11.10 makes it clear that the number we choose as resolution really represents a somewhat arbitrary choice. It refers to the highest spatial frequency at which the contrast is above some given value; for instance, the Rayleigh criterion for bright, point objects on a black background (i.e., stars on a clear night) really assumes that the minimum visible contrast is 26%. The problem with this simplistic attitude to resolution is that it can give one the idea that no matter what their original contrast (i.e., the staining specificity) in the object, all spatial frequencies up to the resolution are equally visible as is implied by the upper dashed line in Figure 11.10. In fact, the most important message from Figure 11.10 is that contrast-in-the-image is equal to contrast-in-the-object as degraded by the CTF of the imaging system, and in particular, *the contrast of small features is much lower than that of larger features.*

Visibility, Resolution, and the Rose Criterion

The reciprocal relationship between contrast and resolution is important because usually what one is actually interested in is not resolution but *visibility*: the ability of an observer to *recognize* two closely spaced features as being separate. Visibility is a much more slippery concept because it depends not just on the calculable aspects of optical theory but on the higher functions of the visual system used to determine how a particular intensity pattern should be interpreted.

Visibility has the advantage of requiring us to talk about the other important parameter of the image data: the signal-to-noise ratio ($R_{S/N}$). The Rose criterion states that to be visible, a black feature that is a single pixel in size on a white ground must have an intensity that differs from that of the background by at least five times the noise level of the background (Rose, 1948). Although the factor 5 is lower for lines and other geometrical features, the main point is that visibility depends on more than geometrical optics and diffraction. Assuming that the contrast is measured in units that are proportional to the statistical precision with which each signal level is known, visibility also depends on the absolute and relative difference between the signal level of the feature and that of the background. In the case of the Rayleigh criterion, the condition states that between the central points of two bright features one must be able to recognize a lower intensity in some intervening pixel that allows them to be seen as separate.

Because of the CTF curve, smaller objects have lower contrast and more photons must be counted in order to reduce the intrinsic noise so that the lower contrast features will become visible above the statistical noise.

This necessity brings together the contrast of the object (as embodied in its staining characteristics), the focusing properties of the optical system as defined by the CTF, and the statistical requirements of the Rose criterion. However, there is still one other process that can limit whether or not a feature is visible in the final image: how the signal is digitized (Stelzer, 1997).

Digitization and the Nyquist Criterion

To convert any continuous, analog signal into a digital representation, it is sampled by measuring its intensity at regular intervals in time (or space). Nyquist's crucial insight was that there is a fixed relationship between the highest temporal (or spatial) frequency present in the data of interest and the minimum rate at which samples must be taken if they are to record accurately all the possible significant variations in that signal. Specifically, for nonperiodic data, the sampling frequency should be at least 2.3 times higher than the highest important frequency in the data. Consequently, to preserve all the information that could be recorded using the CTF shown in Figure 10, it will be necessary to have the pixels smaller than $1/2.3F_c$, where F_c is the cut-off frequency.

The practical problem that arises in CLSM is that, as mentioned above, for a given optical system and wavelength, only one pixel size and one zoom factor match the Nyquist criterion. At zoom settings that provide smaller pixels, the data will be oversampled, with the result that it will be bleached more than necessary and only a smaller field of view can be scanned in a given period. It is now common for manufacturers to display the pixel size as part of the set-up display, but if this is not done, it is not difficult to calculate it by measuring the length of any feature in the image in both pixels and micrometers.

On the other hand, at lower zoom settings, where the pixels are larger than those prescribed by Nyquist, not only may some smaller features be missed entirely, but also features not present in the object may "appear" in the data because of a phenomenon called aliasing.

Proper Nyquist sampling is sometimes neglected because working images that are somewhat oversampled tend to be more pleasant to view on the display screen, and in the absence of standards, the increased bleaching rate is hard to detect. Alternatively, on weakly stained specimens, the pixel intensities involve so few quanta that statistical variations in the signal pose a greater limitation to spatial resolution than does the quantization. Nevertheless, as applications of confocal technology more closely approach the absolute limits, the costs of spatial (and temporal) quantizing will become more apparent.

It should be pointed out that the Nyquist rule also applies to adequate sampling in the z-direction and that the z-step setting used to image a 3D specimen should also be less than half the z-resolution.

Two other matters about sampling should be considered: temporal aliasing and mismatch of probe and pixel shape.

Temporal Aliasing. The accuracy with which temporal changes can be imaged is limited by the frame-scan rate: the "wagon-wheel effect." In the CLSM, temporal aliasing limits not only the ability to follow motion but also the precision in the measurement of the location (or motion) of objects. Because the pixel time is much shorter than the frame time, motion of the specimen during data collection produces distortion rather than the directional blurring that would occur if the whole field were imaged for the same period of time.

Mismatch of Probe and Pixel Shape. There is a mismatch in shape between the circularly symmetrical shape of the Airy disk of the actual probe, the Gaussian shape of the theoretical probe, and the square shape represented by a pixel in a rectangular raster. This is made worse if the probe moves much faster in one direction than the other.

In the case of a signal recorded by a cooled CCD detector on an optimized disk-scanning system, the 2D detector spatially quantizes the signal in a manner entirely determined by the sensor geometry and the total optical magnification. However, because the fast-scanning mirror in the CLSM follows a ballistic rather than stepped trajectory, the output from the PMT of these instruments is effectively continuous in the horizontal direction. As a result, averaged over the pixel time, the effective Airy disk is not round but is slightly blurred in the fast-scan direction.

Practical Considerations Relating Resolution to Distortion

To obtain the theoretical spatial resolution of a confocal microscope, it is necessary, but not sufficient, to have a diffraction-limited optical system. This section discusses the limitations imposed by mechanical stability and the repeatability of the scanning system.

In disk-scanning instruments, the image is real and therefore distortion can originate only from the optics, not from the scanning system. If a CCD sensor is used to detect this image, geometrical distortion in the digitized image can be extremely low, and because of the inherent mechanical stability of both the optics and the sensor, any residual distortion can be corrected by digital image processing. The mechanical problems of this instrument are therefore confined to the effects of vibration and the relative motion of the stage and lens.

In all confocal instruments, the relative motion between the objective and the specimen should be kept to less than 10% of the resolution limit in x, y, and z. In the disk-scanning instruments, both the rotating disk and sometimes the cooling system of the illumination source represent potential internal sources of vibration. It is important to recognize that it is not sufficient to isolate the microscope mechanically on a vibration-isolation table if laser-cooling fans, spinning disks, or scanning galvanometers are permitted to *introduce* vibration into the table.

Less straightforward is the effect of vibration (and possibly stray electrical or magnetic fields either acting directly on the galvanometer mirror suspension or introducing spurious signals into the current amplifiers that control them) on the performance of the scanning mirrors in laser confocal systems. In these instruments, accurate imaging depends on the mirrors causing the beam to scan over the sample in a precise pattern, duplicating the mathematically perfect raster represented by the data locations in the image memory. Failure to duplicate this raster produces image distortion. On a system with a 1000-line raster and a 10 : 1 zooming ratio, keeping beam placement to within 0.1 of a pixel requires a scanning accuracy of 1 part in 10^5. No present system even preserves this level of performance in the electrical signals used to drive their mirror galvanometers, let alone in their actual motion. The electromechanical properties of the galvanometers (e.g., mass, spring constant, frequency response, resonant frequency, overshoot, bearing tolerance, and rigidity) produce additional errors. Image distortions produced by these errors are often masked by the paucity of test samples having an accurately defined geometry (Pawley et al., 1993b) and by the fact that current instruments at high zoom greatly oversample the image so the smallest visible structural features are many pixels wide. Careful measurements will show that the features recorded of a periodic specimen will vary in size by a few percent as the specimen moves across the field of view.

This problem merits mention here because it is possible to measure the x, y, z *position of the centroid* of an object in a digital image to an accuracy that is *much smaller than the spatial resolution limit of the image*. Indeed, WF light microscopy techniques have been used to measure motion on the order of 1 nm (Gelles et al., 1988). However, due to random imprecision in the systems used to position the mirrors, it is unlikely that measurements of similar reliability could be made on any present CLSM. In this context, then, the accuracy and precision with which the mirrors can be controlled is a fundamental limitation on the ability of CLSMs to determine position or motion.

The presence of scan instability can be detected either by visually comparing sequential *single-scan* images of a diagonal knife-edge viewed at the highest possible magnification, contrast, and signal level or, alternatively, by computing the apparent motion of the centroids of two fixed objects, each covering >100 pixels, as they are recorded on a number of sequential scans. The variations measured by the second method should decrease rapidly with increasing illumination intensity because of improved statistical accuracy, although they may then begin to measure stage drift.

Another important and insidious source of distortion is stage drift. Usually the cause is temperature variations in the microscope body, and the solution is to shield the system from drafts. Beyond this, drift can be corrected for digitally as long as there are features visible within the field of view that can be relied upon to not move of their own volition (i.e., not cells).

To summarize:

- Although resolution in the confocal microscope is primarily a function of the NA of the optical system and the λ of the light, it can also be limited if the signal level represents so few quanta that the signal lacks the statistical precision to produce a "visible" feature or if the data are not correctly sampled because the pixels are too large.
- To improve the statistical precision of the signal, every effort should be made to count all photons from the specimen as accurately as possible: reduce optical losses, select your PMT, check alignment, and routinely make, analyze, and compare images of stable, calibrated test objects.
- The effects of image quantization should not be ignored. Only one zoom setting is optimal for each λ, NA, and objective magnification.
- Care should be taken to keep the pinhole diameter close to that of the Airy disk at the half-power points. This setting will change predictably with λ, NA, and lens magnification.

- Special circumstances may occasionally dictate breaking the sampling and pinhole rules, but the limitations inherent in doing so should always be understood and acknowledged.
- It should be possible to operate the system both at its diffraction-limited resolution and at larger spot sizes.
- In laser-scanned instruments, imperfect scanning precision can introduce distortion.
- Fluorescence saturation places unexpected limits on the speed with which images can be formed with laser-scanning microscopes.

CONCLUSION

This chapter attempts to highlight some aspects of confocal instrumentation that must be addressed in order to attain performance limited only by optical and information theory (Table 2). Although the constituent aspects of both photon efficiency and resolution have been addressed separately, it has been evident that the effects of both these factors overlap and interact in a fairly complex manner.

Further complication surrounds the question of single-versus multi-photon fluorescence excitation. At present, because of the substantial cost and the relative complexity of the lasers involved, it seems that the latter will be most useful in applications in which relatively thick (>100-μm) living specimens such as brain slices or embryos must be viewed over an extended period.

TABLE 2. Limitations of Confocal Microscopy

Parameter	Theoretical		Practical
Resolution	Spatial	Y: $\lambda_1, \lambda_2, \alpha_1, \alpha_2$	Alignment, off-axis aberrations
		X: $\lambda_1, \lambda_2, \alpha_1, \alpha_2$	Alignment, off-axis aberrations, bandwidth, ballistic scan
		Z: $\lambda_1, \lambda_2, \alpha_1, \alpha_2$ pinhole diameter	Alignment, off-axis aberrations, Δ tube length
	Temporal: scan speed, signal decay times		Quantization of t
Position	Objective lens distortion, pixellation limitations; sampling time		Mirror accuracy, vibration
Quantitative measurement	Poisson statistics		Mirror and digitizing losses, detector DQE and noise, bleaching/photodamage, saturation, source brightness

ACKNOWLEDGMENTS

This chapter is a modified and updated version of Chapter 2 in the *Handbook of Biological Confocal Microscopy* (2nd ed.). Kind permission to reproduce material from the original chapter was granted by Plenum Press. Like the original, it has benefited greatly from conversations with Sam Wells (then at Cornell), Jon Art (University of Illinois-Chicago), Nick Doe (then with Bio-Rad, Hemel Hempstead, UK, who provided the MRC-1000 data shown in Fig. 11.8), and Ernst Stelzer, who has done much work on the resolution limitations imposed on confocal microscopy by the statistical variations in the small number of photons counted. Victoria Centonze at the Integrated Microscopy Resource in Madison helped obtain the data for Figure 11.8. Cheryle Hughes and Bill Feeney, artists in the Zoology Department, University of Wisconsin, Madison, made the other drawings. This work was supported by grant DIR-90-17534 and DIR 97-24515 from the U.S. National Science Foundation.

REFERENCES

Brakenhoff GL, Blom PJ, Barends PI (1979): Confocal scanning light microscopy with high aperture immersion lenses. J Microsc 117:219.

Buist A, Müller M, Squier J, Brakenhoff GJ (1998): Real time two-photon absorption microscopy using multi point excitation. J Microsc 192(pt2):217–226.

Egger MD, Petran, M (1967): New reflected light microscope for viewing sustained brain and ganglion cells. Science 157:305–307.

Carlsson K, Danielsson PE, Lenz R, Liljeborg L, Majlof L, Aslund N (1985): Three dimensional microscopy using a confocal laser scanning microscope. Optic Lett 10:53.

Centonze V, Pawley JB (1995): Tutorial on practical confocal microscopy and use of the confocal test specimen. In *Handbook of Biological Confocal Microscopy*. Pawley J (ed). New York: Plenum, pp 549–570.

Denk W, Strickler JH, Webb WW (1990): Two-photon laser scanning fluorescence microscopy. Science 248:73–76.

Gelles J, Schnapp BJ, Steur E, Scheetz MP (1988): Nanometer scale motion analysis of microtubule-based motor enzymes. Proc EMSA 46:68–69.

Gunter WD Jr, Grant GR, Shaw S (1970): Optical devices to increase photocathode quantum efficiency. Appl Opt 9(2):251–257.

Hell SW, Bahlmann K, Schrader M, Soini A, Malak H, Gryczynski I, Lakowicz JR (1996): Three-photon excitation in fluorescence microscopy. J Biomed Opt 1(1):71–73.

Minsky M (1988): Memoir on inventing the confocal scanning microscope. Scanning 10:128–138.

Oldenbourg R, Terada H, Tiberio R, Inoué S (1993): Image sharpness and contrast transfer in coherent confocal microscopy. J Microsc 172:31–39.

Pawley JB (1995): Fundamental limits in confocal microscopy. In Pawley J (ed). *Handbook of Biological Confocal Microscopy*. New York: Plenum, pp 19–38.

Pawley JB (1994): The sources of noise in three-dimensional microscopical data sets. In Stevens J, Mills LR, Trogadis JE (eds). *Three Dimensional Confocal Microscopy: Volume Investigations of Biological Specimens*. New York: Academic Press, pp 47–94.

Pawley JB, Centonze V (1994): Practical laser-scanning confocal light microscopy: Obtaining optimal performance from your instrument. In Celis JE (ed). *Cell Biology: A Laboratory Handbook*, 2nd ed. New York: Academic Press.

References

Pawley JB, Blouke M, Janesick J (1996): The CCDiode: The ultimate photodetector for laser confocal microscopy? Proc Soc Photo-Optical Eng 2655–41.

Pawley JB, Hasko D, Cleaver J (1993a): A Standard Test and Calibration Specimen for Confocal Microscopy II. Proc 1993 Int Conf Confocal Microsc 3-D Image Processing, Sydney, Australia, CJR Sheppard (ed), p 35.

Pawley JB, Wright AG, Garrard CC (1993b): Optical Enhancement and Pulse-Counting Improve the Quality of Confocal Data. Proc. 1993 Int Conf Confocal Microsc 3-D Image Processing, Sydney Australia, CJR Sheppard (ed), p 69.

Rose A (1948): Television pickup tubes and the problem of noise. Adv Electron 1:131.

Sanchez EJ, Novotny L, Holtom GR, Xie, XS (1997): Room-temperature fluorescence imaging and spectroscopy of single molecules by two-photon excitation. J Phys Chem A 101:7019.

Sandison DR, Piston DW, Webb WW (1994): Background rejection and optimization of signal to noise ratio in confocal microscopy. In Stevens J, Mills LR, Trogadis JE (eds). *Three Dimensional Confocal Microscopy: Volume Investigations of Biological Systems.* New York: Academic Press.

Sheppard JCR, Choudhury A (1977): Image formation in the scanning microscope. Optical, 24:1051.

Sheppard JR, Gu M, Roy M (1992): Signal-to-noise ratio in confocal microscope systems. J Microsc 168:209–218.

Slomba AF, Wasserman DE, Kaufman GI, Nester (1972): A laser flying spot scanner for use in automated fluorescence antibody instrumentation. J Assoc Ad Med Instrument 6:230–234.

Stelzer EHK (1997): Contrast, resolution, pixelation, dynamic range and signal to noise ratio: Fundamental limits in fluorescent light microscopy. J Microsc 189:15–24.

Straub M, Hell SW (1998): Multifocal multiphoton microscopy: A fast and efficient tool for 3D fluorescence imaging. Bioimaging 6:177–184.

White JG, Amos WB (1995): Direct view confocal imaging systems using slit apertures. In Pawley J (ed). *Handbook of Biological Confocal Microscopy.* New York: Plenum, pp 403–416.

White JG, Amos WB, Fordham M (1987): Evaluation of confocal versus conventional imaging of biological structures by fluorescence light microscopy. J Cell Biol 105:41–48.

Wokosin DL, Amos B, White JG (1998): Detection sensitivity enhancements for fluorescence imaging with multi-photon excitation microscopy. Proc IEEE Eng Med Biol Soc 20:1707–1714.

12

Scanning Near-Field Optical Imaging and Spectroscopy in Cell Biology

Vinod Subramaniam, Achim K. Kirsch, Attila Jenei, and Thomas M. Jovin
Max Planck Institute for Biophysical Chemistry, Goettingen, Germany

INTRODUCTION

Scanning near-field optical microscopy [SNOM; alternatively near-field scanning optical microscopy (NSOM)] combines the enhanced lateral and vertical resolution characteristic of scanning probe microscopies with simultaneous measurements of optical signals, yielding resolutions beyond the limits of conventional diffraction optics (Pohl et al., 1984; Betzig et al., 1991; Betzig and Trautman, 1992). SNOM can simultaneously map topographic and optical properties with extremely high spatial resolution, is noninvasive, and has the potential for operating in an aqueous environment. The fluorescence detection sensitivity extends to the level of single molecules. Consequently, SNOM is developing into an important technique for imaging biological systems. We do not review the relevant literature, but rather present an overview of our applications of SNOM to cellular systems. For a comprehensive review of SNOM, we refer the reader to the monograph by Paesler and Moyer (1996) and the review by Dunn (1999).

SNOM achieves spatial resolution beyond the diffraction limit by scanning a sub–wavelength-sized aperture confining the excitation light in close proximity to

Emerging Tools for Single-Cell Analysis, Edited by Gary Durack and J. Paul Robinson.
ISBN 0-471-31575-3 Copyright © 2000 Wiley-Liss, Inc.

the sample (Pohl et al., 1984; Isaacson et al., 1986; see Fig. 12.1A for a schematic). The resulting resolution is a function only of the aperture size and the probe-to-sample distance, and not of the wavelength. In practice, the most commonly used near-field probe is an aluminum-coated tapered optical fiber with an aperture diameter ≤100 nm. The light transmitted or emitted from the sample is generally collected with an objective lens and processed according to the different contrast mechanisms. In our laboratory, the SNOM was designed as an optical module for operation with a standard commercial scanning probe microscope (SPM) system. It has been operated to date in a shared-aperture mode, that is, with uncoated tips supplying a near-field mode for excitation as well as for detection of fluorescence signals. The spatial resolution achieved in such a system has been estimated to be ≤200 nm, that is, considerably better than the diffraction limit of the conventional optical microscope. In addition, the microscope has the unique capability, not

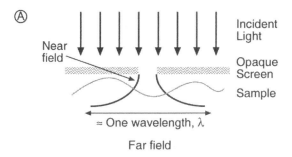

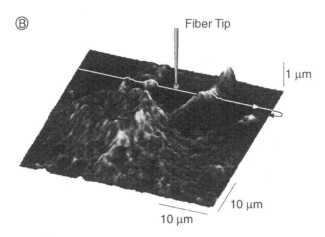

Fig. 12.1. (A) Schematic of the principle behind SNOM. A sub–wavelength-sized aperture (hole in the opaque screen) confines the excitation light and is scanned in close proximity (typically < 10nm) to the sample. (B) Three-dimensional representation (fluorescence overlaid on topography) of a SNOM image of a 3T3 Balb/c cell labeled with fluorescently conjugated concanavalin A bound to cell surface lectin receptors. The fiber tip is raster scanned across the sample to generate the simultaneously acquired topographic and fluorescence images.

achievable in the confocal laser scanning microscope (CLSM), of providing a correlation between high-resolution surface topography and the fluorescence signal(s) (Fig. 12.1B).

BIOLOGICAL APPLICATIONS OF SNOM

The correlation between topography and optical signals is of particular relevance in biological applications. Thus, for example, one can ask questions regarding the distribution of cell surface receptors for external signaling molecules such as growth factors and hormones (see Fig. 12.3 below). While the resolution of SNOM is not sufficient to directly image individual complexes of small proteins or ligands and receptors, in combination with fluorescence resonance energy transfer (FRET) methods, SNOM can be operated with contrast modes dependent on intermolecular separations on the order of 2–10 nm (see below).

Our SNOM was intended primarily for investigations of cellular processes and biological macromolecules engaged in dynamic biochemical processes. The philosophy has been to implement all the available photophysical processes associated with fluorescence, thereby exploiting the specificity and sensitivity of fluorescence probes. The latter can be intrinsic (cellular autofluorescence) or extrinsic (covalent adducts to proteins, nucleic acids, or small molecules) or introduced by cellular transfection and transformation with green fluorescent protein (GFP). The latter protein is unique in its manifestation of a visible fluorophore formed spontaneously by chemical transformation of amino acid residues. The GFP can be targeted to intracellular compartments by genetic fusion to proteins of interest and can thus be expressed in living cells, tissues, and whole organisms.

Some of the biologically relevant applications of our SNOM carried out to date include the study of the following:

- the domain structure of Langmuir–Blodgett films with electron transfer systems (Kirsch et al., 1998a, 1998b);
- GFP expression in bacteria, in fruitfly embryo cells, and in mammalian cells as a fusion protein with the epidermal growth factor (EGF) receptor (Subramaniam et al., 1997, 1998);
- distributions of transmembrane receptors for EGF and platelet-derived growth factors (PDGFs; Vereb et al., 1997);
- distribution and clustering on cell surfaces of the erbB growth factor family of transmembrane receptor tyrosine kinases (particularly erbB2) as a function of activation (Nagy et al., 1998);
- interactions of cell surface proteins via FRET (Subramaniam et al., 1998; Kirsch et al., 1999); and
- multiphoton (two-, three-photon) simultaneous excitation of probes for different intracellular compartments (mitochondrial, nuclear; Kirsch et al., 1998c; Jenei et al., 1999).

TABLE 12.1. Summary of Recently Published Reports on Biological Applications of SNOM

Biological system	References
Cytoskeletal actin	Hecht et al., 1997; Betzig et al., 1993
GFP in bacteria	Muramatsu et al., 1996; Tamiya et al., 1997
DNA	Garcia-Parajo et al., 1998; van Hulst et al., 1997; Wiegräbe et al., 1997
Fluorescence in situ hybridization probes	Moers et al., 1996
Colocalized malarial and host skeletal proteins	Enderle et al., 1997
Fluorescently labeled single DNA molecules	Ha et al., 1996
Phospholipid monolayers	Hwang et al., 1995
Photosynthetic membranes	Dunn et al., 1994, 1995
Rat cortical neurons	Talley et al., 1996; Shiku et al., 1998
Fluorescently labeled myofibrils	Seibel and Pollack, 1997
Cell plasma membranes	Enderle et al., 1998; Hwang et al., 1998

Other groups have used SNOM to probe various biological molecules and systems; a summary of recent references is presented in Table 12.1. The biological applications of SNOM pose several challenges, including the difficulty of imaging soft samples, tracking large topographical changes, accounting for tip–sample interactions, and rationalizing resolution and contrast mechanisms. In particular, operation under water or in physiological buffer solutions, a critical requirement for realistic biological studies, poses problems for implementing the shear-force feedback technique. While some successes have been achieved (Moyer and Kammer, 1996; Keller et al., 1997; Hollricher et al., 1998; Seibel and Pollack, 1997; Hwang et al., 1998; Talley et al., 1998), imaging in an aqueous environment is still not a routine endeavor.

INSTRUMENTAL DETAILS

Our instrument was designed as a module for operation with a standard scanning probe microscope system (Nanoscope IIIa, Digital Instruments, Santa Barbara, CA) and uses uncoated fiber tips as local probes in the shared-aperture mode. In this mode the sample is locally illuminated through the fiber tip and light reflected or emitted by the sample is collected by the same fiber tip. Such tips are formed in a highly reproducible single-step heating and pulling process from telecommunication optical fibers. Although the illumination is not as well localized as with standard apertured probes, the requirement for the excitation light and the collected fluorescence to pass through the fiber tip results in an automatic alignment of the excitation/emission paths with an effective lateral imaging resolution of 150–200 nm (Courjon et al., 1990; Bielefeldt et al., 1994; Kirsch et al., 1996, 1998c). We have implemented multiple photophysical modes in our modular microscope, including (i) multiple laser sources; (ii) dual-

Instrumental Details

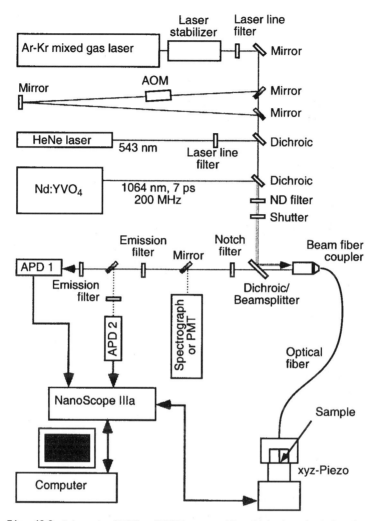

Fig. 12.2. Schematic of MPIbpc SNOM system with multiple photophysical modes.

detection channels; (iii) spectrograph operation for spectral acquisition; (iv) high-intensity continuous wave (cw) or pulsed excitation for multiple-photon (two-photon, three-photon) excitation (MPE); (v) various modalities for detection of FRET, and (vi) fluorescent lifetime detection in the frequency domain. A new design intended for operation in fluids is being implemented. We are also engaged in a collaborative effort for developing microstructure polymer tips for SNOM (Stürmer et al., 1998).

A schematic of the SNOM system depicting the multiple excitation and detection schemes is shown in Figure 12.2. The fiber tip is mounted in a shear-force sensor head mounted above the piezoelectric tube scanner of the Digital Instruments SPM. The sample is laterally scanned beneath the fiber tip, and a distance of 2–5 nm is maintained by monitoring shear-force interactions between the laterally vibrated tip and the sample surface. The scanning process is controlled by the electronics of the SPM system. In addition to one analog-to-digital converter (ADC), the electronics contains two digital pulse counters suitable for dual-channel pulse counting.

Excitation Sources. We have used multiple excitation sources (both cw and pulsed) in the SNOM. Available sources include an Ar–Kr mixed-gas laser (Performa, Spectra Physics, Mountain View, CA), a tunable HeNe laser operated primarily at 543 nm (LSTP-1010, Research Electro-Optics, Boulder, CO), a frequency-doubled cw Nd–YAG diode laser operating at 532 nm (CGL-050-L CrystaLaser, Reno, NV), and for multiphoton imaging purposes (see below) a diode-pumped, pulsed Nd–YVO$_4$ laser (GE 100, Time Bandwidth Products, Zurich, Switzerland). For dual-color imaging and fluorescence resonance energy transfer experiments, we can simultaneously couple two different lasers into the fiber. After stabilization and spectral filtering, the light from the different laser sources is combined with appropriate dichroic mirrors to provide a suitable excitation light source for most commonly used fluorescence dyes.

Fluorescence Imaging. Typical excitation intensities for fluorescence imaging are in the range of 3 nW to 3 μW. The dichroic filter in front of the beam-to-fiber coupler used for separating the fluorescence from the excitation light is easily exchangeable and is chosen according to the dye to be imaged. After additional filtering to remove residual excitation light, the fluorescence is detected with a high-quantum-efficiency single-photon-counting avalanche photodiode (APD; SPCM-AQ 131, EG&G Optoelectronics, Canada).

Emission Spectroscopy. Emission spectra can be collected by directing the fluorescence with a beam-shaping fiber optic into a spectrograph (MS125, Oriel, Stratford, CT) coupled to an intensified charge-coupled device (CCD) system (InstaSpec 5, Oriel).

Dual Detection. The detection of the emission of two dyes in doubly labeled samples is easily accomplished by inserting an additional dichroic mirror in the emission path to reflect the fluorescence from the second dye onto a second APD.

Lifetime Imaging. We have implemented detection of variations in the fluorescence lifetime by the phase modulation technique. In an alternative excitation light path the Ar–Kr laser intensity is modulated with a (standing-wave) acousto-optical modulator (AOM) at a typical frequency of ≈40 MHz. The phase of the modulated fluorescence signal is shifted due to the finite lifetime of the excited state of the fluorophore. The modulated fluorescence signal and the phase shift can be recorded with analog light detectors [e.g., a photomultiplier tube (PMT)] and suitable demodulation (high-frequency lock-in) electronics.

FRET. Fluorescence resonance energy transfer can be detected by changes in the fluorescence lifetime of the donor, by emission spectroscopy, or by donor and acceptor photobleaching methods. The last method requires selective bleaching of the acceptor in some regions of the sample followed by imaging of the donor and acceptor fluorescence and the sensitized emission of the acceptor. The dye pair fluorescein and rhodamine B is imaged quasi-simultaneously by alternating the excitation laser source synchronously with the scanning process and implementing dual detection of the fluorescence of the two dyes. During the trace motion of the sample the 488-nm line of the Ar–Kr laser is used to excite fluorescein donor and sensitized rhodamine B acceptor fluorescence, whereas during the retrace the 543-nm HeNe laser is coupled into the fiber for exciting the acceptor alone.

Multiphoton Excitation of Fluorescence. The optical transmission efficiency of uncoated fiber tips is very high, and they can withstand average light intensities on the order of several hundred milliwatts. Therefore multiphoton excitation of fluorescence is feasible with both cw and pulsed laser sources. The 647-nm laser line of the cw Ar–Kr laser is used to excite near-UV absorbing dyes like DAPI or BBI-342. The diode-pumped solid-state $Nd-YVO_4$ laser operating at 1064 nm and emitting pulses of 7 ps length at 200 MHz repetition rate serves to excite simultaneously green-absorbing dyes like Mitotracker in a two-photon process and near-UV-absorbing dyes like DAPI in a three-photon process (see details below).

SPECIFIC APPLICATIONS

Green Fluorescent Proteins in Cells

Green fluorescent proteins isolated from certain species of jellyfish have attracted enormous attention in recent years due to their use as reporter molecules in cell, developmental, and molecular biology. GFP can be fused to a variety of proteins without affecting their function. These proteins are expressed in vivo and thus act as remarkably versatile indicators of structure and function within cells. The GFP fusion proteins can be visualized and localized in cells and embryos using standard microscopy techniques. We have expressed, imaged, and performed site-specific spectroscopy on various mutants of GFP in bacteria and *Drosophila* Schneider cells and expressed in mammalian cells as a GFP–receptor construct. These mutants of the cloned GFP exhibit red-shifted excitation spectra relative to that of the wild-type protein and are efficiently excited by the 488-nm line of an argon-ion laser. Near-field spectroscopic measurements were performed by positioning the fiber tip over a region of interest and directing the collected light into a spectrograph. The near-field fluorescence spectrum corresponds closely to that obtained in a fluorimeter from a cell suspension, thereby confirming that the optical signal originated from the expressed protein (Subramaniam et al., 1997).

Figure 12.3 shows shear force topography (A), feedback error signal (B), and fluorescence (C) images of a Chinese hamster ovary (CHO) cell stably transfected with a fusion construct of the epidermal growth factor receptor (EGFR) and GFP (Brock et al., 1998), allowing visualization of the distribution of cell surface EGF receptors.

Figure 12.4 shows shear-force topography and fluorescence images of *Drosophila melanogaster* Schneider cells transfected with a GFP mutant optimized for expression in *Drosophila*. Surprisingly, instead of a homogeneous distribution, the cytosolic expressed GFP was distinctly localized in punctate structures within the cell body. The diameters of these features varied from ~150 to ~450 nm.

Photobleaching FRET on Cell Surfaces

Fluorescence resonance energy transfer between excited fluorescent donor and acceptor molecules occurs through the Förster mechanism over the range of 1–10 nm and has been used to assess the proximity of fluorophores in biological samples (for an extensive review, see Clegg, 1996). In the presence of acceptor, energy transfer is manifested by (i) a quenching of the donor emission, (ii) an increased emission of the acceptor excited via the donor (sensitized emission), (iii) a decrease in the donor photobleaching rate, and (iv) a decrease in the donor lifetime. Photobleaching methods, in which either the donor or acceptor molecules are selectively photodestructed, pro-

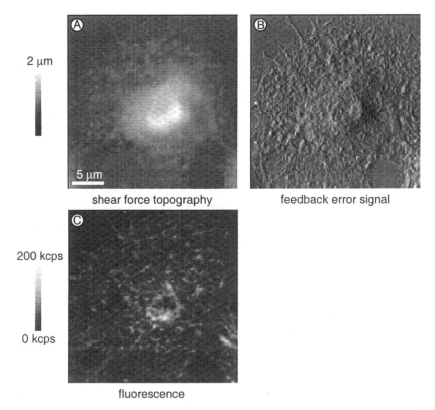

Fig. 12.3. Shear force topography (A), feedback error signal (B), and fluorescence (C) images of CHO cells expressing a fusion construct of the EGFR and GFP; λ_{ex} = 488 nm, 160 nW. *See color plates.*

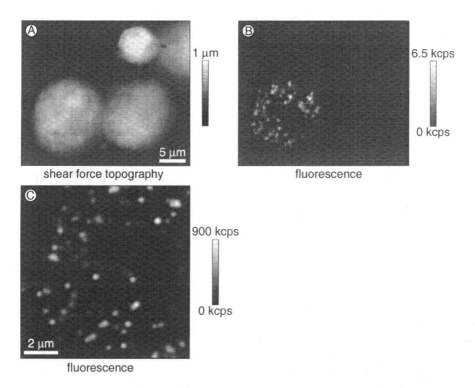

Fig. 12.4. Shear force topography (A) and fluorescence (B, C) images of *D. melanogaster* Schneider cells expressing GFP. Of the three cells seen in the topographic image in (A), only one has been transfected and exhibits GFP fluorescence. (C) Zoomed image of the cell imaged in (B). The area with little fluorescence in the lower right quadrant of the transfected cell in (B) is the cell nucleus, which is devoid of GFP. An analysis of the fluorescence image in (C) reveals that the diameters of these fluorescence features vary between ~150 and ~450 nm.

vide a sensitive means for establishing and quantitating energy transfer efficiencies in cells (Bastiaens and Jovin, 1996; Bastiaens et al., 1996; Jovin and Arndt-Jovin, 1989). The photobleaching time of a fluorophore is inversely proportional to the excited-state lifetime; any process such as energy transfer that shortens the lifetime will decrease the photobleaching rate. Thus, the photobleaching time constant of the donor fluorophore is expected to be longer in the presence of acceptor molecules. In the acceptor photobleaching scenario, destruction of the acceptor leads to a corresponding increase in the donor emission quantum yield (i.e., intensity). The combination of FRET and SNOM makes the length scale from 1 to 100 nm accessible, that is, a domain beyond the limit of standard optical microscopies. To demonstrate the possibility of measuring FRET with our SNOM apparatus using photobleaching approaches, we have used two systems: dye-doped polyvinylalcohol (PVA) thin films and fluorescently conjugated lectins bound to cell surface glycoproteins on 3T3 Balb/c cells (Subramaniam et al., 1998; Kirsch et al., 1999).

Donor Photobleaching Using Labeled 3T3 Cells

The plant lectin concanavalin A (conA) binds the carbohydrate α-methylmannoside, a component of several glycoproteins on mammalian cell surfaces. The 3T3 Balb/c mouse fibroblast cells were grown at 37°C on glass coverslips under standard conditions in Dulbecco's modified Eagle's medium (DMEM) with 10% fetal calf serum. Cells were labeled with fluorescently conjugated conA. Double-labeled cells were incubated with a mixture of fluorescein-conjugated conA (F-conA) and tetramethyl-rhodamine-conjugated conA (R-conA) and had a donor–acceptor (fluorescein–rhodamine) dye ratio of 1 : 2. Control samples (with the appropriate mixture of unlabeled conA and either F-conA or R-conA to yield label densities equivalent to those of double-labeled cells) were also examined.

Figure 12.5 shows topographic (A) and optical (B) images of 3T3 Balb/c cells labeled with a mixture of F-conA and R-conA. For this image, the sample was excited

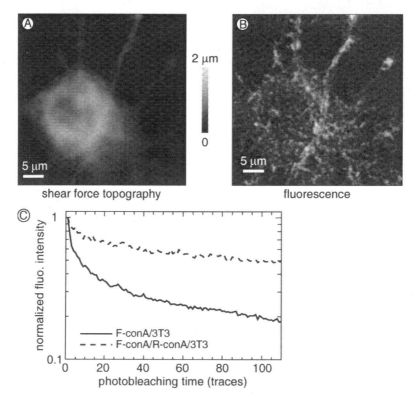

Fig. 12.5. Shear force topography (A) and fluorescence (B) images of a 3T3 Balb/c cell labeled with fluorescein (donor)- and rhodamine (acceptor)-conjugated concanavalin A bound to cell surface receptors; $\lambda_{ex} = 488$ nm, $\lambda_{det} = 530 \pm 15$ nm. Scan parameters: 0.3 Hz line frequency, 128 lines, 4.1 ms counting time/pixel, maximum counts/pixel = 80. The accompanying plot of photobleaching kinetics (C) shows a distinct slowing down of the donor photobleaching in the presence of acceptor, clearly demonstrating energy transfer between the F-conA and R-conA moieties bound to the cell surface. (Adapted from Subramaniam et al., 1998).

at 488 nm and fluorescence was detected with a band-pass filter centered at 530 nm (i.e., the fluorescein emission band).

After locating and imaging an appropriate cell, donor photobleaching was accomplished by disabling the slow-scan axis and repeatedly scanning a single line while delivering a high flux of 488-nm light (fluorescein absorption). Donor photobleaching curves were derived from the optical image acquired during photobleaching. Each curve presented in Figure 12.5 was calculated as an average of the photobleaching curves of six distinct bright features on the scan line. Donor photobleaching was also performed in the same manner on the control samples lacking acceptor. As can be clearly seen in Figure 12.5C, the photobleaching kinetics of the donor in the presence of the acceptor was distinctly slower than in the absence of the acceptor, as expected due to the introduction by FRET of an additional deactivation pathway shortening the lifetime of the excited state.

This demonstration of FRET using photobleaching techniques in the SNOM establishes the possibility of studying on a very localized spatial scale the interactions between a receptor–ligand pair. The capability of combining FRET and SNOM in studies of single molecules has been demonstrated with labeled DNA molecules (Ha et al., 1996). Other applications of photobleaching SNOM are presented elsewhere (Kirsch et al., 1998a). In addition, we have used FRET to assess the cell surface topology of membrane proteins against which antibodies were raised and to measure the intramolecular energy transfer between two chains of the HLA-I protein. We have elaborated a quantitative imaging microscopy technique based on measurement of fluorescence intensities by SNOM and have calculated the energy transfer efficiency on a pixel-by-pixel basis (Jenei et al., manuscript in preparation; Kirsch et al., 1999).

The extension of these techniques to study biologically relevant problems in live cellular systems involves making use of the entire repertoire of reporter molecules (dyes/proteins), including the *Aequoria victoria* GFP (see above). GFP mutants with different spectral properties form excellent FRET pairs; one direction of research in our laboratory involves the study of receptor–ligand interactions using GFP-labeled molecules (Brock et al., 1998).

Multiphoton Excitation and Multicolor Imaging

The excitation spectra of several biologically relevant dyes lie in the near-ultraviolet region, thereby posing serious problems for laser scanning microscopy. High-energy UV light is generally scattered and absorbed strongly by biological tissue, deteriorating image quality and inducing photodamage of living cells. Other major problems include the lack of inexpensive UV laser sources and the difficulty in producing UV-transmissive optics and objectives suitable for high-resolution imaging. Many of these problems can be bypassed by utilizing two-photon excitation (2PE) (Göppert-Mayer, 1931) with a long-wavelength (visible or near-infrared) source, in which a molecule simultaneously absorbs two photons in a single quantum event. The emission induced by this nonlinear excitation process is indistinguishable from the characteristic fluorescence following single-photon excitation (1PE) (Xu and Webb,

1996; Curley et al., 1992). 2PE-based fluorescence microscopy has found widespread use since its introduction in 1990 (Denk et al., 1990, 1995; Bennett et al., 1996; Köhler et al., 1997; Svoboda et al., 1997; Masters et al., 1997; Summers et al., 1996; Potter et al., 1996; König et al., 1996), due to intrinsic advantages such as the ability to excite near-UV dyes, three-dimensional resolution without confocal (pinhole) detection, and suppression of photobleaching and fluorescence background outside the focal plane. These circumstances reflect the quadratic dependence of the excitation, restricting the probe volume to a small region around the focus.

The signal of 2PE and three-photon excitation (3PE) fluorophores is proportional to the time-averaged square and cube, respectively, of the excitation light intensity and is thus enhanced by the use of repetitive short laser pulses. For lasers emitting pulses of length τ at a repetition rate f, the increase of the 2PE fluorescence signal relative to a cw source with the same average power is $\sim 1/\tau f$ (Denk et al., 1995). In 3PE, the corresponding factor scales as $1/(\tau f)^2$ (Schrader et al., 1997; Xu et al., 1996). The 2PE boost factor value can be 5×10^4 ($\tau = 250$ fs, $f = 80$ MHz) for a typical femtosecond Ti–sapphire laser system and has led to the general use of femtosecond laser systems in multiphoton laser scanning microscopy (Denk et al., 1990; Wokosin et al., 1996). Unfortunately, such systems are technically complex and expensive. However, 2PE microscopy can also be implemented with cw lasers, which are in general cheaper and more readily available than femtosecond pulsed devices (Booth and Hell, 1998; Hänninen et al., 1994). To achieve signal levels similar to those attained with femtosecond pulsed lasers, the time-averaged cw intensity has to be increased over the average intensity in the pulsed case by the factor $1/\tau f$, leading to an average excitation intensity of several hundred milliwatts, which can be delivered easily by many cw laser systems. The peak power is decreased by the same factor, thus shifting the potential limitations of the 2PE technique from nonlinear to linear phenomena (Schönle and Hell, 1998). A compromise between such femtosecond Ti–sapphire systems and 2PE based on cw laser systems of higher average power but consequently limited by sample stability is provided by solid-state picosecond lasers. For a picosecond laser with $\tau = 7$ ps and

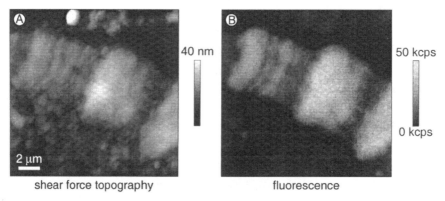

Fig. 12.6. Shear force topopgraphy (A) and cw 2PE-excited fluorescence (B) images of *D. melanogaster* polytene chromosomes labeled with DAPI. Scan parameters: 2.5 s/line, 256 points, 4.1 ms counting time per pixel, excitation: 92 mW at 647 nm. (Adapted from Kirsch et al., 1998c.)

TABLE 12.2. Results of Fit to Power Curve $F = dI^e$ + f for Different Near-UV Absorbing Dyes[a]

Dye	$\langle e \rangle$
DAPI	1.88 ± 0.18
BBI-342	1.90 ± 0.09
EtBr	1.91 ± 0.02

Symbols: [a] F, fluorescence; I, excitation intensity; e, order of the process; f, independent background signal; d, proportionality constant.

f = 200 MHz (characteristic of the laser used in this study), the $1/\tau f$ boost factor is ~700. In order to achieve signal levels in 2PE similar to those attained by femtosecond laser systems, the average power of the excitation in the picosecond case thus has to be increased by the modest factor of ~ 8, resulting in average power levels of ~ 10–40 mW (Bewersdorf and Hell, 1998).

We have implemented both cw and picosecond pulsed multiphoton excitation in our SNOM, permitting the imaging of near-UV dyes and simultaneous, cross-talk-free two- and three-photon imaging, respectively.

Continuous-Wave Two-Photon Imaging. The 647-nm emission of an Ar–Kr mixed-gas laser was coupled into the SNOM and used to excite the UV-absorbing DNA dyes DAPI, the bisbenzimidazole Hoechst 33342 (BBI-342), and the intercalating dye ethidium bromide, which also absorbs in the visible range. Polytene chromosomes of *D. melanogaster* and the nuclei of 3T3 Balb/c cells labeled with these dyes were readily imaged. Figure 12.6 depicts shear force topography and 2PE fluorescence images of a *Drosophila* polytene chromosome. Cellular debris from the squashing procedure is clearly seen in the topographic image, but it does not contribute to the optical image, which is due solely to fluorescence of DNA-bound DAPI.

The multiphoton origin of the detected fluorescence was confirmed by measuring the fluorescence intensity as a function of excitation power, which showed the expected (second-order) dependence on the excitation power in the range of 8–180 mW (Table 12.2).

We also measured the fluorescence intensity as a function of the tip-sample displacement in the direction normal to the sample surface in the single (1PE) and 2PE modes. In 2PE, the fluorescence intensity decayed faster than for 1PE. That is, the thickness of the surface layer contributing to the fluorescence signal is much smaller in the case of 2PE than with 1PE. This result has profound implications for the potential uses of 2PE SNOM.

Picosecond Pulsed Multiphoton Imaging. We have also implemented pulsed picosecond multiphoton excitation in the SNOM using a solid-state Nd–YVO$_4$ laser emitting at 1064 nm. With this system we have achieved simultaneous three-photon excitation of near-UV-absorbing fluorophores and 2PE of dyes excitable in the visible green range in the SNOM. This combination of multiphoton excitations offers the possibility of simultaneous dual-color imaging. Additionally the use of a frequency-doubled cw Nd–YAG laser in this configuration allows us to compare 1PE and 2PE excitation of the same sample directly.

Figure 12.7 shows shear force topography (A), two-photon excitation fluorescence (B), and three-photon excitation fluorescence (C) images of dried MCF7 adenocarcinoma cells labeled with the mitochondrial-specific dye MitoTracker Orange CM-H2TMRos (M7511, Molecular Probes, Leiden, The Netherlands) and the nuclear stain BBI-342 (Calbiochem, Bad Soden, Germany). The contrast was excellent without any crosstalk between the emission signals. The fluorescence intensity showed the expected nonlinear (second- and third-order) dependence on the excitation power in the range of 5–50 mW (see Fig. 12.8). The smallest resolved structures had a width of <200 nm, a value about one-sixth of the fundamental wavelength or one-third the wavelength corresponding to the two-photon energy.

We also measured the fluorescence intensity as a function of the tip-sample displacement in the direction normal to the sample surface in the 1PE and 2PE modes.

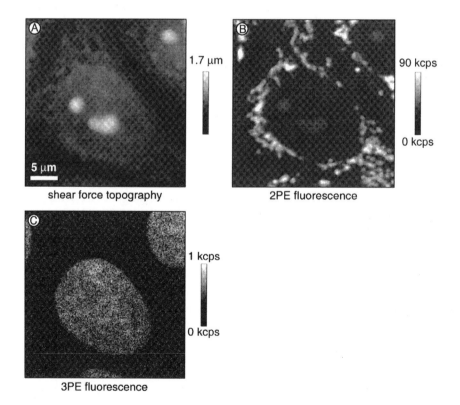

Fig. 12.7. 2PE and 3PE SNOM images of MCF7 cells. (A) Shear force topography, (B) 2PE fluorescence signal of the MitoTracker Orange-labeled mitochondria, and (C) 3PE fluorescence of the BBI-342-labeled nucleus. Scan parameters: 10 s/line, 128 lines, 256 points/line, excitation: 51 mW at 1064 nm. *See color plates.*

Specific Applications

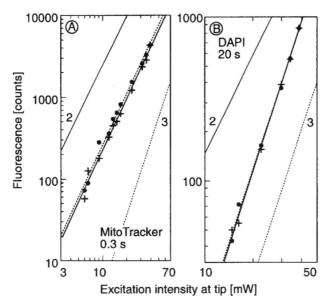

Fig. 12.8. Power curves measured on (A) MitoTracker-labeled MCF7 cell and (B) polytene chromosome labeled with DAPI. The fluorescence intensity values at the fiber tip were measured by varying the excitation intensity using ND filters. Counting times: (A) 0.3 s, (B) 20 s. Data points for increasing excitation intensity (•) and fit to the power curve $F=dI^e$ (—); data for decreasing excitation intensity (+) and fit (. . .). As visual guides, each panel includes lines corresponding to ideal two-photon (—) or three-photon (. . .) processes flanking the data. Fits: 2PE-MitoTracker Orange, $\langle e \rangle = 2.10 \pm 0.11$; 3PE-DAPI, $\langle e \rangle = 2.94 \pm 0.07$.

In the case of 2PE, the fluorescence intensity decayed faster than for 1PE photon excitation as seen from the characteristic decay length in the axial direction for the fluorescence signal (Table 12.3).

A prominent advantage of MPE is the confinement of the excitation to the focal volume, yielding lower collateral damage to the sample due to photobleaching. We characterized the excitation distributions in 1PE and 2PE by photobleaching lines in a rhodamine B–doped PVA film by repeatedly scanning (64 times) a single line with the same light intensity used to subsequently image the bleached pattern, using both 1PE at 532 nm (Fig. 12.9A) and 2PE at 1064 nm (Fig. 12.9B). Both lines were traced and imaged with the same fiber tip on different areas of the same sample. We intentionally overilluminated the sample so as to emphasize any secondary features generated by light leakage from the fiber tip. In the 1PE case the bleached line was surrounded by a larger area exhibiting diffuse bleaching as well as weak secondary bleached lines (Fig. 12.9A inset). With 2PE the fluorescence of the film surrounding the line was uniform, and there were no visible effects of photobleaching in this area (Fig. 12.9B inset). The consequences of the bleaching are seen more clearly in the averaged profiles normal to the bleached lines (Figs. 12.9A and B). The secondary features generated by 1PE are readily evident in the profile shown in panel A.

TABLE 12.3. Results of z Distance Dependence for Single- and Two-Photon Excitation

Excitation mode	$\langle H \rangle^a$
Single-photon	404 ± 62 nm
Two-photon	178 ± 3 nm

[a] Average value of the distance between the probe and the sample surface at which the fluorescence signal dropped to one-half of the initial value at contact.

The secondary photobleaching features were most likely due to coupling of the excitation light out from the conical part of the fiber several micrometers away from the surface. The 1064-nm excitation presumably also coupled out of the fiber, but due to the higher-order dependence of 2PE, the leakage intensity was apparently too weak to yield significant photobleaching. A reduction or elimination of the secondary features due to lateral coupling of light out of the fiber was shown clearly for 2PE. This result has important implications for the microscopy of light-sensitive samples, imaging of which by conventional line-scan 1PE can involve photodamage in neighboring as-yet

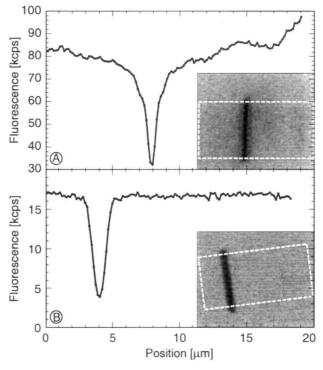

Fig. 12.9. Fluorescence bleaching experiments on rhodamine B-doped PVA film. (A) 1PE bleached line after 64 scans (inset) and mean fluorescence intensity profile perpendicular to the bleached line. (B) 2PE bleached line (inset; note that this image was rotated by ~ 7° due to piezo drift) and mean fluorescence intensity profile perpendicular to bleached line. The dashed boxes indicate the areas over which the mean profiles were calculated.

unscanned regions. The effect is apparently suppressed, or at least diminished significantly, in 2PE SNOM.

The multiphoton operational mode extends the accessible spectral range to the near-UV and also allows multicolor imaging of two different dyes free of crosstalk. In addition, the background signals are significantly reduced, constituting one of the primary advantages of multiphoton excitation. The thickness of the surface layer contributing to the optical signal and the lateral extent of the excitation are also reduced significantly.

PROSPECTS

SNOM in combination with fluorescence methodologies represents a powerful technique for the visualization of cellular systems and biological interactions with a spatial resolution not routinely achievable with far-field optical microscopy. The integration of spectral resolution, lifetime determination, and multiphoton excitation techniques with SNOM has significantly enhanced the capabilities of this microscopy mode. The extension of SNOM to the challenging task of performing near-field microspectroscopy on living cellular systems in aqueous environments promises to reveal new insights into the distribution of cell surface receptors, the mechanisms and dynamics of receptor–ligand interactions, and in general, the physiology of the initial stages of receptor signal-transduction pathways.

ACKNOWLEDGMENTS

We thank Rolando Rivera-Pomar for providing the GFP-transfected *Drosophila* Schneider cells, Roland Brock for providing CHO cells expressing EGFR–GFP fusion proteins, and S. W. Hell for the loan of the Nd–YVO$_4$ laser for the picosecond multiphoton excitation measurements and for motivating discussions. A. J. was supported by a long-term EMBO fellowship and is also affiliated with the Department of Biophysics and Cell Biology, University Medical School of Debrecen, H-4012 Debrecen, Hungary. V. S. was supported by a long-term fellowship from the Human Frontiers Science Program Organization.

REFERENCES

Bastiaens PIH, Jovin TM (1996): Microspectroscopic imaging tracks the intracellular processing of a signal transduction protein: Fluorescent-labeled protein kinase C βI. Proc Natl Acad Sci USA 93:8407–8412.

Bastiaens PIH, Majoul IV, Verveer PJ, Söling H-D, Jovin TM (1996): Imaging the intracellular trafficking and state of the AB_5 quaternary structure of cholera toxin. EMBO J 15:4246–4253.

Bennett BD, Jetton TL, Ying G, Magnuson MA, Piston DW (1996): Quantitative subcellular imaging of glucose metabolism within intact pancreatic islets. J Biol Chem 271:3647–3651.

Betzig E, Trautman JK (1992): Near-field optics—microscopy, spectroscopy, and surface modification beyond the diffraction limit. Science 257:189–195.

Betzig E, Chichester RJ, Lanni F, Taylor DL (1993): Near-field fluorescence imaging of cytoskeletal actin. Bioimaging 1:129–135.

Betzig E, Trautman JK, Harris TD, Weiner JS, Kostelak RL (1991): Breaking the diffraction barrier: Optical microscopy on a nanometric scale. Science 251:1468–1470.

Bewersdorf J, Hell SW (1998): Picosecond pulsed two-photon imaging with a repetition rates of 200 and 400 MHz. J Microsc 191:28–38.

Bielefeldt H, Hörsch I, Krausch G, Lux-Steiner M, Mlynek J, Marti O (1994): Reflection-scanning near-field optical miroscopy and spectroscopy of opaque samples. Appl Phys A Solids Surf 59:103–108.

Booth MJ, Hell SW (1998): Continuous wave excitation two-photon fluorescence microscopy exemplified with the 647 nm ArKr laser line. J Microsc 190:298–304.

Brock R, Hamelers IHL, Jovin TM (1999): Comparison of fixation protocols for adherent cultured cells applied to a GFP fusion proteins of the epidermal growth factor receptor. Cytometry 35:353–362.

Clegg RM (1996): Fluorescence resonance energy transfer (FRET). In Wang XF, Herman B (eds). *Fluorescence Imaging Spectroscopy and Microscopy*. New York: Wiley, pp 179–252.

Courgon D, Vigoureux JM, Spajer M, Sarayeddine K, Leblanc S (1990): Extenral and internal-reflection near-field microscopy–experiments and results. Appl Opt 29:3734–3740.

Curley PF, Ferguson AI, White JG, Amos WB (1992): Application of a femtosecond self-sustaining mode-locked Ti: sapphire laser to the field of laser scanning confocal microscopy. Opt Quant Electron 24:851–859.

Denk W, Piston DW, Webb WW (1995): Two-photon molecular excitation in laser-scanning microscopy. In Pawley JP (ed). *Handbook of Biological Confocal Microscopy*. New York: Plenum Press, pp 445–458.

Denk W, Strickler JH, Webb WW (1990): Two-photon laser scanning fluorescence microscopy. Science 248:73–76.

Dunn RC (1999): Near-field Microscopy. Chem Rev 99:2891–2897.

Dunn RC, Allen EV, Joyce SA, Anderson GA, Xie XS (1995): Near-field fluorescent imaging of single proteins. Ultramicroscopy 57:113–117.

Dunn RC, Holtom GR, Mets L, Xie XS (1994): Near-field fluorescence imaging and fluorescence lifetime measurement of light harvesting complexes in intact photosynethetic membranes. J Phys Chem 98:3094–3098.

Enderle T, Ha T, Chemla DS, Weiss S (1998): Near-field fluorescence microscopy of cells. Ultramicroscopy 71:303–309.

Enderle T, Ha T, Ogletree DF, Chemla DS, Magowan C, Weiss S (1997): Membrane specific mapping and colocalization of malarial and host skeletal proteins in the *Plasmodium falciparum* infected erythrocyte by dual-color near-field scanning optical microscopy. Proc Natl Acad Sci USA 94:520–525.

Garcia-Parajo MF, Veerman JA, Vannoort SJT, Degrooth BG, Greve J, Vanhulst NF (1998): Near-field optical microscopy for dna studies at the single molecular-level. Bioimaging 6:43–53.

Göppert-Mayer M (1931): Über Elementarakte mit zwei Quantensprüngen. Ann Phy 9:273–295.

Ha T, Enderle T, Ogletree DF, Chemla DS, Selvin PR, Weiss S (1996): Probing the interaction between two single molecules: Fluorescence resonance energy transfer between a single donor and a single acceptor. Proc Natl Acad Sci USA 93:6264–6268.

Hänninen PE, Soini E, Hell SW (1994): Continuous wave excitation two-photon fluorescence microscopy. J Microsc 176:222–225.

Hecht B, Bielefeldt H, Inouye Y, Pohl DW, Novotny L (1997): Facts and artifacts in near-field optical microscopy. J Appl Phys 81:2492–2498.

Hollricher O, Brunner R, Marti O (1998): Piezoelectrical shear-force distance control in near-field optical microscopy for biological applications. Ultramicroscopy 71:143–147.

Hwang J, Gheber LA, Margolis L, Edidin M (1998): Domains in cell plasma membranes investigated by near-field scanning optical microscopy. Biophys J 74:2184–2190.

Hwang J, Tamm LK, Bohn C, Ramalingam TS, Betzig E, Edidin M (1995): Nanoscale complexity of phospholipid monolayers investigated by near-field scanning optical microscopy. Science 270:610–614.

Isaacson M, Betzig E, Harootunian A, Lewis A (1986): Scanning optical microscopy at lambda/10 resolution using near-field imaging methods. Science 483:448–456.

Jenei A, Kirsch AK, Subramaniam V, Arndt-Jovin DJ, Jovin TM (1999): Picosecond multi-photon scanning near-field optical microscopy. Biophys J 76:1092–1100.

Jovin TM, Arndt-Jovin DJ (1989): FRET microscopy: Digital imaging of fluorescence resonance energy transfer. Application in cell biology. In Kohen E, Hirschberg JG, Ploem JS (eds). *Cell Structure and Function by Microspectrofluometry* London: Academic Press, pp 99–117.

Keller TH, Rayment T, Klenerman D, Stephenson RJ (1997): Scanning near-field optical microscopy in reflection mode imaging in liquid. Rev Sci Instrum 68:1148–1454.

Kirsch AK, Subramaniam V, Jenei A, Jovin TM (1999): Fluorescence resonance energy transfer (FRET) detected by scanning near-field optical microscopy (SNOM). J Microsc 194:448–454.

Kirsch AK, Meyer CK, Huesmann H, Möbius D, Jovin TM (1998a): Fluorescence SNOM of domain structures of LB films containing electron transfer systems. Ultramicroscopy 71L:295–302.

Kirsch AK, Schaper A, Huesmann H, Rampi MA, Möbius D, Jovin TM (1998b): Scanning force and scanning near-field optical microscopy of organized monolayers incorporating a nonamphiphilic metal dyad. Langmuir 14:3895–3900.

Kirsch AK, Subramaniam V, Striker G, Schnetter C, Arndt-Jovin D, Jovin TM (1998c): Continuous wave two-photon scanning near-field optical microscopy. Biophys J 75:1513–1521.

Kirsch A, Meyer C, Jovin TM (1996): Integration of optical techniques in scanning probe microscopes: the scanning near-field optical microscope (SNOM). In Kohen E, Kirschberg JG (eds). *Proceedings of NATO Advanced Research Workshop: Analytical Use of Fluorescent Probes in Oncology, Miami, Fl. Oct. 14–18 1995.* New York: Plenum Press, pp 317–323.

Köhler RH, Cao J, Zipfel WR, Webb WW, Hanson MR (1997): Exchange of protein molecules through connections between higher plant plastids. Science 276:2039–2042.

König K, Simon U, Halbhuber KJ (1996): 3D resolved two-photon fluorescence microscopy of living cells using a modified confocal laser scanning microscope. Cell Mol Biol 42:1181–1194.

Masters BR, So PTC, Gratton E (1997): Multiphoton excitation fluorescence microscopy and spectroscopy of in vivo human skin. Biophys J 72:2405–2412.

Moers MHP, Kalle WHJ, Ruiter AGT, Wiegant JCAG, Raap AK, Greve J, Degrooth BG, van Hulst NF (1996): Fluorescence in-situ hybridization on human metaphase chromosomes detected by near-field scanning optical microscopy. J Microsc 182:40–45.

Moyer PJ, Kammer SB (1996): High-resolution imaging using near-field scanning optical microscopy and shear force feedback in water. Appl Phys Lett 68:3380–3382.

Muramatsu H, Chiba N, Ataka T, Iwabuchi S, Nagatani N, Tamiya E, Fujihira M (1996): Scanning near-field optical/atomic force microscopy for fluorescence imaging and spectroscopy of biomaterials in air and liquid—observation of recombinant *Escherichia coli* with gene coding to green fluorescent protein. Opt Rev 3:470–474.

Nagy P, Jenei A, Kirsch AK, Szöllösi J, Damjanovich S, Jovin TM (1998): Activation dependent clustering of the erbB2 receptor tyrosine kinase detected by scanning near-field optical microscopy. J Cell Sci 112:1733–1741.

Paesler MA, Moyer PJ (1996): *Near-field Optics.* New York: Wiley.

Phol DW, Denk W, Lanz M (1984): Optical stethoscopy: Image recording with resolution λ/20. Appl Phys Lett 44:651–653.

Potter SM, Wang CM, Garrity PA, Fraser SE (1996): Intravital imaging of green fluorescent protein using two-photon laser-scanning microscopy. Gene 173:25–31.

Schönle A, Hell SW (1998): Heating by linear absorption in the focus of an objective lens. Opt Lett 23:325–327.

Schrader M, Bahlmann K, Hell SW (1997): Three-photon-excitation microscopy: Theory, experiment and appliations. Optik 104:116–124.

Seibel EJ, Pollack GH (1997): Imaging 'intact' myofibrils with a near-field scanning optical microscope. J Microsc 186:221–231.

Shiku H, Hollars CW, Lee MA, Talley CE, Cooksey G, Dunn RC (1998): Probing biological systems with near-field optics. Proc SPIE 3273:156–164.

Stürmer H, Köhler JM, Jovin TM (1998): Microstructured polymer tips for scanning near-field optical microscopy. Ultramicroscopy 71:107–110.

Subramaniam V, Kirsch AK, Jovin TM (1998): Cell biological applications of scanning near-field optical microscopy (SNOM). Cell Mol Biol 44:689–700.

Subramaniam V, Kirsch AK, Rivera-Pomar RV, Jovin TM (1997): Scanning near-field optical imaging and microspectroscopy of green fluorescent protein (GFP) in intact *Escherichia coli* bacteria. J Fluorescence 7:381–385.

Summers RG, Piston DW, Harris KM, Morrill JB (1996): The orientation of first cleavage in the sea urchin embryo, Lytechinus Variegatus, does not specify the axes of bilateral symmetry. Developm Biol 175:177–183.

Svoboda K, Denk W, Kleinfeld D, Tank DW (1997): In vivo dendritic calcium dynamics in neocortical pyramidal neurons. Nature 385:161–165.

Talley CE, Lee MA, Dunn RC (1998): Single-molecule detection and underwater fluorescence imaging with cantilevered near-field fiber optic probes. Appl Phys Lett 72:2954–2956.

Talley CE, Cooksey GA, Dunn RC (1996): High-resolution fluorescence imaging with cantilevered near-field fiber optic probes. Appl Phys Lett 69:3809–3811.

Tamiya E, Iwabuchi S, Nagatani N, Murakami Y, Sakaguchi T, Yokoyama K, Chiba N, Muramatsu H (1997): Simultaneous topographic and fluorescence imagings of recombinant bacterial cells containing a green fluorescent protein gene detected by a scanning near-field optical/atomic force microscope. Anal Chem 69:3697–3701.

van Hulst NF, Garcia-Parajo MF, Moers MHP, Veerman JA, Ruiter AGT (1997): Near-field fluorescence imaging of genetic material—toward the molecular limit. J Struct Biol 119:222–231.

Vereb G, Meyer CK, Jovin TM (1997): Novel microscope-based approaches for the investigation of protein–protein interactions in signal transduction. In: Heilmeyer Jr LMG (eds). *Interacting protein domains, their role in signal and energy transduction. NATO ASI series* H102. New York: Springer, pp 49–52.

Wiegräbe W, Monajembashi S, Dittmar H, Greulich KO, Hafner S, Hildebrandt M, Kittler M, Lochner B, Unger E (1997): Scanning near-field optical microscope—a method for investigating chromosomes. Surf Interf Anal 25:510.

Wokosin DL, Centonze V, White JG, Armstrong D, Robertson G, Ferguson AI (1996): All-solid-state ultrafast lasers facilitate multiphoton excitation fluorescence imaging. IEEE J Sel Top Quant Elec 2:1051–1065.

Xu C, Webb WW (1996): Measurement of two-photon excitation cross sections of molecular fluorophores with data from 690 to 1050 nm. J Opt Soc Am B 13:481–491.

Xu C, Williams RM, Zipfel W, Webb WW (1996): Multiphoton excitation cross-sections of molecular fluorophores. Bioimaging 4:198–207.

13

White-Light Scanning Digital Microscopy

J. Paul Robinson
Purdue University, West Lafayette, Indiana

Ben Gravely
Cosmic Technologies Corporation, Raleigh, North Carolina

INTRODUCTION

Digital microscopy has for many years been a technology difficult to implement. Finally, however, it is clear that new technologies are beginning to reshape the field of microscopy. The dramatic increase in the performance and cost effectiveness of computers, digital storage systems, and telecommunications is fueling rapid growth in the creation and storage of digital microscope images. There are several important issues that must be addressed in dealing with digital images: (1) the image collection system, which is usually a charge-coupled device (CCD) camera; (2) the image capture computer, including necessary boards and software; and (3) image manipulation (image processing) and data transfer from one site to another.

The development of digital microscopy is being driven by the need to make critical decisions based upon the rapid collection and analysis of images. For fifty years a desire to use imaging to define cancerous cells from normal has been a motivating force (Mellors et al., 1952). It has been apparent for many years that these decisions should be made with the aid of imaging technologies in an automated fashion (Bartels et al., 1984). Early attempts at using rapid imaging technologies

---

Emerging Tools for Single-Cell Analysis, Edited by Gary Durack and J. Paul Robinson.
ISBN 0-471-31575-3 Copyright © 2000 Wiley-Liss, Inc.

(video) proved very difficult and were unable to produce high-resolution images (Bibbo et al., 1983; Ingram and Preston, 1964; Prewitt and Mendelsohn, 1966) or simply reached an impasse because of lack of rapid imaging technology (Ingram and Preston, 1970). With the development of more powerful computers and automated microscopes, complete systems for diagnostic use began to take shape (Ploem et al., 1986, 1988). The trend today in medicine, for example, is the move toward electronic data integration of all patient records, including diagnostic images, and this promises to revolutionize not only diagnosis but also teaching with archival databases of images, automated analysis, publication, and local and remote consultation.

Digital microscopy is a key element in a combination of technologies that has made possible the development of telepathology, which involves the electronic transfer of pathological images between geographically distant locations for medical consultation. The remote control of a distant microscope and the transfer of high-quality images over normal telecommunications systems coupled with text and voice will have a significant impact on clinical pathology, resulting in more efficient diagnosis and a greater availability of quality medical services to remote areas.

Telepathology eliminates the need to physically transport critical samples through delivery services, whose availability and reliability may be nonexistent in many areas of the world. Telepathology can be divided into two forms. Still telepathology involves sending "snapshots" of areas of interest from the client to the consultant and works best in a consultation environment between peers. The client should be qualified to select the fields of interest. This technology frequently utilizes high-quality image capture cameras, and images are downloaded in near real time (sometimes several minutes) for remote evaluation, such as that recently reported by Singson et al. (1999). Other less advanced systems have been proposed using regular telephone lines and modems; however, these suffer from loss of high-resolution images and significant and unacceptable time delays (Vazir et al., 1998). Dynamic telepathology involves the robotic operation of a remote microscope by the consulting pathologist, who selects the fields and focuses just as he or she would with a local microscope. A number of publications have appeared recently demonstrating remote microscope control using the internet as a somewhat universal platform (Nagata and Mizushima, 1998; Szymas and Wolf, 1998). The remote microscope responds to the commands and relays images back to the pathologist. Only a qualified technician is required at the remote microscope to prepare slides and place them in the system. Dynamic telepathology is the most demanding of all digital pathology tasks because the high image quality and speed of operation required are very difficult to achieve. Before dynamic telepathology can become common, many complex issues must be solved in image capture and transmission systems.

This chapter discusses one potential solution to the technology problems—a white-light flying-spot scanning microscope system. This microscope is an integrated, fully digital color microscope that utilizes a flying-spot light source, advanced image-processing techniques, and simple straightforward operation. Although the fundamental principles upon which this technology is based were developed nearly a half century ago, only now have technological advancements allowed it to become an economic reality.

SCANNING SPOT

The color scanning microscope (COSMIC) is based on spot-scanning techniques developed in the 1950s for various applications, including film-to-video conversion for television and even black-and-white video microscopes. Figure 13.1 shows the optical diagram of the instrument. A specially designed cathode ray tube (CRT) has a

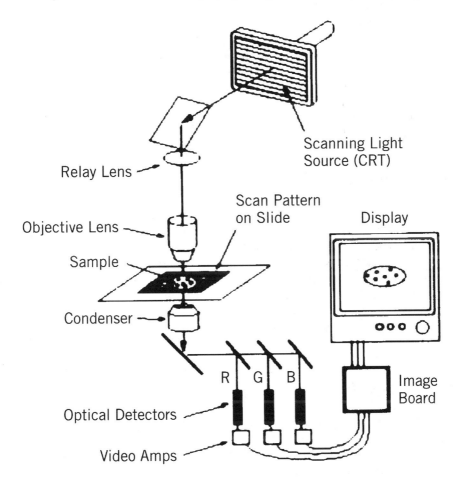

Fig. 13.1. Schematic layout of the COSMIC system showing the light source, optical path, and signal collection components. The spot scans across the CRT create a point of light that illuminates only the Airy disk area on the sample. This reduces bleaching and scattered light on the sample. Since the illumination point is the smallest that can be created by the objective, the system resolution is limited only by the optical resolution of the objective, not the detector components.

white raster pattern on its screen. This pattern is imaged through a microscope relay lens and objective lens onto the sample placed on the microscope stage.

The size of the raster image on the sample is determined by the lens system magnification. For example, an ordinary microscope may have a 40-power objective lens and a 10-power eyepiece lens, giving a combined magnification of 400 from the sample to the eye. Operated in reverse, an object placed outside the eyepiece lens will be demagnified 400 times down to the sample plane.

This one-to-one mapping of the scanning spot from the CRT screen onto the sample plane creates a tiny white probe in a dark field that scans over the sample. The spot characteristics on the sample are completely determined by the microscope objective. As a result, the probing spot has the smallest diffraction-limited diameter possible for each objective lens.

The light transmitted through the sample is divided by dichroic mirrors into red (580–650 nm), green (520–580 nm), and blue (450–520 nm) spectral components and passed to three detectors. A video display reconstructs the image of the sample using the x–y position of the scanning spot and the three color intensities. In the simplest case, the scan frequency of the CRT raster matches the scan frequency of the display, giving a direct correspondence between the spot position on the CRT and the spot position on the display.

DETECTION SYSTEM

The detectors in this digital microscope are not the CCD imaging devices used in most "digital" cameras today. Instead, the detectors are three nonimaging photomultiplier tubes (PMTs) that collect the intensity signals simultaneously in the three spectral bands. Color registration is inherently perfect since there is only one spot scanning over the sample and the PMT detectors read intensity only in the color bands. The PMTs used are ordinary types available from several manufacturers, capable of achieving full color saturation with signals of less than 100 nW of power. More expensive or cooled PMT detectors could be used for special applications.

SIGNAL PROCESSING

Performance can be improved by reducing the scan frequency of the CRT well below the scan frequency of the display. The data are read in at one rate, then presented to the display at a faster rate. The choice of scan rate at the sample is determined by the number of pixels in the image and the desired optical contrast function. From a purely technical standpoint, the number of pixels in the digital image could be chosen as any value, but practical considerations of the data file size, availability of displays, processing and transmission times, and correlation with direct viewing set reasonable boundaries to the image file size.

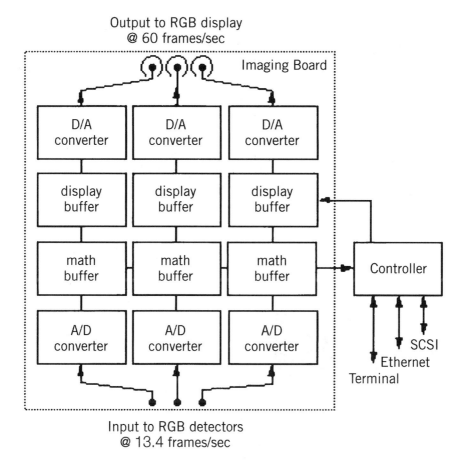

Fig. 13.2. Signal processing path showing how the transmitted light is handled for each spectral channel.

These considerations led to the choice of image size for COSMIC as 1280 × 1024 pixels with 24 bits/pixel color depth. Displays are readily available for this resolution and the data set is manageable for computation, storage, and transmission. The image size is a reasonable match to the number of optical resolution elements in various microscope images as viewed directly by the eye. The selected frame rate of 13.4 Hz is fast enough to observe live cells. The resulting data rate is 420 Mbits/s, which is at the high end of current imaging technologies.

Figure 13.2 illustrates the electronic signal path. The three color signals from the detectors are amplified and sent to a digital imaging board that performs multiple functions. First, the three signals are digitized into parallel 1280 × 1024 math buffers. The data in the three math buffers are transferred to three display buffers, converted into analog signals, and sent to the display at a 60-Hz frame rate. Math functions for averaging up to 256 images and summing (integrating) up to 9999 images are executed in the math buffers by programming built into the imaging board. These functions operate at the full frame rate.

ZOOM MAGNIFICATION

The optical magnification can be changed instantaneously up to 300% (3×) by changing the scanned area on the sample. The principle is illustrated in Figure 13.3. The scan raster overlaying the sample is shown on the left and the resulting image on the display is shown on the right. The image is always digitized into 1280 × 1024 pixels, so the data set remains constant for any magnification, and the pixel size never changes within the field of view.

Operating an objective lens at three times its rated magnification will exceed its classical resolution limit, resulting in so-called empty magnification, which has tra-

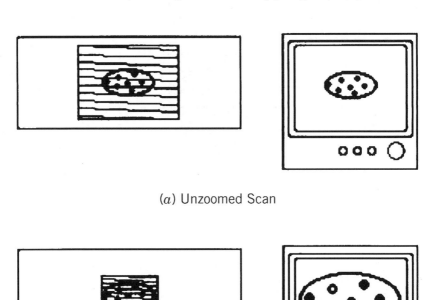

(a) Unzoomed Scan

(b) Zoomed Scan

Fig. 13.3. As the scan area decreases, the specimen fills more of the image window. The display area is constant in size, thus the specimen is enlarged. The image is always digitized at 1280 × 1024 data points, creating a constant file size at all zoom values.

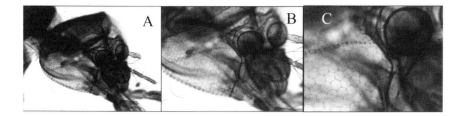

Fig. 13.4. Example of a zoomed image. Here, a mosquito head is imaged using the microscopes 10x objective with no zoom (A), zoom at 1.5x (B), and zoom at 3x (C). *See color plates.*

ditionally been considered a useless exercise. However, what really happens is that each optical resolution element may be digitized into as many as six data points, resulting in a process called digital oversampling. Oversampling tends to push resolution limits beyond standard conventions and results in "superresolution." Oversampling is discussed in more detail below.

The zoom feature also reduces the need for larger digital image files having more pixels since one can instantaneously zoom in to display maximum resolution at reduced field size. COSMIC's constant file size means that transmission times are invariant for all images. Figure 13.4 shows an example of a zoomed image. Here, a mosquito head is imaged using COSMIC's 10x objective with no zoom (A), zoom at $1.5\times$ (B), and zoom at $3\times$ (C).

BRIGHTNESS AND FOCUS CHANGE

Some unexpected characteristics occur with the zoom capability. In an ordinary optical zoom lens two things happen when the field size is changed. The brightness changes because the numerical aperture (NA) has changed (the change in focal length changes the NA), and the focus changes because it is virtually impossible to make a variable-focus lens with moving parts that can maintain constant focus throughout its range. In a point scanning microscope, on the other hand, the zoom feature does not change the focus or the brightness, which are not simply constants but universal constants. The focus does not change because mechanical distances are fixed. There are no alterations in the distance between the lenses and image planes at any time. The changes are related only to the area scanned on the sample. The size of the scanned area has been decreased, but all the mechanical distances remain the same. Another important characteristic of the zoom feature is that the brightness does not change. The illumination energy density on the sample varies with the size of the scanned area. For a large area, the energy per unit area is low. This is because the illumination emitted by the CRT is constant. Only the scanned area changes and therefore the energy density changes linearly with area.

The change in brightness between two different illumination areas can be shown mathematically to be inversely proportional to the ratio of the areas. Since the illumination on the sample changes proportionally to the area ratio, the brightness in the final image does not change with zoom.

CONTRAST ENHANCEMENT AND SUPPRESSION

In other spot-scanning instruments, such as laser scanning microscopes and electron beam microscopes, the intensity of the spot is constant as it scans over the sample. With COSMIC, however, it is possible to modulate the brightness of the spot for every pixel in the sample plane. If the spot is modulated using data from the sample, then several interesting effects are possible. Figure 13.5A illustrates the usual condition of a constant-intensity scan overlaid on the sample and the resultant image on the display. In Figure 13.5B, the CRT raster has a positive image of the sample superimposed on it. When transferred to the sample plane, the spot goes bright when the sample is bright

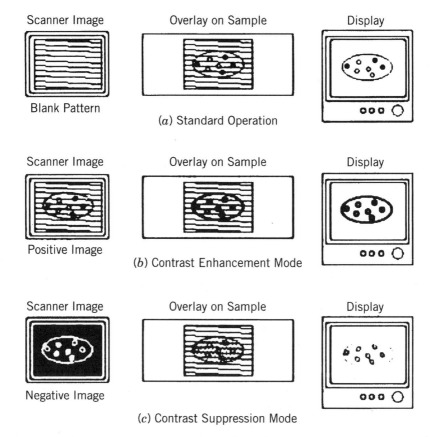

Fig. 13.5. Concepts of spot modulation demonstrated. The brightness of the scanning spot is varied point by point as it traces over the sample in order to increase or decrease the light/dark signal swing. Three effects are shown: standard (A), contrast enhancement (B), and contrast suppression (C).

Fig. 13.6. Image enhancement features of COSMIC. The normal-mode image of Figure 4 (A), the image in contrast suppression mode (B), and the image in color-inverted mode (C). This mode is particularly useful to identify thin structures in cells where the color inversion highlights the objects. *See color plates.*

and dark when the sample is dark, resulting in increased contrast in the image on the display. If the CRT has a negative image superimposed on it (Figure 13.5C), then the spot is dark when the sample is bright and bright when the sample is dark, suppressing contrast in the image. The contrast enhancement feature is very useful for examining unstained samples, and contrast suppression is useful for overstained or dark samples. Many of these samples cannot be successfully imaged with conventional microscopes. Figure 13.6 (A) is the same normal-mode sample as Figure 4. Figure 13.6 (B) demonstrates the contrast enhancement due to spot modulation, and Figure 13.6 (C) shows digital color inversion.

RESPONSE FUNCTION

Most television imaging and display tubes have a nonlinear response represented by an exponential equation using the variable "gamma." Gamma correction circuits are used routinely in video amplifiers to linearize the response.

Although the response function of the digital microscope is inherently linear, a programmable response function has been designed into the imaging board. In terms of gray-level histograms, spot modulation expands or compresses the whole histogram, while the gamma can adjust the midlevel values and not the end points. The gamma response function and spot modulation functions can be used together to fine tune the desired image enhancement effects.

In addition to the modes discussed, digital brightness, contrast, and background correction can be applied. The presence of these functions recognizes the need for simple, real-time image processing, which reduces the need for postprocessing and ensures that the image saved is of the highest quality.

AVERAGING AND SUMMING

In order to boost weak and smooth noisy signals, two digital processing modes have been implemented in hardware using the math buffer. The averaging mode performs a running average of up to 256 images in steps of 2^n, where $n = 1, \ldots, 8$. It is also possible to sum (or integrate) up to 9999 images. The summation mode not only reduces random noise in the final image by the square root of n but also acts as a gain factor up to 9999 for weak signals. The summed image can also be divided by powers of two as high as 256, giving a variable-gain factor. The summing mode is therefore a combination of gain and averaging and is particularly useful for very low signal levels. Both the averaging and summing modes operate at 13.4 frames per second. The new image creation rate is the number of frames processed divided by the system frame rate. For an averaging level of 2, therefore, new images appear at 6.7 Hz, and summing eight frames takes 0.60 s.

COMMUNICATIONS

The light scanning microscope has a SCSI port to save images to any kind of hard drive or removable drive. The file format is TIFF, a universal standard compatible with most image processing programs. An ethernet port is included for connection to networks for storage, archiving, and communication across the Internet or other telecommunication systems.

Files that have been created or modified in TIFF format by image processing or desktop publishing programs can be read back into COSMIC and shown on the display for teaching or conferences. A "slide show" mode can sequentially display all the images on the disk, so this means that text slides or photographs can be displayed along with the microscope images.

OVERSAMPLING AND SUPERRESOLUTION

As mentioned earlier, oversampling is a process of collecting more electronic data points than actually exist in the optical image. If there are 1000 optical resolution elements in a scan line across the image, and that line is digitized into 3000 data points, then that is oversampling. The number of optical resolution elements in a field of view can easily be calculated by using the classical Airy disk formula

$$R_o = \frac{0.61 \lambda}{NA}$$

where λ is the wavelength, typically chosen as 550 nm, and NA is the numerical aperture of the objective. According to the Rayleigh resolution criterion, R_o is the minimum distance resolvable by a lens with a given numerical aperture. It is also the radius of the Airy disk, which is the smallest dot that can be imaged by an objective, and will be referred to as the Airy disk size or the optical resolution element size.

Figure 13.7 shows a 40x objective with a numerical aperture of 0.75 which has an Airy disk size of 0.447 μm that defines its resolution. In a normal field of view of diameter 500 μm, there are approximately 1100 optical resolution elements across the center of the field. The COSMIC CRT raster scans a 3:4 aspect ratio rectangle whose diagonal is 70% of the 500-μm circle diameter, resulting in a scanned area that is approximately 280 × 210 μm, or about 626 × 470 Airy disks.

Oversampling and Superresolution

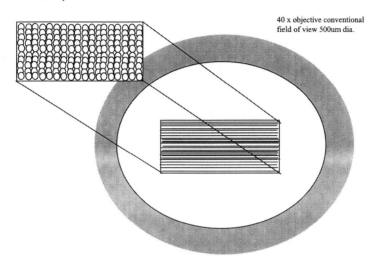

Fig. 13.7. A 40x objective with a NA of 0.75 has an Airy disk size of 0.447 μm that defines its resolution. In a normal field of view of diameter 500 μm, there are approximately 1100 optical resolution elements across the center of the field. The CRT raster scans a 3 : 4 aspect ratio rectangle whose diagonal is 70% of the 500-μm circle diameter, resulting in a scanned area that is approximately 280 × 210 μm, or about 626 × 470 Airy disks.

In order to determine the resolving power of the imaging system, it is necessary to take into account the sampling theory of Nyquist, who derived the number of samples needed to define a single point of a data curve. In terms of optics, to represent the information in a single optical resolution element (Airy disk), one must take two samples across the element. This is not really *oversampling* but *adequate sampling,* since 2 : 1 is the ratio needed for accurate sampling. Oversampling would therefore be anything over the 2 : 1 ratio. Examining Figure 13.7 again, it can be seen that for the 40x objective (with no zoom) 1280 points are digitized across the 626 Airy disks. The sampling ratio is right at 2:1, and it is reasonable to say that the data from the 40x objective are "just resolved."

When the 40x field of view is zoomed by 3x, the number of optical resolution elements in the scanned area is reduced to 218 × 156, but the data are still digitized into 1280 × 1024 data points. This is almost a 6 : 1 oversampling (Plasek et al., 1998). In reality, oversampling does give an improved image over "adequate sampling"; that is, the modulation transfer function improves up to a certain point, after which there is no improvement. The image detail increases with magnification up to a point, after which the image gets larger but no more detail is seen ("empty magnification").

Each objective power has a different number of optical resolution elements within the field of view. Interestingly, lower power objectives have more elements across the field of view, posing a greater imaging challenge than high-power objectives. For example, a 2.5x objective has a scanned area of 4.300 × 3.225 mm that contains

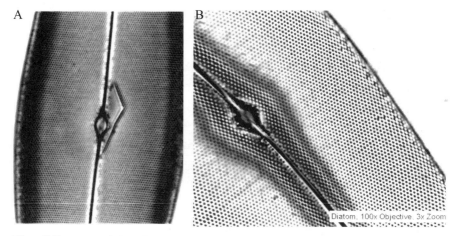

Fig. 13.8. Superresolution effect is demonstrated with an image of the diatom *Pleurasigma angulatum*. The holes in the diatoms are approximately 0.25 μm in diameter and hexagonal in shape. The full-color images were taken in white light. Obtaining such high-resolution images is very difficult with most other systems. Figure 13.8 (A) was taken with a 40x (0.75NA) objective. Figure 13.8 (B) was taken with a 100x (1.3NA oil) objective. However, the sample is in air under the cover slip, so the maximum usable NA for the 100x objective is less than 1.0. The condensor had an NA of 1.4 (oil). *See color plates.*

961 × 721 Airy disks. The 1280 × 1024 electronic data points are less than the 2 : 1 ratio needed to fully resolve the smallest element in the image. However, if the image is zoomed by 1.6, the scanned area covers 600 × 400 Airy disks and all points in the field are fully resolved. The zoom control provides a convenient mechanism to trade field of view for full resolution whenever desired. Because the zoom factor will carry any objective into true empty magnification, the system resolution is limited by the objective's ability to create an Airy disk, not the ability to take the data from it. Therefore, it is reasonable to say that the limit of resolution of this digital microscope is the optics, not the electronics.

This color scanning microscope can also exhibit superresolution, where image detail far smaller than the calculated limit has been observed. The technique uses the oversampling feature coupled with COSMIC's summing, averaging, and background correction functions. Diatom images taken with a 40x objective (0.75 NA) have shown features as small as 0.05 μm, well below the Airy disk size of 0.447 μm. This superresolution effect is demonstrated in Figure 13.8, where an image is shown of *Pleurasigma angulatum*. The holes in these diatoms are approximately 0.25 μm in diameter and are hexagonal in shape.

BRIGHTNESS AND COLOR CALIBRATION

The light transmitted through different objectives varies with the power of the objective. Significantly more light is transmitted through the 2.5x objective than through the 100x objective. In addition, the color balance of each objective is different, so the variation in red/green/blue balance must also be corrected. Calibration of brightness and color balance to standard values is done by adjusting the gain setting of each PMT detector for each objective. A sensor embedded in the objective nose piece automatically selects the proper setup conditions when an objective is selected. These presets can be changed by the user if desired for special effects or if a new objective is placed on the instrument. The calibration procedure establishes a standard color balance and brightness for white, ensuring accurate color reproduction for stained samples.

COMPARISON WITH CAMERA SYSTEMS

Matching a CCD camera detector to a microscope image to ensure maximum resolution requires a different design procedure from COSMIC. CCD camera detectors consist of an x–y array of tiny photodiodes. Each photodiode represents one element of the picture and is called a pixel. Light striking the photodiode surface creates electrons that are read out as the signal. Since the pixels on a CCD detector are fixed, the size and spacing of the pixels must be compared with the Airy disk size to determine performance. For example, to evaluate a camera with a 12-μm square pixel, a simple calculation of the size of the Airy disk image on the detector will indicate resolution. A 100x objective (1.3 NA) has an Airy disk size of 0.256 μm in the sample plane. At the detector, the image of the Airy disk is 100 × 0.256 μm, or 25.6 μm. Since the Airy disk covers two 12-μm electronic pixels, the image will be fully resolved. If the objective is a 40x (0.85 NA), with an Airy disk of 0.394 μm, the image of the Airy disk is 40 × 0.394 μm, or 15.8 μm, which is not fully resolved by the detector, since the sampling ratio is only 1.31 : 1. On a CCD detector, there is no zoom feature available to trade field of view for resolution or to create oversampling. The situation is even worse for a 2.5x objective (0.75 NA), where the Airy disk size is 4.47 μm at the sample plane and 11.18 μm at the detector plane. The sampling ratio is only 0.93 : 1 for this example. To achieve good resolution, CCD camera systems must carefully take into account the pixel size on the CCD detector as well as the Airy disk size of each objective. The number of pixels does not determine resolution: it defines the field size. It is the size of the pixel compared to the size of the Airy disk on the image plane that determines resoluton. Thus a camera with 3500 × 2700 pixels cannot claim a higher resolution than a camera with 1500 × 1200 unless each individual pixel is smaller.

The issues of field size and pixel size are very important in understanding the differences between the point scanning technique of COSMIC and CCD detectors. As with all semiconductor technologies, the size of CCD arrays has become smaller with improved manufacturing techniques. Sizes as commonly designated include 1 in. (12.8 × 9.6 mm), $^2/_3$ in. (8.8 × 6.6 mm), $^1/_2$ in. (6.4 × 4.8 mm), $^1/_3$ in. (4.8 × 3.6 mm), and $^1/_4$ in. (3.2 × 2.4 mm). The signal from a single photodiode pixel is directly proportional to the amount of light striking its surface. If the photodiode pixel is made smaller to get higher resolution (more pixels in the same area), then the signal decreases. A longer time is required to accumulate sufficient signal electrons, and the frame rate goes down accordingly. When light is divided between three CCD detectors to create color images, the loss of sensitivity is even greater. Frame rates for high-resolution color CCD cameras are typically below 2 Hz, which is too slow for scanning around on a sample and makes focusing difficult. With these cameras it is necessary to focus visually through the microscope, then switch the light path to the camera and refocus and readjust the illumination. The process requires operating two instruments to get a digital image. The large image sizes of some cameras (viz. 3200 × 2400 pixels) creates an additional problem. The amount of data in one image can be so large that the time to transfer those data from the camera to a computer for display is significant, taking up to 1 min in some cases. Strategies to increase the frame rate for scanning and focusing include transmitting only a small fraction of the total image. The resulting tiny, low-resolution black-and-white images are not ideal for scanning or focusing. Therefore, the time required to establish the best focus, color levels, and intensity levels and store the image can be very long indeed.

In this scanning microscope the image has been standardized at 1280 × 1024 pixels, which means that the resolution is limited by the Airy disk size of the objective and not by the electronics or pixel size of a detector. With the scan rate of 13.4 Hz (frames per second), it is easy to focus and move the image. Indeed, one can obtain a large number of good images in a short time using this system—a major advantage from a user perspective. There is no reason why a larger image size could not be used, with the caveat that higher file size and slower image frequency would result. Of course, these factors are frequently based upon current computational capabilities, a factor that changes constantly.

There are several other significant differences between COSMIC and traditional camera-based systems. Most, if not all, of the following issues are not applicable to the white-light scanning microscope but may be significant problems for camera systems under certain operating conditions.

First, the regular array pattern of pixels in a CCD camera can cause the generation of false data. When the sample has a periodic structure whose spacing is about the same as that of the detector pixel spacing, a beat frequency can be generated from slight misalignments between the image and the detectors. The result is wavelike patterns appearing in the picture. These patterns, called aliases, or Moiré patterns, are not real in the image but are caused by the interaction of the image with the detector. These aliasing patterns can sometimes be eliminated by rotating the image against the detector. Aliasing can be a significant problem for certain kinds of images, particularly for automated image analysis. Aliasing problems can be greater for color cameras that have three CCD detectors, one each for the red, green, and blue portions of the light spectrum. The three detectors must be carefully aligned to give the same size image for each color and to prevent aliasing between the detectors.

Second, if a portion of a CCD detector is saturated with light, the excess electrons or the excess light may cause a signal from adjacent pixels. The visual effect may look like a star pattern with perpendicular arms and is called blooming. Electronics to detect and minimize blooming are usually built into the camera; however, the result is that the operation of the camera is temporarily halted.

Third, in an ordinary microscope, all points in the image field are illuminated simultaneously, which is called full-field illumination. In this scanning microscope, only a single point is illuminated at a time since illumination is provided by a flying spot. When all the pixels are simultaneously illuminated, scattered light can reduce contrast from pixel to pixel and contribute to blooming.

The fourth issue is interpolation. Medical systems require better resolution than can be obtained from inexpensive mass-produced television-format CCD cameras. Some camera manufacturers have chosen to create artificial resolution by calculation. Several versions have been implemented. One method is to increase resolution simply by interpolation between pixels. In another case, one of the three color detectors is displaced one-half pixel horizontally and one-half pixel vertically. The color data for that physical location are computed from the nearest neighbors. In this technique, there are no physical locations with true information in any part of the image since all data must be computed. It is not clear how interpolation impacts the use of these detector systems, but the issue will have to be addressed eventually. Finally, because the sensitivity of PMTs is at least 1000 times greater than typical CCD detectors, the illumination intensity of the white-light scanning microscope can be 1000 times less than traditional light sources, reducing bleaching or photodamage to the specimen, an issue frequently overlooked in biological imaging.

OPERATIONAL CONSIDERATIONS

The most significant problems to be dealt with in a CRT-based spot-scanning instrument are the cost of the instrument and the aging characteristics of the CRT, which will decrease in light output over time. CRT lifetimes in the 10,000-h range are common, and replacement is not difficult or expensive, particularly when compared with most lasers used in microscopy. However, extended use of the zoom feature can cause preferential aging of the center of the CRT screen. Therefore, zoom time is limited by a timer circuit that disables the zoom after a preset interval (2 to 10 min.), thereby preventing the zoom feature from aging the center of the screen. The brightness and color balance calibration procedure continually corrects for CRT aging until the overall light level is unacceptable and a replacement is needed. Changing or adding different objectives requires calibration for that objective, but once performed, the calibration becomes an instrument preset.

A troublesome issue for those who must deal with microscope and camera systems is the difficulty of obtaining rapid and reproducible images from camera/computer systems. Combining the camera, imaging boards, and computer software is, in itself, an interesting challenge. In this system there are fewer variables and therefore a more consistent and stable image platform.

The clearest departure from traditional microscopy in this scanning microscope is the lack of a viewing ocular. This is a radical difference for microscopists who may feel uncomfortable not being able to view the specimen, except electronically. However, the format is most likely consistent with the future of digital imaging applications. Clearly, the whole concept of telepathology or remote evaluation of images implies that at least one participant will not have access to a traditional viewing ocular. Thus, the complete removal of the ocular may actually be advantageous for the originator of the images to know exactly what the remote viewer is seeing.

The COSMIC instrument is clearly the first fully integrated digital microscope system to offer real-time, full-color, high-resolution digital images. The unique features of zoom magnification, contrast enhancement and suppression, averaging and summing, background correction, and superresolution capabilities are major advantages. The lack of moving parts and alignment protocols is a powerful incentive for use in service roles such as pathology, where a robust, easy-to-operate, high-resolution instrument is desired.

FUTURE APPLICATIONS

Because of the nature of the optical pathway, transmitted fluorescence detection is likely to be a viable option using this technology. Currently, no commercial microscopes use transmitted fluorescence; they are all epifluorescent systems. Because bright-field and fluorescence in a simple, easy-to-use system is a desirable combination, addition of fluorescence to this scanning microscope will significantly enhance its potential in pathology.

REFERENCES

Bartels PH, Layton J, Shoemaker RL (1984): Digital microscopy. Monogr Clin Cytol 9:28–61.

Bibbo M, Bartels PH, Dytch HE, Puls JH, Pishotta FT, Wied GL (1983): High-resolution color video cytophotometry. Cell Biophys 5:61–69.

Ingram M, Preston K, Jr (1970): Automatic analysis of blood cells. Sci Am 223:72–82.

Ingram M, Preston K, Jr (1964): Importance of automatic pattern recognition techniques in the early detection of altered blood cell production. Ann NY Acad Sci 113:1066–1066.

Mellors RC, Keane JF Jr, Papanicolaou GN (1952): Nucleic acid content of the squamous cancer cell. Science 116:265–265.

Nagata H, Mizushima H (1998): World wide microscope: new concept of internet telepathology microscope and implementation of the prototype. Medinfo 9 Pt 1:286–289.

Plasek J, Reischig J (1998): Transmitted-light microscopy for biology: A physicist's point of view. Proc RMS 33:196–205.

Ploem JS (1988): Modern image analysis methods in hematology. Nouv Rev Fr Hematol 30:45–49.

Ploem JS, van Driel-Kulker AMJ, Goyarts-Veldstra L, Ploem-Zaaijer JJ, Verwoerd NP, van der Zwan M (1986): Image analysis combined with quantitative cytochemistry. Results and instrumental developments for cancer diagnosis. Histochemistry 84:549–555.

Prewitt JM, Mendelsohn ML (1996): The analysis of cell images. Ann NY Acad Sci 128:1035–1053.

Singson RP, Natarajan S, Greenson JK, Marchevsky AM (1999): Virtual microscopy and the Internet as telepathology consultation tools. A study of gastrointestinal biopsy specimens. Am J Clin Pathol 111:792–795.

Szymas J, Wolf G (1998): Telepathology by the internet. Adv Clin Path 2:133–135.

Vazir MH, Loane MA, Wootton R (1998): A pilot study of low-cost dynamic telepathology using the public telephone network. J Telemed Telecare 4:168–171.

14

Illumination Sources

Howard M. Shapiro
Howard M. Shapiro, M.D., P.C., West Newton, Massachusetts

REQUIREMENTS FOR A LIGHT SOURCE FOR CYTOMETRY

Cytometric measurements involve the collection of light from small regions of space for short periods of time. In a flow cytometer, measurements of light scattering, transmission, and fluorescence are made in a few microseconds; while longer, the measurement interval in imaging or scanning static cytometers is still short compared to the time needed to examine cells under a microscope. Since even the most efficient light collection optics capture photons over a relatively small solid angle, and there are typically substantial losses associated with wavelength selection and detection, light sources used for cytometry must emit a relatively large number of photons in a desired wavelength region per unit time, and it must be possible to direct a substantial fraction of this photon flux into a small volume of the sample. Two classes of light sources best meet these criteria; they are so-called short or compact arc lamps and lasers.

RADIOMETRIC UNITS AND LIGHT SOURCE CHARACTERISTICS: EXITANCE, RADIANCE, AND THROUGHPUT

The process by which radiant energy or radiation is measured, with all wavelengths given equal weight, is defined as radiometry; in the International System of Units (SI

Emerging Tools for Single-Cell Analysis, Edited by Gary Durack and J. Paul Robinson.
ISBN 0-471-31575-3   Copyright © 2000 Wiley-Liss, Inc.

Units), radiant energy is expressed in joules and radiant flux (energy per unit time) in watts, where one watt equals one joule per second.

Radiant areance, or exitance, expresses the power or radiant flux (ϕ) emitted per unit area (A), in units of watts per square meter. A true point source could be described in terms of its intensity, or pointance, that is, the power emitted per unit solid angle (ω), in units of watts per steradian. (A sphere of radius r has a surface area of $4\pi r^2$; one steradian is defined as that solid angle intercepting an area equal to r^2 on the surface of the sphere.) Most of the light sources encountered in the real world are extended sources such as arc and filament lamps and the sun and other stars, which, while they may be capable of emitting substantial amounts of power, disperse their emission in all directions, that is, over a solid angle of 4π steradians. Their behavior is characterized in terms of radiance, which is the power emitted from, transmitted through, or reflected by a surface per unit of its area per unit solid angle. The units of radiance are watts per square meter per steradian.

The optical system that must be placed between an extended source and a specimen must, at a minimum, include a lens to collect light from the source and another to converge the collected light on the specimen. Both lenses are generally referred to as condensers; for purposes of this discussion, they will be distinguished as the lamp condenser and the microscope condenser. To add to the confusion, it should be noted that, in the epi-illumination systems generally used in fluorescence microscopes, the microscope objective serves as the microscope condenser.

The most efficient illumination, that is, that which produces maximum photon flux through the specimen, can be obtained by making an image of some portion of the emitting surface in the plane of the specimen. In order to maximize light collection and specimen illumination, the condenser lenses should be of high numerical aperture (NA) (equivalent to a low f number) to make the collection and illumination angles as large as possible. Consider, for the moment, that a single convex lens serves as both lamp condenser and microscope condenser. If the emitting surface of the lamp is located two focal lengths away from the lens on one side, and the specimen plane is located two focal lengths away on the other, the lens will form a 1 : 1 image of the source in the specimen plane, and the solid angles for collection and illumination will be the same. If a magnified image of the source is to be formed, the lamp must be placed closer to the lens, and the specimen moved further away; the lens can then collect light over a larger solid angle, but the photon flux per unit area in the specimen plane will not be increased. If a minified image of the source is formed, the lamp must be placed more than two focal lengths away from the source and will therefore collect light over a smaller solid angle; again, the photon flux through the specimen will not be increased. A quantity called optical throughput is said to be conserved; no matter how a real optical system is arranged, there is a point beyond which photons can be lost but not gained.

The largest possible collection angle obtainable using a convex collecting lens will be achieved if the lens is placed one focal length from the source; it will then collect a collimated beam of light, that is, one in which all the "rays" are parallel, requiring that a second lens be used to converge light on the specimen. It is also possible to collect light from a substantial fraction of the emitting surface of a source using ellipsoidal or parabolic reflectors. However, while this can allow substantial amounts of power to be directed through surface areas larger than the emitting surface of the source, it is still not possible to put more photon flux per unit area of the specimen than can be collected per unit area from the source.

LAMP AND LASER CHARACTERISTICS COMPARED

Figure 14.1, which originally appeared in a catalog from Oriel Corporation (Stratford, CT; www.oriel.com), a major supplier of light sources, compares the irradiance of mercury (Hg) and xenon (Xe) arc and quartz tungsten halogen (QTH) filament lamps. The irradiance values are given in milliwatts per square meter area per nanometer wavelength at a distance of 0.5 m; the corresponding radiance is calculated by determining the solid angle represented by a square meter at this distance and by integrating over an appropriate wavelength region.

The 100-W Hg arc lamp has an irradiance of about 120 mW $m^{-2} nm^{-1}$ in a 10-nm band around the strong ultraviolet (UV) line at 366 nm, so total irradiance in this spectral band is approximately 1.2 W m^{-2}. The surface area ($4\pi r^2$) of a sphere with a radius of 0.5 m is $4\pi (0.25)$ m^2. Since this is the surface area subtended by a solid angle of 4π steradians, 0.25 m^2 is the surface area occupied by 1 sr; the radiance is thus about 0.3 W $m^{-2} sr^{-1}$, or 300 mW $m^{-2} sr^{-1}$.

The arc in a 100-W Hg lamp is 0.25 mm (250 μm) in diameter; its surface area is $4\pi (0.000125)^2$ m^2; therefore the surface area of a 1-sr segment of the arc is $(0.000125)^2$ m^2, or 15625 $μm^2$. The power radiated through 1 sr is 300 mW. A spot 10 μm in diameter on the surface has an area of $\pi (5)^2$ $μm^2$, or 78.5 $μm^2$; thus, about 1.5 mW [(78.5/15625) × 300] of UV power in the bandwidth discussed would be emitted through this surface. This represents the maximum amount of power collected from the arc that can be directed back through the same area; thus, if the arc is used to illuminate a cell 10 μm in diameter, no more than 1.5 mW can impinge on the cell at any given time, no matter how efficiently light is collected from the arc and transmitted through the optics.

Mercury arc lamps have strong emission lines at several wavelengths other than 366 nm; peak radiance near that of the 366-nm line is also obtainable at 313 nm, farther in the UV, at 405 nm (violet), 436 nm (blue-violet), 546 nm (green), and 578 nm (yellow). Under the best conditions, neglecting transmission losses in lenses and in the filters used for wavelength selection, one would expect to be able to direct at most 1–1.5 mW in any of these wavelength regions from an Hg arc lamp through a 10-μm cell at any time.

A xenon arc lamp, as can be seen from Figure 14.1, has a relatively flat emission spectrum between the near ultraviolet and the near infrared (350–750 nm); this makes the Xe arc a desirable illumination source for spectrophotometers and spectrofluorometers. However, since the radiance of the Xe arc over this range is only about $1/10$ the radiance of the Hg arc at its strong emission lines, the Xe arc is less desirable as a source for cytometry, except possibly in the region between 450 and 500 nm, where the radiance is slightly higher than that of Hg lamps.

The radiance of a QTH filament lamp is substantially lower than that of a Xe arc lamp below 600 nm and slightly higher between 600 and 800 nm. However, the area of the emitting surface of the filament lamp is much larger than the area of the arc in an arc lamp; thus, much less power can be collected from and directed through a small area. This makes filament lamps poorly suited for fluorescence excitation in flow cytometry, although they have been used quite successfully in flow cytometers that measure absorption and light scattering. For these purposes, they have the advantage that it is relatively easy to achieve precise regulation of output power using a well-regulated dc supply.

Although higher power Hg and Xe arc lamps emit more radiant energy than does a 100-W Hg arc lamp, they are less desirable as sources for cytometry because their arcs have larger surface areas, with the result that their emission per unit area is lower; such lamps may, however, offer an advantage in apparatus in which an area substantially larger than a millimeter or so in diameter must be illuminated, for example, a static cytometer with a charge-coupled device (CCD) or other imaging detector and a large field of view.

The effective power available from a 100-W Hg arc has been increased experimentally by Steen and Sørensen (1993), who modified the front-end electronics and arc-lamp power supply of a flow cytometer to increase the lamp current by a factor of 10 for a few microseconds after the trigger signal indicates the presence of a cell in the observation region. The modification is inexpensive, increases measurement sensitivity, and does not substantially decrease the life of the lamp but has not, to date, been incorporated into a commercial apparatus.

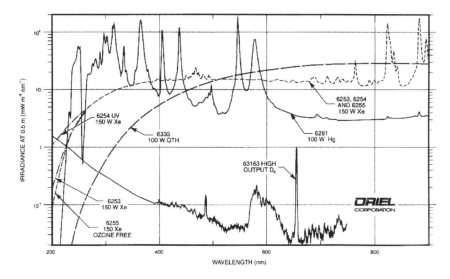

Fig. 14.1. Irradiance of mercury and xenon arc and TOH filament lamps. (From Shapiro HM (1994): Practical Flow Cytometry, 3rd ed. New York: Wiley-Liss. Courtesy of Oriel Corporation.)

Laser sources differ radically from extended sources in several respects; one is that the emission from lasers is confined to a very small solid angle; it is generally possible (neglecting transmission losses) to focus all the energy in the beam to a circular or elliptical spot. Most lasers used for flow cytometry emit a beam in which the energy distribution is normal, or Gaussian. Spot "diameters," in the case of a circular spot, or "width" and "height," in the case of an elliptical spot, define the "$1/e^2$ points," the region within which approximately 87.5%, or $1 - (1/e^2)$, of the total emission is contained. The area of this region corresponds to the area of a central elliptical region of a bivariate normal distribution within two standard deviations of the bivariate mean.

If a laser beam is focused to an elliptical spot 100 μm wide by 20 μm high, a not unreasonable size to use in a flow cytometer, it can be calculated that about 8.8% of the total power in the beam would illuminate a cell 10 μm in diameter at or near the center of the focal spot. If the 1.5 mW already shown to be the maximum available from a 100-W Hg arc lamp were to come instead from a laser, the total laser power would therefore be 1.5/.088 mW, or 17 mW. Assuming that wavelength selection filters, which must be used with an arc lamp but are generally not necessary when laser illumination is used, reduce light transmission by 25–30%, the arc lamp is, at best, equivalent to a laser emitting about 12 mW.

To summarize: If lamp illumination is to be used, the highest intensities, comparable to the output from a laser emitting no more than 12 mW, can be obtained from a 100-W Hg arc lamp (arc diameter approximately 0.25 mm), in spectral regions around 313, 366, 405, 436, 546, and 578 nm. In the region between 450 and 500 nm, slightly more power may be available from a 150-W Xe arc lamp than from a 100-W Hg arc lamp, but this will be equivalent to no more than 1–1.5 mW from a laser (e.g., an argon-ion laser emitting at 488 nm). Even less power is available from filament lamps; while these are usable and may even be desirable for measurements of absorption and light scattering because it is easy to regulate their output, they are poorly suited for fluorescence excitation. Lamps are generally preferable to lasers when uniform illumination over a relatively large area is needed, as is the case in many imaging cytometers.

NOISE IN LIGHT SOURCES AND ITS EFFECTS ON MEASUREMENTS

The output of a light source used for cytometry should, ideally, be constant; since, in the real world, there will be short- and long-term variations in power output, it is desirable to know the magnitudes of such variations, which will allow determination of their effects on measurements. In some cases, it is possible to achieve acceptable measurement precision using an otherwise unacceptably noisy light source by measuring temporal and spatial fluctuations in output and compensating for them using analog and/or digital electronics.

Both lamps and lasers typically exhibit drift in output power over periods of hours to days and power loss over a period of weeks to years due to aging and wear of various components of the source. Such long-term variation generally does not affect measurements made within a few hours of one another; if it is pronounced, it is generally dealt with either by interspersing standard samples at intervals during a day's run or by repair or replacement of the offending source.

Shorter-term fluctuations, ranging in frequency from a few hertz to a few megahertz, can interfere with both measurement precision and measurement sensitivity. The noise of a light source is usually specified in terms of a percentage that can represent either a peak-to-peak or an RMS (root-mean-square) value. The RMS noise value is equivalent to the coefficient of variation (CV) of the fluctuations of power output about the mean. The RMS noise can be estimated from the peak-to-peak noise level by assuming that the peak-to-peak value represents a range of six standard deviations about the mean, the rationale being that more than 99% of the area of a normal distribution lies within three standard deviations of the mean. Thus, dividing the percentage of peak-to-peak noise by 6 provides an approximation to the RMS noise percentage, which, since it represents the CV of power fluctuations, is 100 times the standard deviation of the output power divided by the mean output power.

In flow cytometers, the dc component and slowly varying components of the noise are typically removed by baseline restoration circuits in the preamplifier electronics; the higher frequency components of the noise are superimposed on the output signal. When light scattering is measured, most of the background comes from stray illuminating light; since this is at the same wavelength as the light scattered from the cell, it cannot be reduced or removed by optical filters. When fluorescence is measured, the source-dependent component of the background typically comes from fluorescence of dye in the sample stream and from Raman scattering by water in the sheath and sample streams; potential contributions from illumination light leakage and filter fluorescence can and should be avoided by proper choices of dichroics and detector filters. The amplitude of the fluctuations contributes to the variances of both light scattering and fluorescence measurements, with proportionally greater effects on smaller signals; when the amplitudes of noise fluctuations and signals from particles become comparable, signals from the particles become undetectable, and signals from particles may become unreliable for triggering well before this point is reached.

Under the best of circumstances, the CV of a light scattering or fluorescence measurement cannot be lower than the RMS noise of the light source, except that the effects of source noise on fluorescence signals may be mitigated if the illumination power level is high enough to produce saturation in the relevant fluorophores (van den Engh and Farmer, 1992). For example, chromosome sorters that use relatively high (hundreds of milliwatts) UV and blue-violet high excitation power from ion lasers often achieve fluorescence measurement CVs on the order of 1% when RMS source noise is somewhat higher.

Because cells, even when stained, do not absorb more than a small fraction of the light passing through them, illumination fluctuations have larger effects on absorption and extinction measurements than on measurements of scattered light and fluorescence. If an extinction measurement is made of a particle that removes 10% of the light from the incident beam, a 1% increase in illumination intensity occurring while the particle passes through the beam results in an apparent light loss of 9%, while a

1% decrease in illumination intensity results in an apparent light loss of 11%. Thus, a 1% change in illumination intensity produces a 10% change in the amplitude of an absorption or extinction signal; the less light the particle removes from the beam, the larger the effect of source intensity fluctuations. While the 1% RMS noise level of the air-cooled argon and He–Ne laser sources most commonly used in fluorescence flow cytometers does not significantly compromise the precision of scattering and fluorescence measurements, extinction measurements made using these sources are unacceptably imprecise.

In many cases, the noise level of both lamp and laser sources is determined primarily by the degree of regulation built into the power supply by the manufacturer and is therefore not controllable by the user. Although it is possible, at some sacrifice in power, to decrease output fluctuations using devices incorporating acousto-optic or electro-optic modulators and appropriate feedback circuitry, these typically cost several thousand dollars and require some technical proficiency to adjust. It is less expensive to add noise compensation circuits that monitor source output and adjust amplifier gain using analog difference or ratio circuits; these are not typically implemented in commercial flow cytometers. In the near future, it can be expected that both baseline restoration and source noise compensation will be facilitated by the use of digital signal processing, allowing pulse characteristics to be determined rapidly and precisely after signals from the detectors are digitized at frequencies of several megahertz.

Both low- and high-power ion lasers can generally be operated in either a current control mode or a light control mode. In the current control mode, the laser power supply is regulated to deliver a constant current, and light output remains constant only if the mechanical and optical characteristics of the laser do not change during operation. If if the laser becomes slightly misaligned, light output decreases, even though power supply current remains the same. In the light control mode, power supply output is regulated by a feedback circuit that samples the energy in the beam and adjusts the laser current to maintain constant light output. This generally works well when the laser is emitting at a single wavelength; however, when several lines are emitted simultaneously, it is more difficult to keep power constant. The air-cooled argon lasers in benchtop flow cytometers are operated in the light control mode; so are diode lasers, which are built with a light-sensing photodiode and need a light control feedback loop in the power supply to prevent thermal runaway from destroying the laser itself when it is turned on. Helium–neon and helium–cadmium lasers typically do not incorporate light control circuits. Operation in the light control mode does not guarantee protection against noise if the light output drops substantially enough for the supply current to rise to its maximum value; under these conditions, any further mechanical or optical deterioration cannot be compensated for by either light control or current control circuitry.

Arc lamps exhibit a phenomenon known as arc wander, in which the position of the arc between the electrodes moves over time; this results in changes of illumination intensity in the field of observation. In the author's experience, arc wander has been associated with inadequate cooling of the lamp and housing more than with any other characteristic of the lamp or power supply. To compensate for arc wander, it is necessary to monitor illumination intensity in the region in which cells are measured; the optics and electronics necessary to do this in an arc source flow cytometer have been described by Steen (1980).

LAMP SOURCES: PRACTICAL CONSIDERATIONS

Two basic arrangements are used for specimen illumination by lamps. In Köhler illumination, an image of the light source is formed in the back focal plane of the condenser lens; light rays from the source image leave the condenser lens and pass through the specimen as parallel bundles. Each point in the specimen plane is illuminated by light coming from all points of the source image; this provides uniform illumination of a relatively large area, which is an advantage in imaging systems, particularly those with relatively low magnification and large fields of view. In critical illumination, an image of the source is formed in the specimen plane; intensity variations in the source are replicated in the specimen plane, but, if the intensity of a small area of the source is reasonably uniform, a higher photon flux through the specimen can be achieved than is possible using Köhler illumination; this may be advantageous in flow cytometry.

Arc lamps require relatively elaborate power supplies, including circuitry to generate a high-voltage pulse or RF (radio frequency) pulse train to ionize the gas, producing a conductive medium in which an arc can be started and maintained. Although arc lamps intended for spectrophotometry and microscopy are sold with power supplies that are supposed to provide light output regulated to better than 1%, the author has measured 6–10% variations in light intensity, synchronized with the power line frequency, in arc lamp systems made by several manufacturers. A source with such intensity fluctuations is apt to be unacceptable for flow cytometry unless either improved power supply regulation or intensity compensation in the detector electronics or both are added by the user.

Arc lamps themselves have relatively short operating lifetimes; a 100-W Hg arc lamp bulb typically requires replacement after approximately 200 h of operation at a cost between $200 and $250 (as of late 1998). An arc lamp system itself, including a housing, power supply, and pyrex aspheric condenser, now sells for at least $4500; the pyrex lamp condenser lens is adequate for work at 366 nm and longer wavelengths, while the silica or quartz condenser needed for shorter wavelengths increases the price of the system by at least $1000.

LASER SOURCES FOR CYTOMETRY

Basic Laser Physics

Laser operation depends on stimulated emission in a lasing medium; while the acronym "laser" derives from "light amplification by stimulated emission of radiation," a working laser is actually an optical amplifier driven into oscillation by positive feedback. An input of electrical or optical pump energy from an external source produces a population inversion, a condition in which electronically excited atoms or molecules in the medium outnumber atoms or molecules in lower electronic energy states, favoring stimulated emission; the emission is sustained and provided with directional characteristics by confining the medium in a volume with an appropriate geometry (usually cylindrical, with emission along the axis) and placing it in a resonator, a rigid structure with mirrors, separated by an even number of wavelengths, at opposite ends. One of the mirrors, called the output coupler, transmits a small fraction of photon flux at the laser wavelength. In multiline lasers, a (Littrow) prism is placed between the mirrors; as a result of refraction by the prism, different wavelengths have different optical path distances between the mirrors, and laser operation is sustained only at that wavelength for which the distance represents an even number of wavelengths. Laser output beams may be polarized by placing (Brewster) windows between the lasing medium and the mirrors; these favor emission in one plane of polarization over emission in the perpendicular plane. Further details about the physics of lasers are available elsewhere (Shapiro, 1997a; Hecht, 1992, 1994; Harbison and Nahory, 1997).

Focusing Laser Beams

The divergence of laser beams is sufficiently small that they can, for most practical purposes, be considered as collimated, or composed of parallel rays. According to geometric optics, parallel rays entering a convex lens are brought to a focus of infinitesimal diameter at the focal point of the lens. The actual dimensions of a circular or elliptical spot formed from a laser beam by a convex lens cannot be calculated from the formulas of geometric optics but can be estimated using diffraction theory.

The "diameter" of a Gaussian (TEM_{00}) beam represents the distance between the $1/e^2$ points, within which approximately 87.5%, or $1 - (1/e^2)$, of the total emission is contained. The calculation described here assumes that this beam is collimated, with a diameter of D millimeters, and is brought to a focus by a convex lens of focal length F millimeters with its axis coincident with the beam axis. According to the formula most commonly used for estimating spot sizes (Shapiro, 1997a), the diameter d (in micrometers) of the spot formed at the focal distance F from the lens is given as:

$$d \cong \left(\frac{4}{\pi}\right)\left(\frac{\lambda F}{D}\right) \cong 1.27\left(\frac{\lambda F}{D}\right)$$

with the laser wavelength λ expressed in micrometers rather than nanometers.

The above formula explicitly neglects lens aberrations and assumes that the beam is focused to a diffraction-limited spot. It can be applied to circular spots formed by a single spherical lens or to elliptical spots formed by crossed cylindrical lenses of different focal lengths placed at different points along the beam axis so that their foci coincide. However, because the calculated spot dimensions represent the distances between $1/e^2$ points in the focal spot, the illumination intensity varies substantially from the center of the spot to its "edges" as defined by the formula.

In laser scanning cytometry, it is customary to focus the illuminating laser beam to a circular spot no larger than a few micrometers in diameter and to overscan the specimen to produce an effective pixel diameter somewhat smaller than the spot diameter. Thus, although there are illumination intensity variations within the spot, all pixels in the image are uniformly illuminated.

In a typical laser-source flow cytometer, the illuminating laser beam is brought to a focus on the sample or core stream. As a cell or particle in the core traverses the height (i.e., the dimension in the direction parallel to the direction of sample flow) of the focal spot, it produces a time-varying optical (e.g., scatter or fluorescence) signal pulse, the shape of which represents a convolution of the cell shape and the Gaussian intensity profile of the beam. If the spot height is substantially smaller than the particle diameter, the signal pulse represents a "slit scan" (Wheeless, 1990), and its shape is primarily determined by cell size and internal structure; if the spot height is substantially larger than the particle diameter, the shape of the pulse reflects the Gaussian profile of the spot, and information about, for example, cell size and content of fluorescent material is derived from pulse amplitude or area. Since each particle passes through the entire beam profile in the direction of flow, the variation in illumination intensity along this direction is effectively compensated for, much as was previously described in the case of an overscanning static cytometer.

In a typical flow cytometer, the core diameter may be between 5 and 20 μm. If the focal spot is centered on the core, with the axis representing its height along the axis of the core, a cell in the core may be located as much as 10 μm from the center of the spot along the axis of the spot perpendicular to the direction of flow. The dimension of this axis represents the spot width. If the spot width is equal to the core width, a cell at the edge of the core is at the $1/e^2$ point of the spot and receives substantially less illumination than a cell in the center of the core; this will result in an unacceptably high variance (and CV) of measured signal intensities. While when high laser powers are used saturation effects (van den Engh and Farmer, 1980) may lower the CVs of fluorescence signals measured under these conditions, scatter signal variances will remain high.

The above considerations lead designers of flow cytometers to use spot widths large enough to keep illumination uniform to an acceptable degree. Since one is dealing with the Gaussian distribution, it is possible to use widely available tables to calculate the spot size needed to keep illumination intensity variation within a desired percentage for any given core width. From the tables, it can be seen that illumination intensity remains at over 99% of the peak value within the central 8% of the beam width; maintaining this degree of illumination uniformity for a 10-μm core therefore requires a beam width of approximately 120 μm.

Several approaches may be taken when using more than one laser beam for excitation in flow cytometry. Combinations of mirrors and dichroics can be used to put two or three beams through a single set of crossed cylindrical lenses. Beams may also be placed at small angles to one another and directed through different sets of lenses; this arrangement provides more precise control over focus and spot size and has been used for high-precision analyses of chromosomes. The author has discussed focusing of laser beams in flow cytometers elsewhere at greater length (Shapiro, 1997b).

SPECIFIC LASER TYPES

Ion Lasers

Argon-Ion Lasers. Argon-ion lasers can emit at 275, 305, 351, and 363 nm (UV), 454 and 457 nm (violet-blue), 465 nm (blue), 472, 476, and 488 nm (blue-green), and 496, 501, 515, and 528 nm (green). The 488- and 515-nm lines have the highest gain; air-cooled Ar lasers emitting 10–25 mW at 488 nm are the sources most commonly used in fluorescence flow cytometers. Higher-power (100–300 mW at the high-gain lines) air-cooled systems can emit tens of milliwatts at 457 nm; water-cooled systems are required to obtain UV output and higher power at 457 nm. Air-cooled lasers cost in the neighborhood of $7000 and up; water-cooled systems start at over $20,000. Plasma tube lifetimes are in the thousands of hours. Most fluorescence flow cytometers use air-cooled argon-ion lasers emitting 10–25 mW at 488 nm as their sole light source; these lasers require little or no adjustment by the user. Water-cooled lasers used in larger instruments typically require mirror changes and alignment when switched between UV and visible operation and use Littrow prisms for wavelength selection in the visible; some skill on the user's part is needed to keep them operating properly. Water-cooled ion lasers must be supplied with water at a rate of several gallons per minute; various local regulations usually force users to install a recirculating chiller for the cooling water, which adds significantly to the installation cost. The water-cooled lasers also typically require three-phase electrical service providing at least 208 V at at least 60 A; this also increases installation and operating costs. The output noise of water-cooled argon-ion lasers is typically specified as less than 0.2% RMS, while that of air-cooled systems is specified as less than 1% RMS; the difference reflects the fact that power supply regulation is typically better in the water-cooled systems.

Krypton and Mixed-Gas Ion Lasers. Krypton-ion lasers have lower gain and lower efficiency than argon lasers but operate over a much broader spectral range; they can emit at 337, 350, and 356 nm (UV), 406, 415, and 422 nm (violet), 468, 476, and 482 nm (blue-green), 520 and 530 nm (green), 568 nm (yellow), 647 and 676 nm (red), and 752 and 799 nm (IR). The 647-nm line has the highest gain. When equipped with broadband mirrors, emission can be obtained simultaneously at all lines from 468 to 676 nm; Kr lasers are operated in this mode for light shows. Air-cooled krypton lasers emit tens of milliwatts at these lines; water-cooled lasers are needed for higher power and for UV, violet, and IR output. Optimum values for gas pressure and solenoid magnetic field for krypton laser operation are different for different lines; since the gains at most lines are low, these parameters need to be well controlled to maintain laser action, especially in the UV and violet. There is competition among the visible krypton lines; for example, running in light control mode with optics that allow simultaneous yellow and red emission may result in alternating fluctuations in the yellow and red power outputs. Costs of krypton lasers are similar to those of argon lasers; both types normally produce TEM_{00} (Gaussian) output beams. Plasma tube lifetimes may be shorter for krypton and mixed-gas lasers because the lower gain of the krypton lines generally requires running the laser at higher current. Mixed-gas lasers filled with argon and krypton and capable of emission at any of the visible output wavelengths available from krypton as well as at the major argon lines are also available; the air-cooled varieties are often used in confocal microscopes because they offer a wide range of wavelengths. Ion laser output is generally vertically polarized.

Helium–Neon Lasers

The most common He–Ne lasers emit at 633 nm (red); other available visible wavelengths include 543 nm (green), 594 nm (yellow-orange), and 611 nm (orange). Helium–neon lasers typically require only air cooling by convection and use fixed mirrors; they are more efficient than ion lasers, yielding 1–50 mW optical power from less than 200 W of electrical power. A 1.0% RMS noise level is typically specified. Lasers operating at 633 nm are typically polarized; 543-nm systems are most often not polarized because low gain at this wavelength, which limits power output to a few milliwatts, makes it difficult to introduce Brewster windows into the system without unacceptable light loss. He–Ne lasers cost between a few hundred and a few thousand dollars, depending on power output; plasma tube lifetimes are typically 10,000 h or more.

Helium–Cadmium Lasers

Helium–cadmium lasers can emit 1–100 mW at 325 nm or 1–10 mW at 354 nm (UV) and 5–200 mW at 441 nm (blue-violet). They run on standard line current and are convection or fan cooled, need few adjustments, and have plasma tube lifetimes of a few thousand hours. They are also cheaper than UV ion lasers, ranging in price from around $7000 to about $25,000. He–Cd lasers are intermediate in efficiency between argon-ion and He–Ne lasers; they typically emit Gaussian beams in the blue-violet and "donut-shaped" (TEM_{01*}) or multimode beams in the UV and may be linearly or randomly polarized. Helium–cadmium lasers emitting 325 and 441 nm simultaneously are available, with either wavelength selectable by switching filters; there is not significant competition between the UV and blue-violet lines. The principal problem with He–Cd lasers is noise at frequencies between 300 and 400 kHz. It is difficult to keep RMS noise levels much below 1.5% even when the laser is new; noise levels tend to increase thereafter as helium pressure builds in the plasma tube. This effect can be minimized by running the laser for several hours a week. The effects of noise have been reduced in laboratory-built systems using noise compensation electronics and electro-optic modulators. Noise notwithstanding, He–Cd lasers make convenient UV sources for flow cytometry; however, since the 325-nm emission does not pass through glass, it is difficult to use it for scatter measurements or microscopy without expensive quartz or silica optics.

Semiconductor Diode Lasers

A diode laser is basically a light-emitting diode, the geometry of which is tailored to provide a resonator structure that will support stimulated emission. The emitting surface of a typical diode laser is a stripe approximately 1 µm high and a few micrometers wide; this produces a beam that diverges more in the direction perpendicular to the long axis of the stripe than in the direction parallel to it. When a spherical or aspheric lens is used to collect light from the laser, the collimated beam is asymmetric and when focused to a small spot may show substantial intensity variations along one dimension or another. Improved beam quality is now offered by firms such as Blue Sky Research (San Jose, CA), which mounts a small cylindrical lens assembly inside the case of a laser diode, providing a radially symmetric, near-Gaussian output beam well suited to cytometric applications.

The first diode lasers were made of gallium aluminum arsenide (GaAlAs); GaAlAs lasers now available in quantity emit at 750–780 nm (IR) and are used in compact disc players, CD-ROM readers, and laser printers. Gallium indium phosphide (GaInP) lasers emit between 670 and 690 nm (far red), and aluminum gallium indium phosphide (AlGaInP) devices are now available with emission wavelengths between 630 and 650 nm (red) and power outputs of 3–15 mW. The output wavelength of a diode laser may vary by several nanometers as a function of small differences in semiconductor composition; wavelength variation also occurs with changes in temperature.

A photodiode that senses laser output is built into the case of most laser diodes; this is connected to a feedback-controlled, power supply regulator that converts a 5–6-V DC input to the lower voltage required by the diode and regulates current to prevent the laser from self-destructing. In this obligatory light control mode, noise is less than 0.1% RMS. Diode lasers are extremely efficient (typically 20–30%) and thus consume very little electrical energy. Since hundreds of them can be produced from a single semiconductor wafer, they are also much less expensive than any other type of laser; a 15-mW, 635-nm diode laser system with a circularized beam costs a few hundred dollars, while a 50-mW, 785-nm diode alone costs less than $50. Lifetimes should be thousands of hours.

Nakamura and co-workers at Nichia Corporation (Anan, Japan) have recently developed blue-violet laser diodes made of indium gallium nitride (InGaN) emitting near 405 nm; wavelengths as low as 390 nm should be obtainable from this material (Nakamura and Fasol, 1997). Power outputs of 4–5 mW are now available; the devices, packaged with regulated power supplies and thermoelectric coolers, can now be bought for around $4000. Athough not optimal for the purpose, a blue-violet diode can be used to excite DAPI, certain Hoechst dyes, and several fluorescent labels.

Neodymium-Based Solid-State Lasers

A number of crystalline materials exhibit laser action when optically pumped; the first working laser employed a flashlamp to induce pulsed stimulated emission in a ruby crystal. At present, continuous-wave (CW) and pulsed lasers employing neodymium-doped yttrium aluminum garnet (YAG) or yttrium vanadate (YVO$_4$) rods are widely used, with flashlamps and laser diode arrays used as pump sources. The principal laser transition in these materials is at 1064 nm (IR); frequency doubling, tripling, or mixing crystals can be used to produce emission at 355 nm (UV), 457 nm (blue), 473 nm (blue-green), and 532 nm (green). The UV output is obtainable only from pulsed lasers, and blue-green devices made thus far have been relatively low powered and noisy. However, green frequency-doubled diode-pumped YAG lasers are now available with power outputs ranging from a few milliwatts (for under $1000) to 10 W (for tens of thousands of dollars). Single longitudinal-mode green YAG lasers have very stable power outputs and low noise but are expensive ($6000 and up for 10 mW output); the less expensive devices may be made usable for cytometry by a combination of temperature stabilization and noise compensation. Efficiency is high; a system emitting 2.5 W at 532 nm draws less than 75 W from a wall plug. Improved 473-nm devices could replace 488-nm argon ion lasers for most cytometric applications.

Dye and Vibronic Lasers

Optical pumping can induce laser action in a dye solution; since the electronic transitions involved are in complex molecules, rather than atoms, it is possible to tune the output of dye lasers over a wide range (tens of nanometers) using a prism. CW dye lasers used for cytometry recirculate dye to prevent bleaching; almost all employ rhodamine 6G, which provides output between 570 and 620 nm. Some skill is required to keep both the mechanics and optics of dye lasers in good running condition; a further disadvantage has been the requirement for several hundred milliwatts of pump power in the blue-green or green spectral region. Until recently, water-cooled argon lasers have been the customary pump sources; it is now feasible to use high-powered green YAG lasers.

Vibronic crystalline laser materials behave similarly to laser dyes; laser transitions may occur between an excited electronic state and any vibrational state associated with a lower electronic energy state, allowing output to be tuned over a broad range. Titanium-doped sapphire lasers can operate in pulsed or CW mode between 660 and 1180 nm. Pump energy can be supplied by a high-power green YAG laser. Pulsed Ti–sapphire lasers are used for multiple-photon excitation of fluorescent dyes and intrinsically fluorescent cellular constituents; to date their high cost (typically over $100,000) has limited their use.

Alexandrite (beryllium aluminum oxide containing chromium) emits between 700 and 850 nm, and can be pumped by 635–670-nm diode lasers. UV emission at approximately 380 nm from a frequency-doubled, diode-pumped alexandrite laser has been demonstrated by one company (Light Age, Somerset, NJ); thus far, power output has been only a few milliwatts, and long-term stability and noise are issues. Alexandrite lasers have, to date, been relatively expensive but may become less so if produced in quantity. The author has used a Light Age UV laser for measurements of Hoechst 33342–stained nuclei with good results.

CONCLUSIONS

If present trends continue, it can be expected that more lasers and fewer lamps will be used as illumination sources for cytometry. Diode and diode-pumped solid-state lasers are beginning to replace He–Ne and ion lasers in commercial instruments, reducing size, power consumption, noise, and, in some cases, cost. While ideal UV sources still elude us, it appears that better UV sources will be available before visible-excited substitutes are found for presently unique UV-excited probes such as the Hoechst dyes. For now, the biologists can sit back and watch the race between the chemists and the physicists.

REFERENCES

Harbison JP, Nahory RE (1997): *Lasers: Harnessing the Atom's Light.* New York: Scientific American Library.

Hecht J (1994): *Understanding Lasers: An Entry-Level Guide.* New York: IEEE Press.

Hecht J (1992): *The Laser Guidebook*, 2d ed. Blue Ridge Summit, PA: Tab Books (McGraw-Hill).

Nakamura S, Fasol G (1997): *The Blue Laser Diode. GaN Based Light Emitters and Lasers.* Berlin: Springer.

Shapiro HM (1997a): Lasers for flow cytometry. Unit 1.9. In Robinson JP, Darzynkiewicz Z, Dean P, Orfao A, Rabinovitch P, Stewart C, Tanke H, Wheeless L (eds). *Current Protocols in Cytometry.* New York: Wiley, 1.9.1–1.9.13.

Shapiro HM (1997b): Laser beam shaping and spot size. Unit 1.6. In Robinson JP, Darzynkiewicz Z, Dean P, Orfao A, Rabinovitch P, Stewart C, Tanke H, Wheeless L (eds). *Current Protocols in Cytometry.* New York: Wiley, 1.6.1–1.6.5.

Shapiro HM (1994): Practical Flow Cytometry, 3rd ed. New York: Wiley-Liss. Courtesy of Oreil Corporation.

Steen HB (1980): Further developments of a microscope-based flow cytometer: Light scatter detection and excitation intensity compensation. Cytometry 1:26–31.

Steen H, Sørensen OI (1993): Pulse modulation of the excitation light source boosts the sensitivity of an arc lamp-based flow cytometer. Cytometry 14:115–122.

van den Engh G, Farmer C (1992): Photo-bleaching and photon saturation in flow cytometry. Cytometry 13:669–677.

Wheeless LL Jr. (1990): Slit-scanning. In Melamed MR, Lindmo T, Mendelsohn ML (eds). *Flow Cytometry and Sorting* 2nd ed. New York: Wiley-Liss, pp 109–125.

15

Camera Technologies for Cytometry Applications

Kenneth Castleman
Perceptive Scientific Instruments, Inc., League City, Texas

INTRODUCTION

One of the most critical components of an image cytometry system is the image sensing camera. This is the primary component that converts the optical image into numerical form suitable for subsequent analysis. Any image degradation introduced at this stage will affect all subsequent steps, reducing the quality and accuracy of the system's output.

Compared to those of the recent past, modern cameras offer a wide choice of reasonably inexpensive options. The proper choice of camera depends, in large part, on the intended usage of the instrument. The camera should be well matched to the other system components as well as to the problems that will be addressed.

The camera must convert a two-dimensional spatial distribution of light intensity into a corresponding electrical signal that can then be digitized. The resulting numerical array should be a faithful representation of the specimen. How accurately the camera can conduct this transformation affects the quality of results one can obtain from the instrument.

Image sensing is a complex technology encompassing a variety of physical phenomena. There are numerous sources of noise, distortion, and loss of resolution in the process. Since any particular camera design represents a series of trade-offs and com-

Emerging Tools for Single-Cell Analysis, Edited by Gary Durack and J. Paul Robinson.
ISBN 0-471-31575-3 Copyright © 2000 Wiley-Liss, Inc.

promises, one unit's performance may be better in some areas and worse in others, when compared with similar instruments.

A poor choice of camera can severely limit the accuracy and usefulness of an image cytometry system. The camera with the best specifications or the highest price is not always the right one for a particular job. It is necessary to understand the fundamentals of camera phenomena in order to design an image cytometry system or to use it to best advantage. Fortunately, many of the image degradations commonly introduced by camera shortcomings can be reduced or eliminated by subsequent image processing.

Definitions

Image sensing involves three basic steps: sampling, scanning, and quantization. *Sampling* is the process of dividing the image up into an array (usually a rectangular grid) of small regions, called "picture elements" or "*pixels*," and measuring the light intensity at each. It requires a *sampling aperture,* which is basically the size and shape of a single pixel (usually circular, square, or rectangular). *Scanning* is the process of selectively addressing the picture elements, one at a time, in a logical order. This creates the data stream that represents the image. *Quantization* is the process of generating an integer that reflects the brightness of the image at a particular pixel location. This is accomplished with an analog-to-digital converter (ADC) circuit. *Digitization* refers to the overall process of scanning, sampling, and quantizing an image.

Resolution refers generally to the accuracy with which an imaging system can reproduce small objects in the image. *Optical resolution* relates to the system's ability to resolve (separate) nearby point sources in the image. This is determined primarily by the point spread function (psf) of the microscope objective lens. *Photometric resolution* is specified by the number of gray levels in the gray scale. *Spatial resolution* is specified by the pixel spacing at the specimen plane. This is a function of magnification and pixel spacing at the sensor. *Image size* is the number of pixels per row and per column of the digital image.

Distortion refers to a warping of the objects in the image. Any undesirable additive components in the image are called "*noise.*" *Binning* is the technique of combining sets of adjacent pixels in a sensor array to produce the effect of larger pixels. For example, using 2×2 binning on a 1024×1024 sensor array with 6×6-µm pixels would produce a 512×512 image where the pixels were effectively 12×12 µm.

IMAGE SENSING

Light Sensing

Light-sensing devices produce an electrical signal proportional to the intensity of incident light falling upon them. Semiconductor devices are made from pure silicon crystals that have been "doped" with tiny amounts of specific impurities. Incident photons free up some of the electrons. These "photoelectrons" can be collected and

Image Sensing

measured as an indication of the amount of light hitting the sensor. This phenomenon is harnessed in solid-state image sensor chips.

Photometry

Photometry is the technology of quantifying light intensity, and there are many ways to do this. Photon flux, for example, is one way to specify this. Photons of different wavelength, however, have different energy. Thus, if incident light energy flux is measured, the spectrum of the light affects the intensity. Commonly used image sensors, however, merely count photons, so wavelength considerations do not directly affect the measured intensity. Although different sensors have different sensitivities at different wavelengths, this can be accounted for separately. A quality image digitizer will produce a numerical array wherein each gray level is proportional to the number of photons that landed on that pixel during the exposure time.

The *linearity* of an image sensor specifies how accurately its output reflects the incident photon flux. Modern charge-coupled device (CCD) image sensor chips are quite linear over their entire range, and thus linearity is seldom a problem, as long as saturation (overload) of the sensor is avoided.

Video Scanning Conventions

Part of the camera design exercise is to establish the format of the output signal. This encompasses the number of scan lines, the number of pixels per line, the rate at which pixels are read out of the camera, and whether or not interlaced scanning is used.

Broadcast Scanning Conventions

Figure 15.1 illustrates the Electronic Industries Association (EIA) RS-170 scanning convention, which is the standard for monochrome broadcast television in the United States (Fink and Christiansen, 1989; Hutson et al., 1990; Castleman, 1996). The beam scans the entire image in 525 horizontal scan lines, 30 times each second.

Each frame is made up to two interlaced fields, each consisting of 262.5 lines. The first field of each frame scans all the odd-numbered lines, while the intervening even-numbered lines comprise the second field. Interlacing yields a 60/s field rate to minimize perceived flicker, while the 30/s frame rate keeps the frequency bandwidth within the limits of broadcast television channels.

Each horizontal line scan requires 63.5 μs, of which approximately 50 μs (83%) contains image information (the "active" line time). Of the 525 lines per frame, 16 are lost in the vertical retrace of each field, leaving about 483 active lines per frame. The bandwidth of the standard video signal extends up to 4.5 MHz, and this allows 225 cycles, or about 550 pixels worth of information, assuming two pixels per cycle, across the active portion of each line. RS-170 video is commonly sampled at 480 lines per frame and 640 pixels per line.

The NTSC (National Television Standards Committee) timing standard for color television in the United States differs very slightly from the RS-170 convention. It

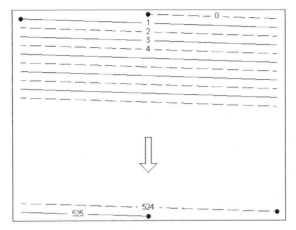

Fig. 15.1. The RS-170 scanning convention, used for monochrome broadcast television transmission in the United States. The CCIR convention, used in much of Europe, is similar, but it employs 625 lines per frame and operates at 25 frames per second.

was designed to accommodate color transmission while maintaining backward compatibility with existing monochrome receivers. Different scanning conventions are used in other countries. For example, the CCIR (Comité Consultatif International des Radiocommunications) standard used in much of Europe runs at 25 frames/s and uses 625 interlaced scan lines per frame. CCIR video signals are commonly sampled at 576 lines per frame and 768 pixels per line.

While there is no reason that the camera on an imaging cytometer should conform to any broadcast television standard, the realities of the marketplace dictate that the majority of mass-produced cameras and video digitizers do so. Besides, if the camera supplies a standard video signal, it facilitates watching the specimen on an ordinary TV monitor.

Other Scanning Conventions

Many cytometry applications require higher resolutions (i.e., more lines, more pixels per line) than what the RS-170 and CCIR broadcast television standards can offer. Different scanning conventions are used in the design of sensor chips intended for scientific cameras. Cameras designed specifically for scientific image sensing can be made to scan by any set of timing rules. Scientific cameras typically have larger image formats (more scan lines and pixels per line), noninterlaced scanning, and slower readout rates (to reduce the noise generated by the readout circuitry on the chip). They also may incorporate variable-length frame integration periods (to increase sensitivity) and sensor-chip cooling to reduce dark current.

Camera Performance

Although cameras differ in the approach they use to sense an image, they can be compared on the basis of their digital-image-producing performance.

Image Size. An important parameter is the size of the image a camera produces. This is specified by the maximum number of scan lines and of pixels per line. A larger image size allows more specimen area to be covered in a single image, assuming fixed resolution.

Pixel Size and Spacing. Two important characteristics are the size of the sampling aperture and the spacing between adjacent pixels. These parameters, when specified at the image plane, are fixed by the design of the image sensor chip. They scale down to the specimen plane by the magnification factor of the microscope.

Magnification. While magnification is a property of the optical system (e.g., microscope), and not of the camera, it effects a trade-off between image size and pixel spacing at the specimen. Higher magnification reduces both the pixel spacing and the area of the specimen that is covered by a single image. Lower magnification produces the opposite effect. Thus the proper magnification for a particular application will be a compromise between resolution and image area.

The magnification is usually the objective power multiplied by any auxiliary magnification that is in place. The eyepieces normally do not contribute to this calculation, but the camera adapter port may do so. The "objective power" number commonly engraved on the lens is a nominal value, and an accurate determination usually requires measurement of the pixel spacing. This can be done with the help of a suitable calibration slide.

Resolution. The word "resolution" is commonly used in three completely different ways in image cytometry. It can refer to the digital image size of a camera or digitizer, that is, the number of scan lines and pixels per line. It can address the pixel spacing at the specimen, or it can relate to the optical resolving power of the microscope, that is, the imaging instrument's ability to reproduce small objects as separate entities. In this chapter we designate these as "digital image size," pixel spacing," and "optical resolution," respectively.

Optical Resolution. According to the Raleigh criterion, one can just resolve (identify as separate) two point-source objects in a microscope image if they are separated by the distance $\delta = 0.61\lambda/NA$, where NA is the numerical aperture of the objective and λ is the wavelength of the narrowband illumination (Castleman, 1996). This resolution is obtained only with a high-quality objective on a microscope that is in proper adjustment. To prevent loss of information, the camera should be able to reproduce detail to this degree.

The psf of a lens is the image produced by a point source in the focal (specimen) plane. To a good approximation, the half-amplitude diameter of the psf in the specimen plane is given by the Abbé distance $r_0 = \lambda/2NA$. The camera's sampling aperture (pixel size and shape) and the microscope's psf combine by convolution to produce the effective scanning spot. To a good approximation, its diameter is the square root of the sum of squares of the diameters of these two components. It is usually most convenient to perform these calculations in the specimen plane.

Linearity. The degree of linearity of the relationship between the input light intensity and the output signal amplitude is another important factor. Although the eye is not particularly critical in this department, the validity of subsequent processing can be jeopardized by a nonlinear camera.

Noise. Finally, one of the most important characteristics of a camera is its noise level. If a uniformly gray image is presented to a camera, its output will show variations in gray level, even though the input brightness is constant across the image. Such noise introduced by the camera is a source of image degradation, and it should be small relative to the contrast of the specimen.

Photon noise. "Photon noise," or "shot noise," results from the quantum nature of light. Even under uniform illumination conditions, the actual number of photons striking a particular pixel in any one exposure period will be random. This random variable has a Poisson distribution, in which case the variance is equal to the mean. Given the large numbers of photons involved, the Poisson distribution is approximately Gaussian. Thus shot noise can be conveniently treated as having a normal distribution with standard deviation equal to the square root of the mean.

In general, the photon noise component is equal to the square root of the number of electrons that accumulate in a well (i.e., photoelectrons plus thermal electrons):

$$N_p = \sqrt{(F \cdot QE + DC)t_e} \qquad (15.1)$$

where F is the incident photon flux, QE is the quantum efficiency, DC is the dark current, t_e is the exposure time, and N_p is the photon noise that results from the statistical nature of light. Shot noise is usually the dominant noise source under high-exposure or high-dark-current conditions.

Task Requirements. Whether or not a particular camera is adequate depends on the specific task at hand. In some applications, digitizing images with relatively few lines, pixels per line, or gray levels or with appreciable noise and nonlinearity may be sufficient. Image cytometry typically requires a high-quality camera that is capable of sensing large images with many gray levels, good linearity, and a low noise level.

TYPES OF CAMERAS

Historically, imaging tubes, such as the vidicon and its relatives, were the backbone of image cytometry. Currently, however, solid-state cameras generally offer more flexibility, better performance, and lower cost.

Tube-Type Cameras

The Vidicon. The vidicon is a common type of television image-sensing tube. It is a cylindrical glass envelope containing an electron gun at one end and a target and light-sensitive faceplate at the other. The tube is surrounded by a yoke containing electromagnetic focus and beam deflection coils. The faceplate is coated on the inside with a thin layer of photoconductor over a thin transparent metal film, forming the *target*. Adjacent to the target is a positively charged fine wire screen called the *mesh*. A smaller positive charge is applied to the target.

A stream of electrons is ejected from the electron gun, focused to a small spot on the target, and steered across the target in a scanning pattern by the time-varying deflection field. The moving electron beam deposits a layer of electrons on the inner surface of the photoconductor to balance the positive charge on the metal coating on the opposite side.

An optical image formed on the target causes the photoconductor to leak electrons until an identical electron image is formed on the back of the target. Electrons will thus remain in dark areas and be absent in light areas of the image.

As the electron beam scans the target, it replaces the lost electrons, causing a current flow in the target's external circuit. This current is proportional to the number of electrons required to restore the charge and thus to the light intensity at that point. Current variations in the target circuit produce the video signal. The electron beam repeatedly scans the surface of the target, replacing the charge that bleeds away. The vidicon target is thus an integrating sensor, with the period of integration equal to the scanning frame rate.

While variations of the vidicon tube have contributed significantly to image cytometry as well as to broadcast television in the past, the present trend is toward solid-state sensors.

CCD Cameras

Silicon Light Sensors. Pure silicon can be grown in large crystals in which each atom is covalently bonded to its six neighbors in a three-dimensional rectangular lattice structure. An incident photon can break one of these bonds, freeing an electron. A thin metal layer deposited on the surface of the silicon and charged with a positive voltage creates a "potential well" that collects and holds the electrons thus freed. Each potential well corresponds to one pixel in an array of sensors. A potential well can hold about 10^6 electrons on typical chips.

Thermal energy also causes random bond breakage, creating "thermal electrons" that are indistinguishable from "photoelectrons." This gives rise to "dark current," current produced in the absence of light. The dark current is temperature sensitive, doubling for each 6°C increase in temperature. At the long integration times that are required for image sensing at low light levels, the wells can fill with thermal electrons before filling with photoelectrons. Cooling is often employed to reduce dark current and thereby extend the usable integration time.

CCD Construction. CCD chips are manufactured on a light-sensitive crystalline silicon chip, as discussed above (Janesick and Elliot, 1992). A rectangular array of photodetector sites (potential wells) is built into the silicon substrate. Photoelectrons produced in the silicon are attracted to and held in the nearest potential well. By controlling the electrode voltages, they can be shifted as a "charge packet" from well to well until they reach an external terminal.

CCD Operation. There are three architectures that can be employed for reading the accumulated charge out of CCD image sensor arrays. These are (1) full-frame architecture, (2) frame transfer architecture, and (3) interline transfer architecture (Fig. 15.2).

Full-Frame CCD. Following exposure, a full-frame CCD must be shuttered to keep it in the dark during the readout process. It then shifts the charge image out of the bottom row of sensor wells, one pixel at a time. After the bottom row is empty, the charge in all rows is shifted down one row, and the bottom row is again shifted out. This process repeats until all rows have been shifted down and out. The device is then ready to integrate another image.

Frame Transfer CCD. A frame transfer CCD chip has a doubly long sensor array. The top half senses the image in the standard manner, while an opaque mask protects the storage array on the bottom from incident light. At the end of the integration period, the charge image that has accumulated in the sensing array is shifted rapidly, row by row, into the storage array. From there it is shifted out, pixel by pixel, in the standard manner, while the sensing array integrates the next image. This technique employs simultaneous integration and readout, making video-rate image sensing possible.

Interline Transfer CCD. In an interline transfer CCD every second column of sensors is covered by an opaque mask. These columns of masked wells are used only in the readout process. After exposure, the charge packet in each exposed well is shifted into the adjacent masked well. This transfer requires very little time because all charge packets are shifted at once. While the exposed wells are accumulating the next image, the charge packets in the masked columns are being shifted out in the same way as in a full-frame CCD. In an interline transfer sensor, the number of pixels per

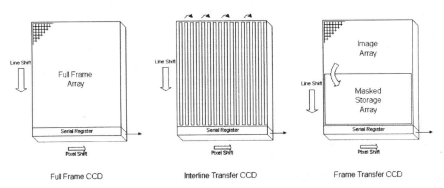

Fig. 15.2. CCD chip operation using full-frame, interline transfer, and frame transfer architecture.

line is half the actual number of wells per row on the chip. No more than 50% of the chip area is light sensitive, because the masked columns cover half the sensing surface. As with frame transfer, this technique employs simultaneous integration and readout.

CCD Performance. Available in a variety of configurations, CCDs give rise to compact and rugged solid-state cameras for both television and image digitizing applications. They are free of geometric distortion and exhibit highly linear response to light. CCDs are therefore emerging as the device of choice for image cytometry.

Scan Rate. The accumulated charge in a CCD chip can be scanned out at television rates (25–30 frames/s, 14 megapixels/sec) or much more slowly, if necessary, to reduce readout noise or to accommodate slower ADCs.

Dark Current. Because they can integrate for periods of seconds to hours to capture low-light-level images, CCDs are often used in fluorescence microscopy. Integration times longer than a few seconds at room temperature would fill the wells with thermal electrons before photoelectrons had a chance to build up. Thus cooling the chip to reduce dark current is a requirement for long integration times. Each 6°C decrease in temperature halves the dark current.

Because of imperfections in the crystal lattice, dark current varies significantly from one pixel to the next, particularly in less expensive chips. In long-exposure images, this leaves a "starfield" of fixed-pattern noise due to the few pixels that have the highest dark current. Because this pattern is stationary, it can be recorded and subtracted out. This effectively removes the pattern, provided the exposure is controlled so that the offending pixel wells are not allowed to saturate.

Dead Pixels. Defects in the crystal lattice can cause "dead pixels," which cannot hold or sometimes cannot shift electrons. This can wipe out a pixel or a column of pixels. CCD sensors are graded on the number of such defects, and the higher grade chips are generally more expensive.

Readout Noise. The operation of the on-chip electronics generates noise that can contaminate the image. It ranges from a few to many electrons per pixel depending on the chip design. It gets worse as the charge is read out at a faster rate. It is usually the dominant noise factor under short-exposure, low-light conditions where the dark current and photon noise components are small.

Charge Transfer Efficiency. The charge developed at a particular pixel may be shifted as many as 2000 times, depending upon its location in the array. The charge transfer efficiency must be extremely high or significant numbers of photoelectrons will be lost in the readout process.

Fill Factor. Half or more of the available area of the sensor chip can be covered by opaque charge transfer circuitry, leaving gaps between the pixels and reducing the "fill factor" below the ideal 100%. The chip can be coated with a microarray of "lenslets," each of which focuses the incoming light it receives onto the sensitive area of one pixel well.

Blooming. Overexposure of a CCD sensor can cause excess photoelectrons to spread to adjacent pixel wells. This produces the appearance of "blooming" in the image.

Spectral Sensitivity. Silicon sensors become less sensitive at the deep blue and ultraviolet end of the wavelength spectrum. This can be overcome by a "lumigen coating" that absorbs the short-wavelength photons and then reemits the energy as longer wavelength photons that the silicon can see.

Dynamic Range. Dynamic range is computed as pixel well capacity divided by readout noise level, both measured in electrons. It is usually expressed in decibels. This parameter is independent of exposure conditions (light level and exposure time). It characterizes the performance of the chip at high light levels, where the wells become filled with photoelectrons (rather than dark current) during a relatively short exposure.

Signal-to-Noise Ratio. The signal-to-noise ratio (SNR) for an image sensor chip can be computed as the number of photoelectrons received by a well divided by the total (photon plus readout) noise level, that is,

$$\text{SNR} = F \cdot \text{QE} \cdot t_e / \sqrt{N_p^2 + N_r^2} \tag{15.2}$$

where F is the incident photon flux, QE is the quantum efficiency, t_e is the exposure time, N_r is the readout noise level, and N_p is the photon noise that results from the statistical nature of light (see above). The SNR is quite dependent on light level and exposure time. Further, one or the other of the separate noise sources will often dominate, depending upon exposure conditions.

Charge Injection Devices

Charge injection device (CID) sensors (Williams and Carta, 1989; Kaplan, 1990) employ the same photoelectronic properties of silicon as CCDs, but they use a different method of readout.

CID Construction. At each pixel site, the CID has two adjacent electrodes (Fig. 15.3) that are insulated from the silicon surface by a thin metal–oxide layer. One electrode in each pixel is connected to all the pixels in the same column, while the other electrode is connected to all the pixels in the same row. Thus, a single pixel can be addressed uniquely by its row and column address. If all rows and columns of electrodes are held at a positive voltage, the entire chip accumulates a photoelectron image.

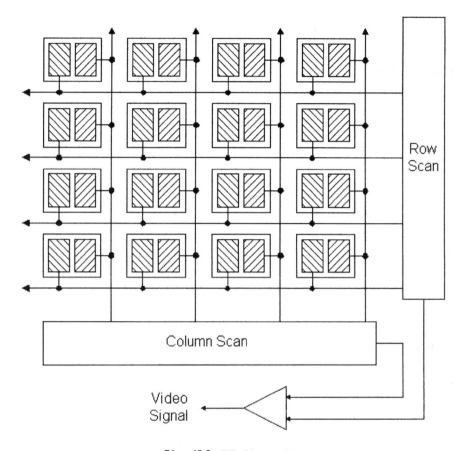

Fig. 15.3. CID chip operation.

CID Operation. When one electrode is driven to 0 V, the accumulated photoelectrons will shift underneath the second electrode, creating a current pulse in the external circuitry. The size of this current pulse reflects the amount of accumulated photoelectronic charge. Because the accumulated photoelectrons remain in the well after the shift, this is a nondestructive type of image readout: the pixel can be read out repeatedly by shifting the charge back and forth between the electrodes.

When both electrodes are driven to 0 V, the accumulated photoelectrons are injected into the underlying substrate, producing a current pulse in the external circuitry. The size of the pulse again reflects the amount of accumulated charge, but this process leaves the well empty. This destructive readout mode is used to prepare the chip for integrating another image.

The circuitry on the chip controls the voltages on the row and column electrodes to effect image integration and destructive and nondestructive readout. Because the CID can address individual pixels in any order, subimages of any size can be read out at any speed. Nondestructive readout allows one to watch the image accumulate on the chip, which is useful when the length of the required integration period is unknown.

CID Performance. CIDs are largely immune to blooming (charge spreading to adjacent pixels) and to radiation damage. Also, with nondestructive readout, the control program can monitor the filling of the wells and selectively flush individual pixels that become full before the integration period is over.

Because CIDs do not shift charge packets across the array, charge transfer efficiency is not a concern. Unlike the CCD, a small defect in the crystal lattice affects only one pixel. Also, essentially the entire surface area is light sensitive, leaving virtually no gaps between pixels. Even with these advantages, CIDs are considerably less light sensitive than similar CCDs, and their use has not become widespread.

APS Cameras

An emerging new solid-state image-sensing technique is *active pixel sensor* (APS) camera technology (Janesick and Elliot, 1992; Fossum, 1993, 1995). Like CCD cameras, these are fabricated on a silicon chip, but they use complementary metal–oxide–semiconductor (CMOS) integrated circuit technology, rather than the NMOS technology more common in CCDs. CMOS allows designers to build into the sensor chip signal-processing circuitry that would normally reside elsewhere in the camera. Indeed, much of the circuitry that traditionally resides on various circuit boards in the camera can be fabricated directly on an APS image sensor chip.

An active program of APS sensor development is being pursued by the NASA Jet Propulsion Laboratory. Experimental APS camera chips have been developed with emitter–follower amplifiers at each pixel, with special noise-reducing readout circuitry and with analog-to-digital converters on the chip. Some APS chips contain circuitry that allows them to read out a specified rectangular subimage continuously and nondestructively. They can incorporate automatic exposure control and dynamic contrast control.

CMOS APS cameras offer up to a 10X reduction in camera mass and volume and a 100X reduction in power requirements, in addition to their inherently high degree of system integration. They can be fabricated by existing CMOS foundry facilities.

Existing APS chips have sensor arrays of up to 1024×1024 10-μm pixels. The near future promises 2048×2048 arrays of 8-μm pixels. ADC resolution, currently at 8–10 bits, will probably increase to 12 bits. Current chips require only 20 mW of power, but 5-mW chips are possible. Image readout can be done at 1 MHz now, with 10-MHz readout a future possibility. A dynamic range of 75 dB is achievable now, and 100 dB is likely to come. Noise levels of six electrons per pixel are attainable, with two electrons per pixel on the horizon. Today's dark current level of 158 pA/cm^2 will probably drop to 50 pA/cm^2. Quantum efficiency is expected to increase from the present 40% to above 60%.

Single-chip cameras have been fabricated. These devices are easily interfaced into a system since they require connections only for power input and video signal output.

They have on-chip circuitry for timing, control, noise reduction, and analog-to-digital conversion. On-chip image compression is planned for the future.

Due to the large amount of on-chip circuitry, the pixel fill factor of these devices can be relatively low. This can be remedied by a microarray of lenslets covering the sensor. These tiny lenses collect light from a larger area and focus it onto the pixel well, bringing the fill factor up near 100%.

APS technology promises to reduce the cost of cameras in the future and perhaps improve their performance as well.

IMAGE DIGITIZATION

Conversion of an image into an array of numbers must be done so as to preserve the specimen content of interest without significant degradation. The noise level must be sufficiently low that specimen apearance and subsequent quantitative image analysis are not compromised. Photometric resolution (number of gray levels in the gray scale), optical resolution (psf diameter), spatial resolution (pixel spacing at the specimen plane), and the number of pixels per row and per column must be adequate for the tasks at hand.

Noise Sources

Quantization Noise. The time-varying electrical signal emerging from the image sensor is sampled and quantized by an ADC circuit. If B is the number of bits used in the quantization, the gray scale goes from zero to $2^B - 1$. Since quantization alters the gray level at each pixel by a small random amount, it can be viewed as a source of noise. The signal-to-noise ratio for quantization (SNR_q) is the (full-scale) signal amplitude divided by the quantization noise level. For images with a Gaussian distribution of gray levels, the SNR, measured in decibels, is

$$SNR_q = 20 \log_{10}\left(\frac{2^B}{\sigma_q}\right) = 6B + 11 \qquad (15.3)$$

where B is the number of bits used in the quantization and σ_q is the standard deviation (root-mean-square value) of the resulting quantization noise. Each 20 dB represents a factor of 10 in the SNR.

The actual ratio of full-scale signal to RMS quantization noise amplitude is

$$SNR = \frac{2^B}{\sigma_q} = 10^{SNR/20} = 10^{(6B+11)/20} \rightarrow \sigma_q = \frac{2^B}{10^{(6B+11)/20}} \qquad (15.4)$$

The commonly used 8-bit gray scale ($B = 8$, $2^B = 256$ gray levels, $SNR = 59$ dB $\rightarrow$ $SNR = 891$) is adequate for many image cytometry applications. Normally this quantization noise level ($\pm 0.11\%$ of full scale) is significantly less than the other two noise sources (photon and readout noise) and is thus tolerable. One should verify this, however, and use 10, 12, or more bits of gray-scale resolution when required by the application. At low light levels (where the CCD wells are not being filled to capacity) and without sufficient signal amplification between the sensor and the ADC, the full B-bit range of the ADC may not be utilized. Then the SNR of Equation (4) will not be realized, since it is based on full-scale signal amplitude.

In summary, the camera introduces three uncorrelated random-noise components: (1) photon noise, which results from the statistical nature of light; (2) readout noise, introduced by the circuitry on the sensor chip; and (3) quantization noise, which results from the conversion of a continuous value into an integer. In general, the three independent random-noise sources combine in such a way that the overall noise level is the square root of the sum of the squares of their individual amplitudes.

$$\sigma_n^2 = \sigma_r^2 + \sigma_p^2 + \sigma_q^2 \tag{15.5}$$

It often occurs that one of these sources is the dominant one. Which source dominates depends upon the imaging conditions in use at the time. For example, if the chip is read out slowly, readout noise is reduced; and if a 10- or 12-bit ADC is used, quantization noise is reduced. Then photon noise may be the dominant source.

Spatial Resolution

The well-known Shannon sampling theorem states that one can reconstruct, by proper interpolation, a sinusoidal signal from equally spaced sample points if there are no fewer than two sample points per cycle of the sine wave (Castleman, 1996). If the sampling is done more sparsely, one can encounter the phenomenon of aliasing, which introduces Moiré patterns into the image.

A microscope objective cannot pass image detail at frequencies higher than the optical cutoff frequency of $f_c = 2NA/\lambda$, where NA is its numerical aperture and λ is the wavelength of narrowband illumination (Goodman, 1988; Born and Wolf, 1980; Castleman, 1996). Thus, aliasing can be avoided completely if the pixel spacing at the specimen is no larger than $\lambda/4NA$. This is about 1/8 μm for an objective with NA = 1.0 operating in green (λ = 500 nm) light.

For applications in which the specimen does not contain detail at the resolution limit of the objective lens, larger pixel spacing will suffice. However, even smaller pixel spacing may be required for accurate measurement of objects in the image or for optimal display of the image. In these cases, reduced pixel spacing can be achieved by interpolation ("resampling") of the image after it has been digitized (Castleman, 1996).

SELECTING CCD CHIPS AND CAMERAS

An impressive array of solid-state cameras, incorporating a variety of different CCD chips, is commercially available. These cover a wide range of cost and performance, and the CCD camera situation is subject to rapid change.

The performance of a particular CCD camera depends on two major design factors: the choice of the CCD sensor itself and the design of the supporting electronics in the camera. Overall camera performance cannot exceed the limitations of either the chip or the electronics. A poor-quality chip in a well-designed camera and a good chip embedded in poorly matched circuitry will be equally disappointing. The circuitry in a well-designed camera will tend to exploit the best characteristics of the sensor chip and cover for its weaknesses. Thus a particular CCD camera must be evaluated as a complete system.

CCD Chips

Comparing CCD specifications is difficult since each manufacturer chooses to specify chip characteristics differently. Well capacity and RMS readout noise are usually given in electrons and dark current in electrons per second for a single pixel at 0°C. Dark current doubles for each 6°C increase in temperature, and vice versa. Dynamic range, computed as well capacity divided by readout noise, often appears in the data sheets. The SNR is quite dependent on exposure conditions (light level and exposure time) and thus is seldom listed in a readily usable form.

CCD Cameras

The vast majority of CCD cameras produced have been designed with the human eye (i.e., television applications) in mind. Quantitative electronic imaging imposes a significantly different set of constraints. In some applications, a well-designed television camera can be pressed into service for image cytometry to great advantage. In other applications, however, only a specially designed scientific-grade camera can perform satisfactorily.

The scientific cameras from *Photometrics*, for example, have three gain settings. The 1× setting matches the full-scale range of the ADC to the well capacity of a single pixel. The 4x gain mode, where one-quarter of full-well capacity saturates the ADC, achieves greater sensitivity for use at low light levels. The 0.5x gain mode, when used with binning, increases the effective well size to improve the SNR at high light levels. Photometrics product literature (see http://www.photomet.com) provides useful specifications and other helpful information about CCD camera operation and performance.

CONCLUSION

Each image cytometry application deserves its own analysis of camera and digitizing requirements. When selecting a specific camera, the camera characteristics should be evaluated in light of the requirements of the planned experiments. In any case, the camera should be well matched to the problem and to the other system components as well. Although an inadequate camera can forestall success, camera overkill can waste resources that might be better applied elsewhere.

Modern cameras are quite good in performance and reasonable in cost compared to the past and will undoubtedly continue to improve. In addition, some of the image detail lost to camera-induced noise, distortion, and lack of resolution can be recovered with digital image processing.

REFERENCES

Born E, Wolf E (1980): *Principles of Optics*, 6th ed. Oxford: Pergamon Press.

Castleman KR (1996): *Digital Image Processing.* Englewood Cliffs, NJ: Prentice Hall.

Fink DG, Christiansen D (1989): *Electronics Engineers Handbook.* New York: McGraw-Hill.

Fossum ER (1993): Active pixel sensors—are CCDs dinosaurs? *Proc SPIE* 1900:2–14.

Fossum ER (1995): CMOS image sensors: electronic camera on a chip. *IEEE International Electron Devices Meeting Technical Digest,* Dec. 10–13, 1995, Washington, DC.

Goodman JW (1988): *Introduction to Fourier Optics.* New York: McGraw-Hill.

Hutson G, Shepherd P, Brice J (1990): *Colour Television Theory: System Principles, Engineering Practice & Applied Technology.* New York: McGraw-Hill.

Janesick J, Elliot T (1992): History and advancements of large area array scientific CCD imagers. *Astronom Soc Pacific Con Series* 23:1–67.

Kaplan H (1990): New jobs for charge-transfer devices. *Photonics Spectra,* November issue.

Williams B, Carta D (1989): CID cameras: More than an alternative to CCDs. *Adv Imag*, January issue.

Index

Abbé equation, for resolution, 245
Acoustic devices, for cell transport and manipulation, 108
Acoustic pollution, 38–39
Acoustic signal, in droplet cell sorters, 34–39, 38f
Acoustic-optical modulator (AOM), 154–155
Active pixel sensor (APS) camera, 334–335
Acute myelogenous leukemia, CD34$^+$ hematopoietic stem cells for, 77
ADC. *See* Analog-to-digital converter
Adequate sampling, 301
Air-cooled lasers, 313, 317
Air-handling units, standards for, 83
Airy disk size
   in charge-coupled device (CCD) cameras, 303–304
   in digital microscopy, 300–302
Alexandrite lasers, 320
Aliasing
   in camera systems, 304, 336
   temporal, in confocal microscopy, 264
Altra cell sorter, 8, 8f, 85
Aluminum gallium indium phosphide (AlGaInP) lasers, 319
Amplitude modulation, in measuring fluorescence lifetimes, 179–181
Analog-to-digital converter (ADC), 255, 276, 324

Anticoincidence systems, for rare-cell sorting, 69f–70f, 69–71
Antigen presentation, monitoring of, flow cytometry for, 204–215
Antigen processing, 197–199, 198f
   MHC II-peptide complexes in, 215–217, 216f
   protease inhibitors and, 207, 208f
   study/monitoring of
      comparison of methods, 207–208, 215
      flow cytometry for, 204–215
      fluorescence intensity measurements for, 204–208, 205f, 207f–209f
      fluorescence lifetime measurements for, 213–215, 214t
      fluorescence polarization measurements for, 209–213, 210f–212f
      fluorescence-derivatized bovine serum albumin for, 200–201, 201f
      two-photon fluorescence microscopy for, 202–204, 203f
APS camera. *See* Active pixel sensor
Arc lamps, 309–310, 310f
   arc wander with, 313–314
   costs of, 314
   operating lifetimes of, 314
   power supplies of, 314
Arc wander, 313–314
Areance, radiant, 308
Argon-ion lasers, 313, 317

Note: Page numbers followed by letters f and t indicate figures and tables, respectively.

Artifacts
  in confocal microscopy, 247
  in high-speed cell sorting, 29–31
  in rare-cell sorting, 29
Autoclaving, 85
Autofluorescence, tissue, 225
Autoimmune disease, CD34$^+$ hematopoietic stem cells for, 78
Automation
  in cell collection, 45
  in digital microscopy, 291–292
Avalanche photodiode (APD)
  cost of, 130
  count rates of, 130
  dead time of, 118, 130
    correction of, 127, 128f
  for DNA fragment analysis, 115, 118–119, 123–127, 130, 135–136
  use of two, 135–136
Averaging mode, of digital microscopy, 299–300

Background rejection, in fluorescent lifetime imaging microscopy, 168–169
Backscattered light, 247
Beckman-Coulter cell sorters
  droplet, 8, 8f
  for research samples, 80
  sterilization of, 85
Becton Dickinson cell sorters
  fluid-switching, 16f–17f, 17
  for research samples, 80
  sterilization of, 85
Binning, definition of, 324
Bio-Rad MRC-600, 255–257, 257f, 261
Bio-Rad MRC-1000, 256–257, 257f
Biosafety
  of fluid-switching cell sorters, 3, 17, 55
  standards for, 83
Blank field test, for confocal microscopy, 240–241
Blind deconvolution algorithm, for deep-tissue two-photon fluorescence microscopy, 229–231, 230f
Blooming effect, 304–305, 332
Bone marrow cells
  CD34$^+$, 73–93
  high-speed sorting of, 22, 73–93
  tagged data for classifier systems of, 58f–59f
Bovine serum albumin, fluorescence-derivatized
  as exogenous antigen, 200–201, 201f
  in study of antigen processing, 200–217
Break-off point, of droplet cell sorters, 34–36, 35f, 40

piezosignal amplitude and, 37f
  stability of, 43–44, 55
Breast cancer, CD34$^+$ hematopoietic stem cells for, 76, 84, 88t, 88–89
Breast cancer cells
  ROC analysis of, 57, 60f
  tagged data for classifier systems of, 58f–59f
Brightness, in white-light scanning digital microscopy, 297–298, 302–303
Broadcast scanning conventions, 325–326, 326f
Brownian motion, in microfluidics, 96, 98
Burst, in DNA fragment detection, 116, 118, 119f, 132–134, 133f

Camera technologies
  accuracy of, 323
  active pixel sensor (APS), 334–335
  aliasing in, 304, 336
  charge injection device (CID), 332–334
    construction of, 332, 333f
    operation of, 333–334
    performance of, 334
  charge-coupled device (CCD), 329–332
    blooming effect in, 304–305, 332
    charge transfer efficiency of, 331
    construction of, 329–330
    dark current of, 331
    dead pixels in, 331
    dynamic range of, 332
    false data with, 304
    fill factor of, 331–332
    in fluorescence lifetime imaging microscopy, 141, 154f, 154–156
    frame transfer, 330, 330f
    full-frame, 330, 330f
    interline transfer, 330f, 330–331
    operation of, 330f, 330–331
    performance of, 331–332, 337
    readout noise in, 331, 336
    resolution of, 303–305
    scan rate of, 331
    selection of, 336–337
    signal-to-noise ratio of, 332
    silicon light sensors in, 329
    spectral sensitivity of, 332
    in video rate two-photon fluorescence microscopy, 231
    white-light scanning digital microscopy versus, 303–305
  for cytometry applications, 323–338
  definitions in, 324
  image digitization in, 335–336
  image size of, 327

interpolation in, 305, 336
light-sensing, 324–325
linearity of, 325, 328
magnification in, 327
noise in, 328, 331
  photon, 328, 336
  quantization, 335–336
  sources of, 335–336
performance of, 326–328, 331–332, 334, 337
photometric, 325
pixel size and spacing in, 327, 336
resolution of, 303–305, 327
  optical, 327
  spatial, 336
scanning conventions for, 325–326
  broadcast, 325–326, 326f
  video, 325
selection of, 323–324
task requirements of, 328
tube-type, 328–329
types of, 328–335
vidicon, 328–329
Cathepsins, in antigen processing, 197–199, 207
Cathode ray tube, in digital microscopy, 293–294, 305
CCD. *See* Charge-coupled device
CD34 antigen, 74–76, 197–199
CD34$^+$ hematopoietic stem cells
  clinical applications of
    potential, 76–80
    results with, 88t–89t, 88–90
  engraftment data, 89, 89t
  FDA regulations for, 74, 80, 82
  high-speed sorting of, 73–93
  immunomagnetic separation in, 79, 81–82
  in-process and final product testing in, 86–88, 87f
  pre-enrichment for, 79–81
  SyStemix process for, 76, 79–90, 84f, 88t–89t
  technical challenges in, 79
  manufacturing practices for, 80, 82
  purified hematopoietic stem cells versus, 77–78
  quality assurance for, 80, 82–86
  sources of, 75
  subfractionation of, 75
  Thy-1$^+$ subset of, 76–77, 80–82, 84
Cell assays, in microfabricated devices, 107–108, 108f–109f
Cell biology, scanning near-field optical imaging in, 271–290
Cell diameter, and transit time, 26–27, 28f

Cell fusion, 104, 106f
Cell separation, cell sorting versus, 2
Cell sorter(s)
  for clinical-scale tissue, 80
  cycle time of, 3, 13–14, 14f, 23, 29
  dead time of, 13, 23
  droplet. *See* Droplet cell sorters
  duty cycle of
    50 percent, 23
    full, 23
    high, 23, 46
  effective rate of, 22–23
  fluid-switching, 2–3, 16f–17f, 17–18
  high-speed. *See* High-speed cell sorting
  parallel, 47
  peak rate of, 22–23
  performance of, defining, 14–15
  photodamage, 3
  processing rates of, 22–23
  quality assurance for, 82–86
  recovery of, 14–15, 23
  for research samples, 80
  robustness of, 23, 46
  sterilization of, 85–86
  types of, 2–3
  well-tempered, 23
  yield of, 14–15, 23
Cell sorting
  versus cell separation, 2
  emerging tool in, 18
  high-speed. *See* High-speed cell sorting
  indexed, 53
  push-button approach to, 16f–17f, 17
  rare. *See* Rare-cell sorting
  as real-time data analysis, 55, 57
  technology
    lack of understanding of, 50
    review of, 1–19
    smart, 50
    utilization of, 50
Cell therapy
  current Good Manufacturing Practices for (cGMP), 80, 82
  FDA regulations for, 74, 80, 82
  quality assurance for, 80, 82–86
Cell viability
  in confocal microscopy, 243–244, 244f
  in high-speed sorting, 46, 80
  in rare-cell sorting, 66
Centrifugal elutriation, 79, 81
Ceprate-LC, 79
cGMP. *See* Good Manufacturing Practices, current

Charge, in cell sorters, 6–7, 11–12, 12f
  adjusting synchronization of, 41–43
  and drop deflection, 44–45
  stages of, 40
  timing of, 39–40
Charge injection device (CID), 332–334
  construction of, 332, 333f
  operation of, 333–334
  performance of, 334
Charge transfer efficiency, of CCD cameras, 331
Charge-coupled device (CCD) cameras, 329–332
  aliasing in, 304
  blooming effect in, 304–305, 332
  charge transfer efficiency of, 331
  construction of, 329–330
  dark current of, 331
  dead pixels in, 331
  dynamic range of, 332
  false data with, 304
  fill factor of, 331–332
  in fluorescence lifetime imaging microscopy, 141, 154f, 154–156
  frame transfer, 330, 330f
  full-frame, 330, 330f
  interline transfer, 330f, 330–331
  interpolation in, 305
  linearity of, 325, 328
  noise in, 331
    readout, 331, 336
  operation of, 330f, 330–331
  performance of, 331–332, 337
  resolution of, 303–305
  scan rate of, 331
  selection of, 336–337
  signal-to-noise ratio of, 332
  silicon light sensors in, 329
  spectral sensitivity of, 332
  in video rate two-photon fluorescence microscopy, 231
  white-light scanning digital microscopy versus, 303–305
Charge-coupled device (CCD) chips, 329–330, 336–337
Charged-coupled devices (CCD), cooled
  in confocal microscopy, 254–255, 265
  quantum efficiency of, 254f, 254–255
Charge-to-mass ratio, in droplet cell sorters, 44–45
Chinese hamster ovary cells. *See* CHO cells
Chloride concentrations, fluorescent lifetime imaging microscopy of, 163f–165f, 163–166

CHO cells
  fluorescence lifetime flow cytometry of, 185–186, 188f
  green fluorescent protein in, scanning near-field optical microscopy of, 277–278, 278f
Chromosome(s)
  fluorescence lifetime flow cytometry of, 189f, 190
  sorting
    eight-way, 45
    high-speed, 22, 45
    photodamage technique for, 3
Chronic myelogenous leukemia, CD34[+] hematopoietic stem cells for, 77
CID. *See* Charge injection device
Class II vesicle (CIIV), 199
Clinical trials
  cell sorting for, 74, 83–86
  regulations for, 74, 80, 82
Coaxial flow system, of droplet cell sorters, 3–5, 5f, 9–10
Coincidence
  in cell sorters, 12–13, 23, 32
  in rare-cell sorting, 62–63, 63f–64f
    minimal cell-cell in excitation source, while maintaining sufficient overall sensitivity, 66–67
Color scanning microscope (COSMIC)
  averaging mode of, 299–300
  brightness and color calibration in, 302–303
  brightness and focus change in, 297–298
  communications of, 300
  comparison with camera systems, 303–305
  contrast enhancement and suppression in, 298–299, 298f–299f
  detection system in, 294
  future applications of, 306
  image size for, 295
  operational considerations with, 305–306
  oversampling in, 297, 300–302, 301f
  resolution of, super, 297, 300–302, 302f
  response function of, 299
  scan rate of, 294
  scanning spot in, 293–294
  schematic layout of, 293f
  signal processing in, 294–296, 295f
  spot modulation in, 298f, 298–299
  summing mode of, 299–300
  zoom magnification in, 296–297, 296f–297f, 302
Complementary metal-oxide-semiconductor APS camera, 334–335
Computers, for DNA data collection, 124–125

Confocal microscopy
  basic concept of, 222
  biological reliability versus physical accuracy in, 243–244
  blank field test for, 240–241
  contrast transfer function of, 262, 262f
    and resolution, 263
  cooled CCD in, 254f, 254–255, 265
  counting statistics in, 240–243
  deep-tissue, 222–225
  detection and measurement losses in, 249–261
  detection geometry of, 223, 224f
  detector in
    quantum efficiency of, 249–255, 250f–251f, 253f–254f
    testing, 249–255
  digitization in, 255, 256f, 264–265
  fluorescence saturation in, 240, 243–244
    singlet-state, 243
  interaction problems in, 243–246, 244f
  lifetime imaging with, 141
  limits of, 239–269, 267t
    imposed by spatial and temporal quantization, 261
    parameters related to, 240, 241f
  mechanical stability of, 265–267
  mirrors in, 247–248, 258, 266
  noise in, 240–243
    extrinsic, 240
    fixed-pattern, 243
    intrinsic, 240–243
    level of, 250
  Nyquist criterion for, 245–246, 261, 264–265
  objective transmission in, 246–247, 247t
  optical system losses in, 246–249, 247t
  penetration depth of, 222
  photomultiplier tube in, 239, 250–252, 251f, 253f, 256, 258–261
  photon counting in, 255–258, 256f–257f
  photon efficiency of, 246
    measuring, 258–261, 260f
    practical tests of, 246–261
  pinholes of, testing, 248
  pixels in
    shape, mismatch with probe, 264–265
    size of, 245–246, 264
  position and motion in, determination of, 266
  pulse heights in, 252, 253f
  repeatability of scanning system in, 265–267
  resolution of, 245–246, 261–267
    practical considerations in relating to distortion, 265–267
    spatial, 222, 246
  temporal, 246, 264
  Rose criterion for, 263
  sampling theory for, 261–267
  sectioning ability of, 239–240
  signal-to-noise ratio in, 263
  spatial frequencies of, 262
  specimen response to dye in, 243–244
  stage drift in, 266
  task of, 240
  temporal aliasing in, 264
  versus two-photon fluorescence, 222–225, 224f
  visibility in, 263
  wagon-wheel effect in, 264
  zoom settings for, 245–246, 261, 264
Continuous wave (CW) lasers, 320
Continuous-wave two-photon imaging, with scanning near-field optical microscopy, 282f, 283, 283t
Contrast enhancement and suppression, in white-light scanning digital microscopy, 298–299, 298f–299f
Contrast transfer function, of confocal microscopy, 262f, 262–263
Cooled CCD
  in confocal microscopy, 254f, 254–255, 265
  quantum efficiency of, 254f, 254–255
Cooling, for lasers
  air, 313, 317
  water, 317
Cornea
  anatomy of, 225–226, 226f
  as deep-tissue imaging model, 225–226
  two-photon fluorescence microscopy of, 225–226, 234f, 234–235
Correlation error, 29–30, 30f
COSMIC. *See* Color scanning microscope
Coulter particle counter device, microchip, 99–100, 100f
Coulter volume instrument, 3–4
Critical illumination, 314
Cuvette systems, 28
CW. *See* Continuous wave
Cycle time
  definition of, 23
  determination of, 29
  of droplet cell sorters, 13–14, 14f
  of fluid-switching cell sorters, 3
Cytomation sorters
  droplet cell, 8, 8f
  high-speed, 22, 80
Cytometry
  flow. *See* Flow cytometry

Cytometry (*continued*)
  lasers for, 314–317
  light sources for, 314–317
    requirements for, 307
  microfabricated, 98–100, 99*f*
  microfluidics and, 96–100

Dark current, 331
Dead pixels, 331
Dead time, in equipment, 13, 23
  calculation of, 64–65
  in DNA fragment detection, 118, 130
    correction of, 127, 128*f*
  functional, 64
  in rare-cell sorting, 63–65, 65*f*
    minimal, 67
Decompression, and cell viability, 46
Deconvolution, blind algorithm, for deep-tissue two-photon fluorescence microscopy, 229–231, 230*f*
Deep-tissue imaging
  confocal microscopy for, 222–225
  of mouse ear, 228–229, 229*f*
  skin and cornea as models of, 225–226
  two-photon fluorescence microscopy for, 221–237
    blind deconvolution algorithm for, 229–231, 230*f*
    video rate, 231–233
  two-photon fluorescence spectroscopy for, 233–235
    biomedical applications of information, 234–235
    importance of information, 233–234
Delay time, of droplet cell sorters, 41–43
  accuracy of, 41–43
  visual method for determining, 41, 42*f*
Density gradient separation, 79
Detection time, of droplet cell sorters, 13–14, 14*f*
Dielectrophoresis, 103–104, 105*f*–106*f*
  negative, 103
  positive, 103
Differential counting, 100, 100*f*
Diffraction theory, 315–317
Diffusion, between sheath and core, 98
Digital microscopy, 291–293
  development of, 291–292
  important issues in, 291
  remote-control, 292
  in telepathology, 292, 305
  white-light scanning, 291–306
Digital oversampling, 297, 300–302, 301*f*
Digitization
  in camera systems, 335–336
  in confocal microscopy, 255, 256*f*, 264–265
  definition of, 324
Diode lasers, 319
Discriminant function analysis
  assumptions of, 58–59
  in rare-cell sorting, 57–59, 61*f*, 69
Distortion, definition of, 324
DNA fragment(s)
  analysis and sizing, 115–137
    avalanche photodiode for, 115, 118–119, 123–127, 130, 135–136
    beam-shaping optics for, 121, 121*f*
    burst data analysis in, 118, 119*f*, 132–134, 133*f*
    capillary and pressurized sample delivery for, 122–123, 123*f*
    versus conventional flow cytometry, 116
    critical aspects of system, 128–131
    data collection system for, 115, 124–125
    data processing software for, 125–127, 126*f*
    detection volume in, 130–131
    DNA samples for, 120
    dye-DNA interactions and, 128–129, 129*t*, 134
    flow cell and fluidics for, 121*f*, 121–122
    future directions in, 135–136
    instrument dead time in, 118, 127, 128*f*
    instrument development for, 115, 117–118
    instrumentation for, 120–127, 121*f*
    of large fragments, 134–135, 135*f*
    laser for, 120–121, 121*f*
    light collection path for, 123–124
    materials and methods for, 120–127
    MiniSizer for, 116–136
    nucleic acid stain for, 120
    results of, 131–136
    sample delivery and flow rate in, 129–130
    of small fragments, 131, 132*f*
  physical length of
    calculation of, 134
    extension in flow, 134–135
    staining and, 128–129, 129*t*
  single, detection by flow cytometry, 115–137
DNA injection, 101
DNA-binding dyes
  development of, 136
  interactions with DNA, 128–129, 129*t*, 134
  POPO-3, 117–118, 120, 128–129, 134
  TOTO-1, 116
DNA-binding fluorochromes, 188*f*, 189
DNASizer, 124
Droplet cell sorters, 2–16
  acoustic signal in, 34–39, 38*f*

Index

acoustically polluted, 38–39
automated cell collection in, 45
biohazards with, 2
break-off point of, 34–36, 35f, 40
 piezosignal amplitude and, 37f
 stability of, 43–44, 55
brief history of, 3–4
charge of, 6–7, 11–12, 12f
 adjusting synchronization of, 41–43
 and drop deflection, 44–45
 stages of, 40
 timing of, 39–40
charge-to-mass ratio in, 44–45
coaxial flow system of, 3–5, 5f, 9–10
coincidence in, 12–13, 32
components of, 4f, 4–8
contaminating cells in, 12–13
cycle time of, 31
delay time of, 41–43, 42f
droplet generation of, 5–6, 8–10
 assemblies for, 8, 8f
 frequency of, 8–10, 31–34, 33f–34f
 high-speed, 31–34
 optimal, 32–34, 34f
 physics of, 32–34, 33f
 stability of, 2, 5, 8
efficiency of, 14–15, 23–26, 25f–26f
eight-way configuration of, 45
electronic sorting system of
 cycle time of, 13–14, 14f
 dead time of, 13
 detection time of, 13–14, 14f
 processing of cell measurement in, 13–14, 14f
evaporation and, 45
exclusion zone of, 32, 43
field strength of, 44–45
flow cells of, 10–11, 11f
four-way sorting configuration of, 6f–7f, 7
frequency coupling of, 34–36, 35f
 numerical example of, 36–37, 37f
high-speed, 21–48
hydrodynamic properties of, 4–5, 9
kinetic energy of, 32–34
laser intersection of, location of, 10–11, 11f
microdroplets of, 5–6
multiple sort directions in, 6f–7f, 7, 45
nozzle of, 10–11
 assemblies of, 10–11, 11f, 37–39, 38f
 diameter of, 8–10
 geometry of, 5
 jet-in-air, 10–11, 11f, 28
 quartz, 10–11, 11f
 resonance frequency of, 37–38
performance of

 defining, 14–15
 predicting, 15–16
 variables in, 14
Poisson statistics in assessment of, 23–26, 25f–26f
pressure of, 31–34
purity of, 14–15, 24, 26f
queuing in, 12–13
for rare cells, 55
rates of, 2
recovery of, 14–15
sequence of operations in, 7–8
sheath pressure of, 8–10
for sorting into microwell plate, 6f–7f, 7
stream deflection in, variations of, 6f–7f, 7
stream undulations in, 5–9, 12–13, 34–36, 35f, 39–40
temperature stability of, 38–39
timing and drop deflection synchronization in, 39–40
typical, schematic of, 4f
velocity of, 32–34, 33f–34f, 39–40
 parabolic profile of, 9–10
yield of, 14–15
Drosophila melanogaster Schneider cells, green fluorescent protein in, scanning near-field optical microscopy of, 277–278, 279f
Duty cycle
 50 percent, 23
 full, 23
 of high-speed cell sorters, 23, 46
Dye lasers, 320
Dyes, DNA-binding, 116–118, 120, 128–129, 129t, 134, 136
Dynamic telepathology, 292

Ear, mouse, two-photon microscopy of, 228–229, 229f
E-beam radiation, for equipment sterilization, 85–86
Effective rate, definition of, 22–23
Electrical impedance, 104
Electrical transport, sorting and characterization, in microfabricated devices, 103–104, 105f–106f
Electrokinetic microchip, 107–108, 108f–109f
Electronic(s)
 in confocal microscopy, 255
 defocusing, 151
 for DNA analysis and sizing, 116
 of droplet cell sorters, 13–14, 14f
 in fluorescence lifetime flow cytometry, 182–183, 185f
 of high-speed cell sorters, 29–31, 80

Electronic(s) (*Continued*)
  errors in, 29–31, 30*f*
  parameters for, 29
  quality tests of, 30–31
  of light sources, 311–314
  in white-light scanning digital microscopy, 294–296, 295*f*
Electrophoretic separations
  continuous, 108, 109*f*
  in microfabricated devices, 107–108, 108*f*–109*f*
Electrostatic field, strength of, in droplet cell sorters, 44–45
Embryo manipulation
  microfabricated devices for, 102–103, 104*f*
  traditional methods of, 101
Empty magnification, 296–297, 301
Endocytic system, antigen processing in, 197–219
Epidermal growth factor receptor (EGFR), green fluorescent protein in, scanning near-field optical microscopy of, 277–278, 278*f*
Epi-illumination system, 308
Error(s)
  in charge synchronization, 41–43
  correlation, 29–30, 30*f*
  in high-speed cell sorting, 29–31, 30*f*
Evaporation, and droplet cell sorters, 45
Event(s)
  definition of, 22
  in DNA fragment detection, 116
  ghost, 29–30, 30*f*
  interval distribution, 24–26, 25*f*, 27*f*
  Poisson statistics for, 23–26, 25*f*–26*f*
  processed per second, 22–23
  rare, detection of, 49–72
Exclusion zone, in cell sorters, 32, 43
Exitance, of light sources, 307–309
Exponential delay, in fluorescent lifetime imaging microscopy, 170
Extrinsic noise, in confocal microscopy, 240

FACS Vantage, 85
FACSCalibur, 86–87, 87*f*
Facsort, 16*f*–17*f*, 17
Fast Fourier transform
  for fluorescence lifetime flow cytometry, 175
  for fluorescent lifetime imaging microscopy, 153, 155
FDA. *See* Food and Drug Administration
Fill factor, of CCD camera, 331–332
Fixed-pattern noise, in confocal microscopy, 243
FLIM. *See* Fluorescence lifetime imaging microscopy

Flow
  laminar, 96–97
  sheath, 97*f*, 97–98, 99*f*
  steady, 96–97
  turbulent, 96–97
  unsteady, 96–97
Flow cells
  for DNA fragment analysis and sizing, 121*f*, 121–122
  for droplet cell sorters, 10–11, 11*f*
Flow cytometry
  of CD34$^+$ hematopoietic stem cells, 73–93
  in clinical diagnostic medicine, 175
  conventional, 95, 175
  fluorescence lifetime. *See* Fluorescence lifetime flow cytometry
  high-speed, 21–48. *See also* High-speed cell sorting
  kinetics of, 207–208, 209*f*
  lasers for, 316–317
  light sources for, requirements for, 307
  microfabrication in, for single-cell handling and analysis, 95–113
  for monitoring antigen processing and presentation, 204–215
  quality standards for, 82–86
  of rare cells, 49–72. *See also* Rare-cell sorting
  for single DNA fragment detection, 115–137
  technology, 1–19, 50
Fluid(s)
  classifications of, 96
  flow
    laminar, 96–97
    sheath, 97*f*, 97–98, 99*f*
    steady, 96–97
    turbulent, 96–97
    unsteady, 96–97
  Newtonian, 96
  non-Newtonian, 96
Fluidic(s)
  devices, microfabricated, for single-cell handling and analysis, 95–113
  for DNA fragment analysis and sizing, 121*f*, 121–122
  in high-speed cell sorting, 27–28, 80
  microsystems
    basics of, 96–98
    and cytometry, 96–100
  in rare-cell sorting, 55
  sterilization of, 85
Fluid-switching cell sorters, 2–3, 16*f*–17*f*, 17–18
  biosafety of, 3, 17, 55
  cycle time of, 3
  for rare cells, 55

rates of, 3, 17–18
stability of, 3
Fluorescein isothiocyanate (FTIC)
  conjugated to bovine serum albumin
    as exogenous antigen, 200–201, 201f
    in study of antigen processing, 200–217
  in fluorescent lifetime imaging microscopy, 178
  fluorescent properties of, 200–201
  immunogenicity and antigenicity of, 200–201
  pH dependence of, 200, 202f
  spectral properties of, 200, 204
Fluorescence decay, 140, 176
Fluorescence decay time, methods of measuring, 140
Fluorescence emission signals, separation of
  fluorescence lifetimes for, 177–178
  phase-resolved, 181–182
Fluorescence intensity
  measurements, of antigen processing and presentation, 204–208, 205f, 207f–209f
  in scanning near-field optical microscopy
    as function of excitation power, 283t, 283–284
    as function of tip-sample displacement, 284–285, 286t
Fluorescence lifetime(s)
  of fluorochromes used for labeling cellular complexes and cells, 176, 177t
  measurement of, 179–181
    amplitude modulation method of, 179–181
    in fluorescence lifetime flow cytometry, 179–181, 184–187
    phase-shift method of, 179–181
    in steady state, 179–180
    two-phase method of, 179
  for separating overlapping emissions, 177–178
  unique properties of, 141–142
Fluorescence lifetime flow cytometry, 175–196
  of antigen processing, 213–215, 214t
  cell preparation and staining in, 184–187
  fluorescence lifetimes in
    of markers for cellular complexes and cells, 176, 177t
    measurement of, 178–181, 184–187
  in frequency domain, 176–177, 182
  instrumentation for, 183–184, 184f–185f
  materials and methods for, 178–187
  multifrequency, 181
    conceptual diagram of, 181f
  phase-resolved separation of emission signals in, 181–182
  results of, 187–192, 187f–192f
  signal processing in, 182–183, 185f

single-frequency, 180–181
  conceptual diagram of, 180f
theory of operation, 178–183
in time domain, 176
Fluorescence lifetime imaging microscopy (FLIM), 139–173
  background rejection and noise immunity in, 168–169
  of chloride concentrations, 163f–165f, 163–166
  comparison of techniques, 166–170
  examples of, 162–166
  frequency-domain heterodyning method of, 152–162, 153f
    background of, 152–154
    calculation of ac, dc, phase and modulation for, 153, 153f
    instrumentation for, 154f, 154–159, 156f–158f
  frequency-domain homodyning method of, 145–152, 147f–149f
    background of, 145–150
    instrument for, 150–152, 151f
    modulation of excitation light source in, 150–151
    modulation of microchannel plate image intensifier in, 151
    signal detection in, 152
  historical perspective on, 139–140
  intensified camera-based, 154f, 154–156
    modulation of laser in, 154–155
    signal detection and data acquisition system in, 155–156, 156f
  laser scanning versus wide-field illumination in, 167–168
  lifetime resolution of, 166
  microchannel plate-based, 154–156, 156f
  optical mixing methods of, 141, 159–162
    background of, 159–160
    instrument for, 160–162
  photobleaching and, 169–170
  photon and time efficiency of, 166–167
  pump-probe method of, 159–162, 160f–161f
    laser sources for, 161–162
    optics, signal detection and data acquisition in, 162
  recent developments in, overview of, 140–141
  sequential laser scanning based, 156–159
    excitation and optics in, 157f, 157–158
    synchronization of instrument and data acquisition in, 158f, 158–159
  single- or multiple-exponential delay in, 170
  stimulated emission technique of, 159–162
  in time domain, 142–145, 143f

Fluorescence lifetime imaging microscopy
(*Continued*)
  background of, 142–144
  instrument for, 144–145, 145*f*
  pulsed light source in, 144–145
  time gating of detector in, 145
  time-resolved, 141–142, 154*f*, 154–162, 157*f*, 160*f*–161*f*
  two-photon. *See* Fluorescence microscopy, two-photon
Fluorescence microscopy, two-photon, 142–143, 156–157, 227*f*
  advantages of, 223–225
  of antigen processing, 202–204, 203*f*
  axial depth discrimination of, 223
  biomedical applications of fluorescence spectroscopic information in, 234–235
  blind deconvolution algorithm for, 229–231, 230*f*
  challenges in clinical setting, 231
  versus confocal, 222–225, 224*f*
  of cornea, 225–226, 234*f*, 234–235
  deep-tissue, 221–237
  detection geometry of, 223, 224*f*
  excitation wavelengths of, 223–225
  of mouse ear, 228–229, 229*f*
  penetration depth of, 223
  photodamage with, 223
  photon collection with, 223–225
  setting up, 227*f*, 227–228
  of skin, 225–226, 230*f*, 231, 232*f*, 233
  three-dimensional imaging capabilities of, 202–203
  tissue thermal damage with, 222–223
  video rate, 231–233, 232*f*, 232–233
Fluorescence polarization measurements, of antigen processing, 209–213, 210*f*–212*f*
Fluorescence resonance energy transfer (FRET), 140
  with scanning near-field optical microscopy, 273, 276–281
  for photobleaching on cell surfaces, 278–279
Fluorescence resonance energy transfer, with scanning near-field optical microscopy, for donor photobleaching using labeled 3T3 cells in, 280*f*, 280–281
Fluorescence saturation
  in confocal microscopy, 240, 243–244
  singlet-state, 243
Fluorescence spectroscopy, two-photon, in deep tissue, 233–235
Fluorescence-derivatized bovine serum albumin

  as exogenous antigen, 200–201, 201*f*
  in study of antigen processing, 200–217
Food and Drug Administration (FDA), cell therapy regulations of, 74, 80, 82
Fourier transform
  for fluorescence lifetime flow cytometry, 175
  for fluorescent lifetime imaging microscopy, 146, 153, 155
Frame transfer CCD, 330, 330*f*
Frequency coupling, 34–36, 35*f*
  numerical example of, 36–37, 37*f*
Frequency domain
  fluorescence lifetime flow cytometry in, 176–177, 182
  fluorescent lifetime imaging microscopy in, 145–162
    background rejection and noise immunity in, 168–169
    exponential delay in, 170
    lifetime resolution of, 166
    photobleaching and, 169–170
    photon and time efficiency of, 166–167
  for measuring fluorescence decay time, 140
  spatial, confocal microscopy in, 262
Frequency-domain heterodyning method, of fluorescent lifetime imaging microscopy, 152–159, 153*f*
  background of, 152–154
  calculation of ac, dc, phase and modulation for, 153, 153*f*
  instrumentation for, 154*f*, 154–159, 156*f*–158*f*
  pump-probe, 159–162, 160*f*–161*f*
Frequency-domain homodyning method, of fluorescent lifetime imaging microscopy, 145–152, 147*f*–149*f*
  background of, 145–150
  instrument for, 150–152, 151*f*
  modulation of excitation light source in, 150–151
  modulation of microchannel plate image intensifier in, 151
  signal detection in, 152
FRET. *See* Fluorescence resonance energy transfer
FTIC. *See* Fluorescein isothiocyanate
Full duty cycle, 23
Full-field illumination, 304
Full-frame CCD, 330, 330*f*
Functional dead time, 64
Functional genomics, 95, 101
Fused-silica device, 98, 99*f*
Fused-silica microcoils, 105
Fusion, cell, 104, 106*f*

Fuzzy logic, in rare-cell sorting, 57

Gallium aluminum arsenide (GaAlA) lasers, 319
Gallium indium phosphide (GaInP) lasers, 319
Gases, dissolved, and cell viability, 46
Ghost events, 29–30, 30*f*
Good Manufacturing Practices, current (cGMP), 80, 82
Graft-versus-host disease, purified hematopoietic stem cells and, 78
Granulocyte colony-stimulating factor-mobilized peripheral blood, CD34$^+$ hematopoietic stem cells from, 75, 81
Green fluorescent protein
  in cells, 277–278, 278*f*–279*f*
  with scanning near-field optical microscopy, 277–278, 278*f*–279*f*

Helium-cadmium lasers, 313, 318
Helium-neon lasers, 313, 318
Hematopoietic rescue, following myeloablative therapy, 75–76
Hematopoietic stem cells
  CD34^{+} high-speed sorting of, 73–93
  clinical applications of
    potential, 76–80
    results with, 88*t*–89*t*, 88–90
  definition of, 74
  engraftment data, 89, 89*t*
  FDA regulations for, 74, 80, 82
  functions of, 74–75
  genetic modification of, 78
  homologous, 82
  in-process and final product testing of, 86–88, 87*f*
  manufacturing practices for, 80, 82
  non-homologous, 82
  phenotype of, 74–76
  purified
    applications of, 77–78
    versus CD34$^+$ hematopoietic stem cells, 77–78
    for in utero transplantation, 78
  quality assurance for, 80, 82–86
  SyStemix sorter for, 76, 79–90, 84*f*, 88*t*–89*t*
Heterodyning detection
  in fluorescence lifetime flow cytometry, 182
  in fluorescent lifetime imaging microscopy, 141, 152–159
High-speed cell sorting, 1, 21–48
  adjusting charge synchronization in, 41–43
  applications of, for CD34$^+$ hematopoietic stem cells, 73–93

automated cell collection in, 45
average event rate of, 29
break-off point in, 34–36, 35*f*, 40
  piezosignal amplitude and, 37*f*
  stability of, 43–44, 55
and cell viability, 46, 80
for clinical-scale tissue, 80
criteria for, 22
definitions in, 22–23
delay time in, 41–43, 42*f*
drop deflection, drop charge and field strength in, 44–45
drop generation in, 30*f*, 31–34
electronics, 29–31, 80
  parameters for, 29
  quality tests of, 30–31
errors in, 29–31, 30*f*
exclusion zone in, 32, 43
fluidics and optics in, 27–28, 80
frequency coupling in, 34–36, 35*f*
  numerical example of, 36–37, 37*f*
laser power in, 28
maximum analysis rate in, 26–27
multiple sort directions in, 45
nozzle assembly of, construction of, 37–39, 38*f*
parallel approach in, 47
peak rate of, 22–23
Poisson statistics in, 23–26, 25*f*–26*f*
practical limits of, 46–47
processing rate of, 22–23
of rare cells, 54–55, 60–71, 70*f*
robustness of, 23, 46
speed versus precision in, 21–22, 46–47
  workable compromise for, 47
timing and drop deflection synchronization in, 39–40
well-tempered sorter for, 23
HLFs. *See* Human lung fibroblasts
Homodyning detection
  in fluorescence lifetime flow cytometry, 182
  in fluorescent lifetime imaging microscopy, 141, 145–152, 147*f*–149*f*
Human Genome Project, 49
Human immunodeficiency virus (HIV), 73, 78
Human lung fibroblasts (HLFs), fluorescence lifetime flow cytometry of, 184–187, 187*f*, 187–188
Hydrodynamic focusing, 97–98
Hydrodynamic orientation, of cells, 97–98
Hydrodynamic properties, of droplet cell sorters, 4–5, 9
Hydrogen peroxide, for equipment sterilization, 85–86

Illumination
  critical, 314
  full-field, 304
  Köhler, 314
  in scanning microscope, 304–305
  sources, 307–321
    characteristics of, 307–309
    comparison of lamp and laser
      characteristics, 309–311, 321
    for cytometry, 314–317
      requirements for, 307
    exitance of, 307–309
    lamps, practical considerations with, 314
    lasers, 314–321
      types of, 317–320
    noise in
      and effects of measurements, 311–314
      peak-to-peak, 312
      root-mean-square (RMS), 312–313
    output variations in, 311–314
    radiance of, 307–309
    radiometric units of, 307–309
    requirements for, 307
    throughput of, 307–309
  wide-field, in fluorescent lifetime imaging microscopy, 167–168
Image sensing, 323–338
  cameras. *See also* Camera technologies
    performance of, 326–328, 331–332, 334, 337
    types of, 328–335
  definitions in, 324
  digitization of, 335–336
  light-sensing, 324–325
  linearity of, 325, 328
  performance of, 326–328
  photometric, 325
  scanning conventions for, 325–326
    broadcast, 325–326, 326*f*
    video, 325
  steps in, 324
Image size
  of camera systems, 327
  definition of, 324
  for white-light scanning digital microscopy, 295
Immune response. *See also* Antigen processing
  initiation of, 199
Immunomagnetic separation, of CD34⁺
  hematopoietic stem cells, 79, 81–82
Indexed cell sorting, 53
Indium gallium nitride (InGaN) lasers, 319
Inertial forces, 96

Intensity
  fluorescence
    of antigen processing and presentation, 204–208, 205*f*, 207*f*–209*f*
    in scanning near-field optical microscopy, 283*t*, 283–285, 286*t*
  of light source, 308
Interactive Data Language (IDL), 124–127
Interline transfer CCD, 330*f*, 330–331
International System of Units, 307–308
Interpolation, 305, 336
Interval distribution, of events, 24–26, 25*f*, 27*f*
Intrinsic noise, in confocal microscopy, 240–243
Ion lasers, 317–318
Iris effect, 151
Isolex device, 79

Jet-in-air nozzles, 10–11, 11*f*, 28
Joules, 307–308

Kinetic(s)
  of droplet cell sorters, 32–34
  of flow cytometry, 207–208, 209*f*
Köhler illumination, 314
Krypton-ion lasers, 317–318

Laminar flow, 96–97
Lamp(s)
  arc, 309–310, 310*f*
    arc wander with, 313–314
    costs of, 314
    operating lifetimes of, 314
    power supplies of, 314
  basic arrangements of, 314
  versus lasers, 309–311, 321
  noise levels of, 313
  practical considerations with, 314
  quartz tungsten halogen (QTH) filament, 309–310, 310*f*
Laser(s). *See also specific applications*
  air-cooled, 313, 317
  alexandrite, 320
  aluminum gallium indium phosphide (AlGaInP), 319
  argon-ion, 313
  basic physics of, 314–315
  beams
    diffraction theory for, 315–317
    focusing, 315–317
  continuous wave, 320
  current control mode of, 313
  dye, 320
  explanation of acronym, 314

gallium aluminum arsenide (GaAlA), 319
gallium indium phosphide (GaInP), 319
helium-cadmium, 313, 318
helium-neon, 313, 318
in high-speed cell sorting, 27–28
indium gallium nitride (InGaN), 319
ion, 317–318
krypton-ion, 317–318
light control mode of, 313
as light sources
 for cytometry, 314–317
 versus lamps, 309–311, 321
mixed-gas ion, 317–318
multiline, 315
neodymium-based solid-state, 319–320
noise levels of, 313
population inversion with, 315
pulsed, 320
semiconductor diode, 319
Ti-sapphire, 320
types of, 317–320
vibronic, 320
water-cooled, 317
yttrium aluminum garnet (YAG), 320
yttrium vanadate (YVO4), 320
Lee filter, 125
Leukemia, CD34[+] hematopoietic stem cells for, 77
Lifetime(s), fluorescence. *See* Fluorescence lifetime(s)
Lifetime imaging
 with confocal microscopy, 141
 fluorescent, 139–173
 recent developments in, overview of, 140–141
 with scanning near-field optical microscopy, 276
Lifetime resolution, in fluorescent lifetime imaging microscopy, 166
Light
 backscattered, 247
 white. *See* White-light scanning digital microscopy
Light sensing, 324–325
Light source(s), 307–321
 characteristics of, 307–309
 comparison of lamp and laser characteristics, 309–311, 321
 for cytometry, 314–317
 requirements for, 307
 exitance of, 307–309
 extended, 308
 in fluorescent lifetime imaging microscopy, 144–145, 150–151
 lamps, practical considerations with, 314
 lasers, 314–321
 types of, 317–320
 noise in
 and effects of measurements, 311–314
 peak-to-peak, 312
 root-mean-square (RMS), 312–313
 output variations in, 311–314
 pulsed, 144–145
 radiance of, 307–309
 radiometric units of, 307–309
 throughput of, 307–309
Line scanning approach, for video rate two-photon fluorescence microscopy, 231
Linearity
 of camera systems, 325, 328
 of fast photon counting systems, 256–257, 257*f*
Listmode data, for testing classifiers, 56*f*, 56–57, 58*f*–59*f*
Lithographic arrays, 102
Lithographic microcoils, 105
Littrow prisms, 315, 317
Logistic regression, in rare-cell sorting, 59–60, 69
Lookup tables, 55, 60
Lumigen coating, 332

Macrophage(s)
 processing, 198*f*, 201–219
 flow cytometry for monitoring, 204–215
 fluorescence intensity measurements of, 204–208, 205*f*, 207*f*–209*f*
 fluorescence lifetime measurements of, 213–215, 214*t*
 fluorescence polarization measurements of, 209–213, 210*f*–212*f*
 MHC II-peptide complexes in, 215–217, 216*f*
 two-photon fluorescence microscopy of, 202–204, 203*f*
 receptor, 204–206
Magnetic sorting, in microfabricated devices, 104, 106*f*
Magnification
 in camera systems, 327
 empty, 296–297, 301
 zoom
 for confocal microscopy, 245–246, 261, 264
 for white-light scanning digital microscopy, 296–297, 296*f*–297*f*, 302
Major histocompatibility complex class II molecule(s) (MHC II)
 in antigen processing, 197–200, 199*f*, 206, 208, 215–217

Major histocompatibility complex class II
molecule(s) (*Continued*)
synthesis of, 198–199
Major histocompatibility complex class II molecule-peptide complexes, processing and presentation of, model of, 215–217, 216f
MCS-2. *See* Multichannel scaler card
Mechanics, of microfabricated devices, 101–104, 102f–104f
MEMS. *See* MicroElectroMechanical Systems
Mercury arc lamps, 309–310, 310f
MHC class II containing compartment (MIIC), 199
Microchannel plate-based system, for fluorescent lifetime imaging microscopy, 154–156, 156f
Microcoils, 105
Microdroplets, of droplet cell sorters, 5–6
MicroElectroMechanical Systems (MEMS), 1, 18
microfabrication and, 98–100
for single-cell handling and analysis, 100–108
Microfabricated cytometry, 98–100
Microfabricated devices, 101–104
acoustic methods in, 108
cell assays and electrophoretic separations in, 107–108, 108f–109f
cell fusion in, 104, 106f
cytometry, 98–100, 99f
development costs of, 109
development of, 98–100, 99f–100f
dielectrophoresis in, 103–104, 105f–106f
for differential counting, 100, 100f
electrical impedance in, 104
electrical transport, sorting and characterization in, 103–104, 105f–106f
fluidic, for single-cell handling and analysis, 95–113
magnetic sorting in, 104, 106f
mechanical transport, sorting, manipulation and characterization in, 101–103, 102f–104f
micro NMR in, 105–107, 107f
MicroElectroMechanical Systems (MEMS), 18, 98–108
multiple streams in, 99–100, 100f
Microfluidics
basics of, 96–98
and cytometry, 96–100
diffusion (Brownian motion) in, 96, 98
sheath flow in, 97f, 97–98, 99f
MicroGreen laser, 120–121
Microscopy. *See also specific types*
confocal, 141, 222–225, 224f
limits of, 239–269

fluorescence lifetime imaging (FLIM), 139–173
multiphoton multifocal, 232
remote-control, 292
scanning near-field optical, 271–290
time-resolved, 141–142, 154f, 154–162, 157f
two-photon fluorescence, 142–143, 156–157
of antigen processing, 202–204, 203f
versus confocal, 222–225, 224f
deep-tissue, 221–237
white-light scanning digital, 291–306
Microwell plate, droplet cell sorting into, 6f–7f, 7
MIIC. *See* MHC class II containing compartment
Miniaturization technology, 95. *See also* Microfabricated devices
MiniMACS, 79
MiniSizer
avalanche photodiode in, 118–119, 123–127, 130, 135–136
beam-shaping optics for, 121, 121f
burst outline analysis of, 132–134, 133f
capillary and pressurized sample delivery in, 122–123, 123f
critical aspects of, 128–131
data collection system for, 124–125
data processing software for, 125–127, 126f
dead time of, 118
correction of, 127, 128f
detection of phycoerythrin molecules by, 116, 117f
detection volume of, 130–131
detector in, 123, 130
development of, 117–118
for DNA fragment analysis and sizing, 116–136
of large DNA fragments, 134–135, 135f
of small fragments, 131, 132f
DNA-dye interactions in, 128–129, 129t, 134
flow cell and fluidics of, 121f, 121–122
laser of, 120–121, 121f
light collection path in, 123–124
optical schematic of, 121f
results of, 131–136
sample delivery and flow rate in, 129–130
Mirrors, in confocal microscopy, 247–248, 258, 266
Mixed-gas ion lasers, 317–318
Mobilized peripheral blood (MPB), CD34$^+$ hematopoietic stem cells from, 75, 77, 80–82
MoFlo cell sorter, 8, 8f, 22, 68
Moiré patterns, 304, 336

Index   353

Molecular analysis, 116. *See also* DNA fragment(s)
Molecular analyzer, 136
Momentum, of Newtonian fluids, 96
Monoclonal antibody (MAb), in immunomagnetic CD34 selection, 81–82
Mouse ear, two-photon microscopy of, 228–229, 229f
MPB. *See* Mobilized peripheral blood
Multichannel scaler card (MCS-2), 124–125
Multiphoton excitation, 248–249, 258
  microlens system of, 255
  in scanning near-field optical microscopy, 277, 281–287, 282f
  picosecond pulsed, 283–287, 284f–285f
Multiphoton multifocal microscopy, 232
Multiple myeloma
  CD34$^+$ hematopoietic stem cells for, 76–77, 84, 88
  tumor-free graft for, 77
Multivariate statistical classification, of rare cells, 57–60, 69
  real-time, 60–62
Myeloablative therapy, hematopoietic rescue following, 75–76

NA. *See* Numerical aperture
Navier-Stokes equation, 96
Near-field scanning optical microscopy. *See* Scanning near-field optical microscopy
Neodymium-based solid-state lasers, 319–320
Newtonian fluids, 96
  momentum of, equation of, 96
NMR. *See* Nuclear magnetic resonance
Noise
  in camera systems, 328, 331
    sources of, 335–336
  in confocal microscopy, 240–243
    extrinsic, 240
    fixed-pattern, 243
    intrinsic, 240–243
    level of, 250
  definition of, 324
  in fluorescent lifetime imaging microscopy, 168–169
  in light sources
    and effects of measurements, 311–314
    lamps, 313
    lasers, 313
  photon (shot), 328, 336
  quantization, 335–336
  readout, 331, 336
Noise immunity, in fluorescent lifetime imaging microscopy, 168–169

Non-Hodgkin's lymphoma, CD34$^+$ hematopoietic stem cells for, 76–77, 84, 88t, 88–89
Nozzles, 10–11
  assemblies of, 10–11, 11f
  construction of, 37–39, 38f
  diameter of, 8–10
  geometry of, 5
  jet-in-air, 10–11, 11f, 28
  quartz, 10–11, 11f
  resonance frequency of, 37–38
  sterilization of, 85–86
Nuclear magnetic resonance (NMR), micro, 105–107, 107f
Nuclear transplantation, 101
Nucleic acid stain, for DNA fragments, 120
Numerical aperture (NA), and illumination, 308
Nyquist criterion
  for confocal microscopy, 245–246, 261, 264–265
  for digital microscopy, 301

Optical microscopy, scanning near-field. *See* Scanning near-field optical microscopy
Optical mixing methods, of fluorescent lifetime imaging microscopy, 141, 159–162
  background of, 159–160
  instrument for, 160–162
Optical pumping, for lasers, 319–320
Optical resolution
  of camera systems, 327
  definition of, 324
Optical resolution elements
  calculation of, 300
  size of, 300–302
Optical system
  of confocal microscopy
    contrast transfer function of, 262f, 262–263
    losses in, 246–249, 247t
  for DNA sizing and analysis, 121, 121f
  in fluorescent lifetime imaging microscopy, 141, 157–162
  for high-speed cell sorting, 27–28
Optical throughput, 307–309
Oversampling, digital, 297, 300–302, 301f

Parallel operations, 47
Partec fluid-switching sorting system, 16f, 17
Particle analysis, 136
Particle counter, 99–100, 100f
Peak rate, definition of, 22–23
Peak-to-peak noise, in light sources, 312
Penetration depth
  of confocal microscopy, 222
  of two-photon fluorescence microscopy, 223

pH dependence, of fluorescein, 200, 202f
Phase modulation methods, in fluorescent
    lifetime imaging microscopy, 141
Phase-resolved separation, of fluorescence
    emission signals, 181–182
Phase-sensitive detectors (PSDs), in fluorescence
    lifetime flow cytometry, 179, 183
Phase-shift method, of measuring fluorescence
    lifetime, 179–181
Phenylalanine methyl ester (PME) lysis, 81
Photobleaching
    in fluorescent lifetime imaging microscopy,
        169–170
    in scanning near-field optical microscopy,
        276–277
        donor, using labeled 3T3 cells, 280f,
            280–281
        of fluorescence resonance energy transfer,
            on cell surfaces, 278–279
        multiphoton imaging and, 285–287, 286f
Photodamage, 3, 223
Photometric resolution, definition of, 324
Photometry, 325
Photomultiplier tube
    in confocal microscopy, 239, 250–252, 256,
        258
        measuring output signal of, 258–261
        quantum efficiency of, 250–252, 251f, 253f
    in fluorescence lifetime flow cytometry, 183
    in fluorescent lifetime imaging microscopy,
        154–158
    in white-light scanning digital microscopy,
        294
Photon counting
    in confocal microscopy, 255–258, 256f–257f
    detector, for DNA fragments, 115–116, 123,
        130, 135–136
    fast circuitry, 256–257, 257f
Photon efficiency
    of confocal microscopy, 246–261
        measuring, 258–261, 260f
        definition of, 246
    of fluorescent lifetime imaging microscopy,
        166–167
Photon flux, 325
Photon noise, 328, 336
Phycoerythrin molecules, single, detection of,
    116, 117f
PI. See Propidium iodide
Picosecond pulsed multiphoton excitation, in
    scanning near-field optical microscopy,
    283–287, 284f–285f
Picture elements. See Pixels

Piezoelectric device, in droplet cell sorters, 5, 8,
    35–40, 37f–38f
Pinholes, in confocal microscopy, 248
Pixels, 324
    in camera systems
        active pixel sensor (APS), 334–335
        dead, 331
        size and spacing of, 327, 336
    in confocal microscopy
        shape, mismatch with probe, 264–265
        size of, 245–246, 264
PME. See Phenylalanine methyl ester
Pockels cell, 154–155
Pointance, of light source, 308
Poisson statistics, 23–26, 25f–26f
Polymerase chain reaction, in rare-cell sorting,
    52–53, 54f, 66
Population inversion, with lasers, 315
Pressure
    and cell viability, 46
    for DNA fragment analysis and sizing,
        122–123, 123f
    of droplet cell sorters, 8–10, 31–34
Processing rate
    definition of, 22
    types of, 22–23
Progenitor cell sorting. See Stem cell sorting
Propidium iodide (PI), 178, 190
Protease inhibitors, and antigen processing, 207,
    208f
PSDs. See Phase-sensitive detectors
Pulsed lasers, 320
Pump-probe
    definition of, 159
    in fluorescent lifetime imaging microscopy,
        159–162, 160f–161f
        laser sources for, 161–162
        optics, signal detection and data acquisition
            in, 162
Purity
    of cell sorters, 14–15, 24, 26f
    of rare-cell sorting, 51, 55–56, 62–63, 63f

QTH filament lamps. See Quartz tungsten
    halogen
Quality assurance
    for cell product therapy, 80, 82–86
    in facility and process design, 83
    for good tissue practices, 83
Quantization
    definition of, 324
    noise, in camera systems, 335–336
    spatial and temporal, in confocal microscopy,
        261

Quantum efficiency
  in confocal microscopy, 249–255, 250f–251f, 253f–254f
  of cooled CCD, 254f, 254–255
  of photomultiplier tube, 250–252, 251f, 253f
Quartz capillary, 122–123, 123f
Quartz nozzle, 10–11, 11f
Quartz tungsten halogen (QTH) filament lamps, 309–310, 310f
Queuing
  in droplet cell sorters, 12–13
  in rare-cell sorting, 67f, 68
Queuing theory, for defining and measuring instrument performance, in rare-cell sorting, 55–56, 62–64, 63f–65f

Radiance
  definition of, 308
  of light sources, 307–309
  units of, 308
Radiant areance, 308
Radiant energy, units of, 307–308
Radiant flux, 307–308
Radiometric units, 307–309
Radiometry, definition of, 307
Rare cell(s), definition of, 49
Rare-cell sorting, 24, 49–72
  analysis speeds in, 54–55
  anticoincidence systems for, 69f–70f, 69–71
  applications of, 50
    flexible, modular design for optimization to, 70f, 70–71
    of CD34[+] hematopoietic stem cells, 73–93
  cell arrival statistics in, 55–56
  classification/sort decisions in
    acceptable costs of misclassification in, 57–58
    multivariate statistical, 57–60, 69
    need for sophisticated, flexible, 69f–70f, 69–71
    parameters for, 54–55, 69
    real-time multivariate statistical, 60–62
    stages of, 54–55
    standard for, 57
  coincidence in, 62–63, 63f–64f
    minimal cell-cell in excitation source, while maintaining sufficient overall sensitivity, 66–67
  dead time in, 63–65, 65f
    minimal, 67
  discriminant function analysis in, 57–59, 61f, 69
  eliminating not-of-interest cells in, 51
  false positives in, 52–55
  fast sensors and preamplifiers for, 67
  ideal instrument for, designing, 66–69, 70f
  importance of, 49–50
  indexed, 53
  instrument performance for, queuing theory for defining and measuring, 55–56, 62–66, 63f–65f
  limits in sorting speed and purities, 55–56
  linked, transformed parameters for, 68f, 69
  live-cell requirements in, limitations imposed by, 66
  logistic regression in, 59–60, 69
  measurement artifacts in, 29
  modeling for, 52
    importance of, 56–62
    multiparameter, 56–57
    simple one-parameter, 56
  multiple probes for, 51, 53–54
  multistage, multiqueuing signal processing in, 67f, 68
  problems in, 50–51
    fundamental, 52–56
    reasons for, 50–51
    theoretical aspects of, 51–52
  receiver operating characteristic analysis in, 53–54, 57, 60f
  signal-to-noise ratio of cell sample, 50–51
  single-cell, multitube PCR, 52–53, 54f, 66
  sort boundaries in, 60–62
  staining specificity for, 49, 51–52
  statistical sampling limits of, 52–53
  tagged data mixtures for, 56f, 56–57, 58f–59f
  time required to collect desired cells in, as function of sorting rate, 53t
  truth table for, 57
  yield versus purity in, 51, 63
Rare-event detection, 49–72. See also Rare-cell sorting
Rayleigh criterion, 300, 327
Readout noise, 331, 336
Reagents, sterile, 86
Real-time data analysis, cell sorting as, 55, 57
Real-time multivariate statistical classification, of rare cells, 60–62
Receiver operating characteristic (ROC) analysis, in rare-cell sorting, 53–54, 57, 60f
Recovery, of cell sorters, 14–15, 23
Red blood cells, handling and analysis, microfabricated devices for, 101–103, 103f, 107–108
Remote-control digital microscopy, 292
Resolution
  Abbé equation for, 245
  of camera systems, 303–305, 327

Resolution (*Continued*)
  of confocal microscopy, 245–246, 261–267
  contrast transfer function and, 263
  definition of, 324
  lifetime, in fluorescent lifetime imaging microscopy, 166
  Nyquist criterion for, 245–246, 261, 264–265, 301
  optical
    of camera systems, 327
    definition of, 324
  optical elements
    calculation of, 300
    size, 300–302
  photometric, definition of, 324
  Rayleigh criterion for, 300, 327
  Rose criterion for, 263
  spatial
    of camera systems, 336
    of confocal microscopy, 222, 246
    definition of, 324
    of scanning near-field optical microscopy, 271–272
  super, 297, 300–302, 302f
  temporal, in confocal microscopy, 246, 264
  of white-light scanning digital microscopy, 297, 300–302, 302f
Response function, of digital microscopy, 299
Reynolds number
  calculation of, 96
  of microfluidics, 96–97
    for parabolic velocity profile, of droplet cell sorters, 9–10
RMS noise. *See* Root-mean-square noise
Robust design, of cell sorters, 23, 46
ROC analysis. *See* Receiver operating characteristic analysis
Root-mean-square (RMS) noise, in light sources, 312–313, 318
Rose criterion, for confocal microscopy, 263

Sample size, in rare-cell sorting, 52–53
Sampling
  adequate, 301
  aperture, definition of, 324
  definition of, 324
  oversampling, 297, 300–302, 301f
  in rare-cell sorting, limits of, 52–53
  theory
    for confocal microscopy, 261–267
    for digital microscopy, 301
    Shannon, 336

Scan rate
  of CCD camera, 331
  of color scanning microscope, 294
Scanning conventions, 325–326
  broadcast, 325–326, 326f
  video, 325
Scanning, definition of, 324
Scanning near-field optical microscopy (SNOM)
  biological applications of, 273–274
    summary of recently published reports on, 274t
  in cell biology, 271–290
  continuous-wave two-photon imaging with, 282f, 283, 283t
  dual detection of, 276
  emission spectroscopy of, 276
  excitation sources for, 276
  fluorescence imaging of, 276
  fluorescence intensity in
    as function of excitation power in, 283t, 283–284
    as function of tip-sample displacement, 284–285, 286t
  fluorescence resonance energy transfer with, 273, 276–281
  green fluorescent protein with, 273, 277–278, 278f–279f
  instrumental details of, 274–277
  lifetime imaging with, 276
  multicolor imaging in, 281–287
  multiphoton excitation of fluorescence in, 277, 281–287, 282f
    picosecond pulsed, 283–287, 284f–285f
  multiple photophysical modes of, 274–276, 275f
  as optical module for scanning probe microscope, 272
  photobleaching with, 276–277
    on cell surfaces, 278–279
    donor, using labeled 3T3 cells, 280f, 280–281
    multiphoton imaging and, 285–287, 286f
  principle of, schematic of, 272f
  prospects for, 287
  spatial resolution of, 271–272
  specific applications of, 277–287
  topographical properties of, 271–273
Scanning probe microscope, scanning near-field optical microscopy as optical module for, 272, 274. *See also* Scanning near-field optical microscopy

Scanning spot, in digital microscopy, 293–294
Seattle Instruments, 22
Semiconductor diode lasers, 319
Sequential laser scanning based microscopy system, 156–159
　excitation and optics in, 157f, 157–158
　synchronization of instrument and data acquisition in, 158f, 158–159
Shannon sampling theorem, 336
Shear forces, and cell viability, 46
Sheath flow, in microfluidics, 97f, 97–98, 99f
Sheath pressure, of droplet cell sorters, 8–10
Shot noise, 328
Signal-to-noise ratio
　of CCD camera, 332
　of cell sample, in rare-cell sorting, 50–51
　of confocal microscopy, 263
Silicon detector, in confocal microscopy, 254–255
Silicon light sensors, in camera systems, 329
Single-cell handling and analysis, 101–108
　acoustic devices for, 108
　cell assays and electrophoretic separations, 107–108, 108f–109f
　dielectrophoresis for, 103–104, 105f–106f
　electrical impedance, 104
　electrical transport, sorting and characterization, 103–104, 105f–106f
　fusion, 104, 106f
　magnetic sorting, 104, 106f
　mechanical transport, sorting, manipulation and characterization, 101–103, 102f–104f
　micro NMR, 105–107, 107f
　microfabricated fluidic devices for, 95–113
　PCR, in rare-cell sorting, 52–53, 54f, 66
　traditional methods of, 101
Skin
　anatomy of, 225–226, 226f
　as deep-tissue imaging model, 225–226
　two-photon fluorescence microscopy of, 225–226
　　blind deconvolution algorithm for, 230f, 231
　　video rate, 232f, 233
Slit scan, 316
Smart technology, 50
SNOM. See Scanning near-field optical microscopy
Spatial frequency domain, confocal microscopy in, 262
Spatial quantization, in confocal microscopy, 261
Spatial resolution
　of camera systems, 336

　of confocal microscopy, 222, 246
　definition of, 324
　of scanning near-field optical microscopy, 271–272
Spectroscopy
　in cell biology, 271–290
　fluorescence, two-photon, in deep tissue, 233–235
Spot modulation, 298f, 298–299
Spot-scanning techniques, 293–294
Stage drift, in confocal microscopy, 266
Stanford Fluorescence Activated Cell Sorter (FACS II), 3–4
Steady flow, 96–97
Stem cell(s), hematopoietic. See Hematopoietic stem cells
Stem cell sorting
　discriminant function analysis in, 57, 61f
　high-speed, of $CD34^+$ hematopoietic cells, 73–93
　multivariate statistical classification of, 57
　ROC analysis in, 57, 60f
Sterilization, of equipment, 85–86
Stimulated emission technique, of fluorescent lifetime imaging microscopy, 159–162
Summing mode, of digital microscopy, 299–300
Superresolution, 297, 300–302, 302f
Surface-to-volume ratio, in microfluidics, 96–97
SyStemix, 22
　high-speed cell sorter, 80, 83–86, 84f
　　for $CD34^+$ hematopoietic stem cells, 76, 79–90, 88t–89t
　　engraftment data, 89, 89t
　　isolation system for, 86
　　quality assurance for, 83–86
　　sterilization of, 85–86

Tagged parameters, for rare-cell sorting, 56f, 56–57, 58f–59f
T-cell stimulation, antigen processing and, 197–199, 207
Telepathology, 292, 305
　for consultation, 292
　dynamic, 292
Temperature stability, in droplet cell sorters, 38–39
Temporal aliasing, in confocal microscopy, 264
Temporal quantization, in confocal microscopy, 261
Temporal resolution, in confocal microscopy, 246, 264
Three-photon excitation, in scanning near-field optical microscopy, 282, 284, 284f

Throughput, of light sources, 307–309
Thy-1$^+$ subset, of CD34$^+$ hematopoietic stem cells, 76–77, 80–82, 84
Thymocytes, fluorescence lifetime flow cytometry of, 184–192, 187f, 190f–192f
TIFF file format, 300
Time domain
  fluorescence lifetime flow cytometry in, 176
  fluorescent lifetime imaging microscopy in, 142–145, 143f
    background of, 142–144
    background rejection and noise immunity in, 168–169
    exponential delay in, 170
    instrument for, 144–145, 145f
    lifetime resolution of, 166
    photobleaching and, 169–170
    photon and time efficiency of, 166–167
    pulsed light source in, 144–145
    time gating of detector in, 145
  for measuring fluorescence decay time, 140
Time efficiency, in fluorescent lifetime imaging microscopy, 166–167
Time-gating approach
  in fluorescent lifetime imaging microscopy, 141–145, 143f, 145f
  for measuring fluorescence decay time, 140
Time-resolved fluorescence lifetime imaging microscopy, 141–142, 154f, 154–162, 157f, 160f–161f
Time-shifting method, for measuring fluorescence decay time, 140
Ti-sapphire lasers, 320
Tissue autofluorescence, 225
Tissue, deep. *See* Deep-tissue imaging
Topography, in scanning near-field optical microscopy, 271–273
Transit time
  cell diameter and, 26–27, 28f
  as rate-limiting factor, 28
Transplantation
  high-speed sorting of CD34$^+$ hematopoietic stem cells for, 73–93
  nuclear, 101
  tumor cells in, elimination of, 76–80
  in utero, purified hematopoietic stem cells for, 78
Truth table, 57
Tube-type cameras, 328–329
Tumor cell sorting
  CD34 selection in, 76–80
  discriminant function analysis in, 57, 61f
  ROC analysis in, 57, 60f

Turbulent flow, 96–97
Two-phase method, of measuring fluorescence lifetime, 179
Two-photon excitation, 248–249
  in scanning near-field optical microscopy, 281–287, 284f, 286t, 286f
Two-photon fluorescence microscopy, 142–143, 156–157, 227f
  advantages of, 223–225
  of antigen processing, 202–204, 203f
  axial depth discrimination of, 223
  biomedical applications of fluorescence spectroscopic information in, 234–235
  blind deconvolution algorithm for, 229–231, 230f
  challenges in clinical setting, 231
  versus confocal, 222–225, 224f
  of cornea, 225–226, 234f, 234–235
  deep-tissue, 221–237
  detection geometry of, 223, 224f
  excitation wavelengths of, 223–225
  of mouse ear, 228–229, 229f
  penetration depth of, 223
  photodamage with, 223
  photon collection with, 223–225
  setting up, 227f, 227–228
  of skin, 225–226, 230f, 231, 232f, 233
  three-dimensional imaging capabilities of, 202–203
  tissue thermal damage with, 222–223
  video rate, 231–233
    high-rate, point scanning system of, 232f, 232–233
    implementing, 231–232
    line scanning approach for, 231
    multiphoton multifocal microscopy for, 232
Two-photon fluorescence spectroscopy, deep-tissue, 233–235
  biomedical applications of information, 234–235
  importance of information, 233–234

Ultra-rare cells
  definition of, 49
  multiple probes for, 53
Unsteady flow, 96–97

Vibronic lasers, 320
Video rate two-photon fluorescence microscopy, 231–233
  high-rate, point scanning system of, 232f, 232–233
  implementing, 231–232

line scanning approach for, 231
  multiphoton multifocal microscopy for, 232
Video scanning conventions, 325
Vidicon, 328–329
Viscous forces, in microfluidics, 96
Visibility, in confocal microscopy, 263

Wagon-wheel effect, 264
Water-cooled lasers, 317
Watts, 307–308
White blood cells, microfabricated devices for, 102, 103f
White-light scanning digital microscopy, 291–306
  averaging mode of, 299–300
  brightness and color calibration in, 302–303
  brightness and focus change in, 297–298
  communications of, 300
  comparison with camera systems, 303–305
  contrast enhancement and suppression in, 298–299, 298f–299f
  detection system in, 294
  future applications of, 306
  image size for, 295
  operational considerations with, 305–306
  oversampling in, 297, 300–302, 301f
  resolution of, super, 297, 300–302, 302f
  response function of, 299
  scan rate of, 294
  scanning spot in, 293–294
  signal processing in, 294–296, 295f
  spot modulation in, 298f, 298–299
  summing mode of, 299–300
  zoom magnification in, 296–297, 296f–297f, 302
Wide-field illumination, in fluorescent lifetime imaging microscopy, 167–168

Xenon arc lamps, 309–310, 310f

Yield
  of cell sorters, 14–15, 23
  of rare-cell sorting, 51, 63
Yttrium aluminum garnet (YAG) lasers, 320
Yttrium vanadate (YVO4) lasers, 320

Zappers, 3
Zoom
  for confocal microscopy, 245–246, 261, 264
  for white-light scanning digital microscopy, 296–297, 296f–297f, 302

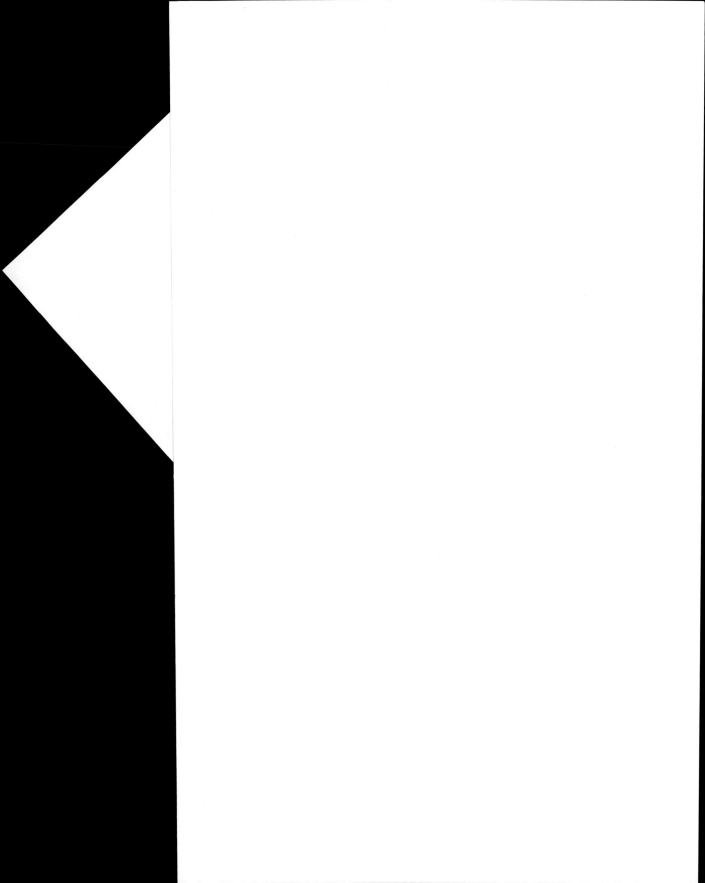

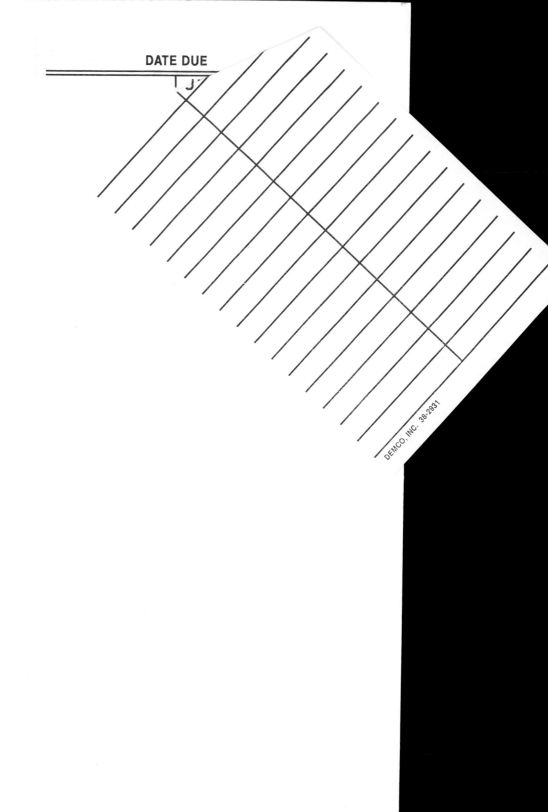